Basic Machine Tool Operations

John E. Neely

Prentice Hall
Upper Saddle River, New Jersey   *Columbus, Ohio*

Library of Congress Cataloging-in-Publication Data

Neely, John
  Basic machine tool operations / John E. Neely.
    p.   cm.
  ISBN 0-13-099677-7
    1. Machine-tools. 2. Machine-shop practice.   I. Title
TJ1185.N365 2000
670.42'3—dc21

99-19572
CIP

Cover photo: FPG International
Editor: Ed Francis
Production Editor: Christine M. Buckendahl
Production Coordination: Clarinda Publication Services
Design Coordinator: Karrie Converse-Jones
Cover Design: Brian Deep
Production Manager: Patricia A. Tonneman
Marketing Manager: Chris Bracken

This book was set in Slimbach and Helvetica by The Clarinda Company, and was printed and bound by Banta/Harrisonburg. The cover was printed by Phoenix Color Corp.

©2000 by Prentice-Hall, Inc.
Pearson Education
Upper Saddle River, New Jersey 07458

Printed in the United States of America

10 9 8 7 6 5 4 3 2 1

ISBN: 0-13-099677-7

Prentice-Hall International (UK) Limited, *London*
Prentice-Hall of Australia Pty. Limited, *Sydney*
Prentice-Hall of Canada, Inc., *Toronto*
Prentice-Hall Hispanoamericana, S.A., *Mexico*
Prentice-Hall of India Private Limited, *New Delhi*
Prentice-Hall of Japan, Inc., *Tokyo*
Prentice-Hall (Singapore) Pte. Ltd., *Singapore*
Editora Prentice-Hall do Brasil, Ltda., *Rio de Janiero*

PREFACE

Basic Machine Tool Operations is written for anyone who wishes to take a one-semester course in manual machining. Those who want to enhance their abilities in any trade and those who plan to continue in their second semester with computer numerical control (CNC) training will find this text easy to read and understand. Although the book is useful for a short course in machine shop, its major purpose is to acquaint students with the machines and processes that have led up to the development of precision numerically controlled machine tools. Thus, this textbook is not intended as a complete two-year manual machining course, but instead can be completed in one term. Its smaller size and consequently lower cost makes it an ideal adjunct for CNC instruction.

It would be very difficult to program CNC machines without having an understanding of the operating principles of manually controlled machine tools. Speeds, feeds, and coolants are important to know, but a hands-on perception for the safe and productive operation of the machine is even more desirable.

With the modern computer and high-precision machinery, parts can be produced within extremely close tolerances (within microinches). This kind of precision is not possible with manual machining processes with the exception of precision grinding machines. Manual machine tools can normally produce parts only within plus or minus .001 in. or more, and the CNC operator requires less skill to maintain such precision than would a manual machinist. The computer makes most of the complex decisions; yet an intuitive machine operator can make sophisticated judgments as to how a tool or machine is performing simply by listening and observing. In order to be such an efficient machine operator, one must have had training first in basic machine tools and their functions. The computer will never replace the operator.

From this text, students will learn how machining processes are used to manufacture mechanical parts, and how to identify and use tooling for manual machines that is the same or similar to that used on CNC machines. Where it is appropriate, the transition from manual machine tools to the CNC machines is discussed.

The book is made up of sections and unit divisions, which include a statement of purpose, objectives, an information section, and questions at the end of each unit. Answers to these test questions are given in Appendix II to help with self-study. *Basic Machine Tool Operations* takes a highly visual approach in which drawings and photographs illustrate and clarify the mechanical principles taught in this book.

An instructor's manual is also available, and includes multiple-choice tests and keys as well as information on using the text in various teaching systems. Some projects and worksheets are also included.

Basic Machine Tool Operations is equally adaptable to lecture-lab or totally competency-based instruction or even a hybrid form of part lecture and part self-paced instruction. This is possible because of its unique format, which provides a flexible, yet complete, learning program. It is my hope that this textbook will help students and others in the metals trades to understand the materials and machines of their trades more fully and to solve more of the problems encountered in their particular fields.

Acknowledgments

My thanks to those educators who gave me some ideas for machine shop preparation for CNC students.

I wish to thank the following for their contribution to this textbook:

Aloris Tool Company, Inc.
Armco, Inc.
Bay State Abrasives Division, Dresser Industries, Inc.
The Bendix Corporation
Boyar-Schultz Corporation
Brown & Sharpe Mfg. Company
Bryant Grinder Corporation
Buck Tool Company
Cincinnati Milacron
Clausing Machine Tools
David Drotar, Macomb Community College
The Desmond-Stephen Manufacturing Company
DoAll Company
El-Jay Inc., Eugene OR
Enco Manufacturing Company
Giddings & Lewis, Inc.
Goran N. Smith, St. Clair College
Haas Automation, Oxnard CA
Hardinge Brothers, Inc.
Harig Manufacturing Corporation
Hitachi Magna-Lock Corporation
Illinois/Eclipse, Division Illinois Tool Works, Inc.
John Allan, Jr.
John Wiley & Sons, Inc.
Kasto-Racine, Inc.
Kennametal, Inc.
Landis Tool Company, Division of Litton Industries
Lane Community College
The Lodge & Shipley Company
Louis Levin & Son, Inc.
Lovejoy Tool Company, Inc.
The L. S. Starrett Company
Mark Drzewiecki, Surface Finishes, Inc.
Mohawk Tools, Inc., Machine Tools Division
Monarch-Sidney, Division of Monarch Machine Tool Company
MTI Corporation National Twist Drill & Tool Division, Lear Siegler, Inc.
PMC Industries
Ralmike's Tool-A-Rama
Sandvik Madison, Inc.
Sellstrom Manufacturing Company
Southwestern Industries, Inc., Surface Finishes, Inc.
Sweetland Archery Products, Eugene, OR
Syclone Products, Inc.
Taper Micrometer Corporation TRW Inc.
Tony Schuls, Elgin Community College
The Warner & Swasey Company, Warren Tool Corporation
The Weldon Tool Company
Whitnon Spindle, Division of Mite Corporation
Wilton Corporation

John E. Neely

CONTENTS

SECTION A
Shop Techniques and Information

Some machinists go through their entire career with no more than a scratch or two, while others are often cut or even lose a finger. The difference is in making a habit of safe actions around machinery. A person must be able to anticipate hazardous situations and avoid them.

This section covers some principles of shop safety, but in each section where the use of machines is described, safety around them is discussed.

In these units you will learn to select metals that are used in the shop. You will learn how to cut off a piece of material for a machining operation.

Unit 1 Shop Safety
Unit 2 Selection and Identification of Metals
Unit 3 Cutoff Machines

Safety is not often considered as you proceed through your daily tasks. Often you expose yourself to needless risk because you have experienced no harmful effects in the past. Unsafe habits become almost automatic. None of us really likes to think about the possible consequences of an unsafe act. However, safety can and does have an important effect on people who make their living in a potentially dangerous environment such as a machine shop. An accident can reduce or end your working career. Years spent in training and gaining experience can be wasted in an instant if you should have an accident, not to mention causing you a possible permanent physical handicap and hardship on your family. Safety is an attitude that should extend far beyond the machine shop and into every facet of your life. You must constantly think about safety in everything you do.

OBJECTIVES

After completing this unit, you should be able to:

1. Identify hazards in the machine shop.
2. Identify and use safety equipment designed for a machinist.
3. Describe safe procedures around the shop and around a machine tool.

PERSONAL SAFETY

Eye Protection

Eye protection is a primary safety consideration around the machine shop. Machine tools produce metal **chips,** and there is always a possibility that these may be ejected from a machine at high velocity. Sometimes they can fly many feet. Furthermore, most cutting tools are made from hard materials. They can occasionally break or shatter from the stress applied to them during a cut. The result can be more flying metal particles.

Eye protection must be worn *at all times* in the machine shop. There are several types of eye protection available. Plain safety glasses are all that is required in most shops. These have shatterproof lenses that may be changed if they become scratched. The lenses have a high resistance to impact. Figure A1.1 shows the high impact resistance of these shatterproof lenses. Probably the most comfortable types are the fixed bow safety glasses (Figure A1.2).

Side shield safety glasses must be worn around any grinding operation. The side shield protects the side of the eye from flying particles. Side shield safety glasses may be of the solid or perforated type (Figure A1.3). The perforated side shield fits closer to the eye.

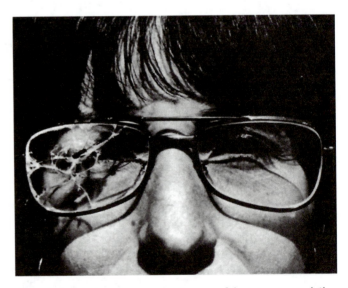

FIGURE A1.1 These shatterproof lenses saved the eye of the wearer when they were struck by a flying piece of metal. (Photo courtesy of John Allan, Jr.)

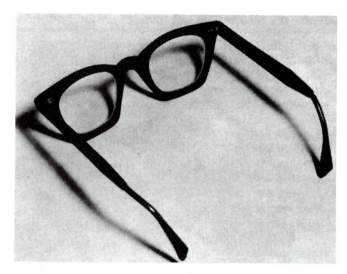

FIGURE A1.2 Fixed bow safety glasses. Their only disadvantage is that particles are able to enter from the sides.

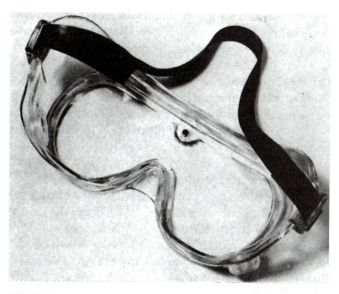

FIGURE A1.4 Safety goggles are adequate except in the vicinity of heavy flying metal objects. The relatively soft plastic on these goggles can be pierced.

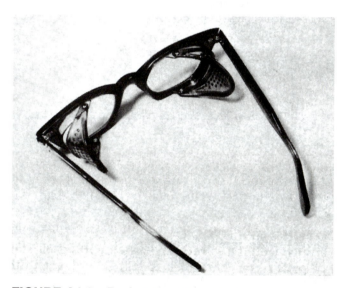

FIGURE A1.3 Perforated side shield safety glasses.

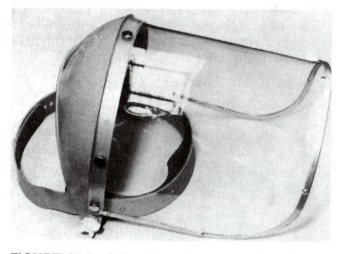

FIGURE A1.5 Safety face shield protects the entire face from flying particles or liquids.

If you wear prescription glasses, you may want to cover them with a safety goggle (Figure A1.4). The full face shield may also be used (Figure A1.5). Prescription glasses can be made as safety glasses. In industry, prescription safety glasses are sometimes provided free to employees.

Foot Protection

Generally, the machine shop does not present too great a hazard to the feet. However, there is always a possibility that you could drop something on your foot. A safety shoe is available. This will have a steel toe shield designed to resist impacts (Figure A1.6). Some safety shoes also have an instep guard. Shoes must be worn at all times in the machine shop. A solid leather shoe is recommended. Tennis shoes and sandals should not be worn. You must never even enter a machine shop with bare feet. Remember that the floor is often covered with razor-sharp metal chips.

Ear Protection

The instructional machine shop usually does not present a noise problem. However, an industrial machine shop may be adjacent to a fabrication or

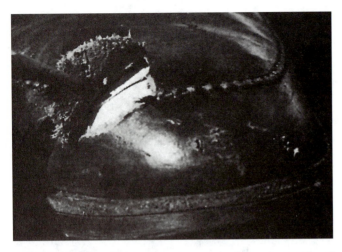

FIGURE A1.6 This safety-toe shoe is an example of the protection it can give your toes from falling objects. The steel liner is exposed. (Photo courtesy of John Allan, Jr.)

punch press facility. New safety regulations are quite strict regarding exposure to noise. Several types of sound suppressors and noise reducing ear plugs may be worn.

FIGURE A1.7 Ear muffs are designed to protect the ears from damage caused by loud noises.

TABLE A1.1 The decibel level of various sounds

130—Painful sounds; jet engine on ground
120—Airplane on ground: reciprocating engine
110—Boiler factory
    —Pneumatic riveter
100—
    —Maximum street noise
    —Roaring lion
90—
    —Loud shout
80—Diesel truck
    —Piano practice
    —Average city street
70—
    —Dog barking
    —Average conversation
60—
    —Average city office
50—
    —Average city residence
40—One typewriter
    —Average country residence
30—Turning page of newspaper
    —Purring cat
20—
    —Rustle of leaves in breeze
    —Human heartbeat
10—
 0—Faintest audible sound

Excess noise can cause a permanent hearing loss. Usually this occurs over a period of time, depending on the intensity of the exposure. Noise is considered an industrial hazard if it is continuously above 85 **decibels,** the units used in measuring sound waves. If it is over 115 decibels for short periods of time, ear protection must be worn (Figure A1.7). Ear muffs or plugs should be used wherever high intensity noise is likely. A considerate workman will not create excessive noise when it is not necessary. Table A1.1 shows the decibel level of various sounds; sudden sharp or high intensity noises are the most disturbing to your eardrums.

CUTTING OFF STOCK

Care in handling is required when cutting off stock in a power saw. The stock should be brought to the saw on a rollcase (Figure A1.8) or a simple rollstand. The pieces being cut off can sometimes be several feet long and should be similarly supported. Sharp burrs left from the cutting should be removed immediately with a file. You can acquire a nasty cut by sliding your hand over one of these burrs.

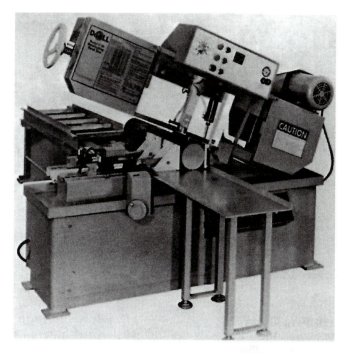

FIGURE A1.8 The material is brought into the saw on the rollcase (opposite side) and, when pieces are cut off, they are supported by the stand (this side of the saw). The stand prevents the part from falling to the floor. (The DoAll Company)

Carrying Objects

Do not carry long stock in the vertical position because of the chance of hitting light fixtures and ceilings. A better way is to have someone carry each end of a long piece of material. Do not carry sharp tools in your pockets. They can injure you or someone else.

HOT METAL SAFETY

Oxyacetylene torches are often used for cutting shapes, circles, and plates in machine shops. Safety when burning them requires proper clothing, gloves, and eye protection. It is also very important that any metal that has been heated by burning or welding be plainly marked, especially if it is left unattended. The common practice is to write the word *HOT* with soapstone on such items. Wherever arc welding is performed in a shop, the arc flash should be shielded from the other workers. *Never* look toward the arc because if the light enters your eye even from the side, the eye can be burned.

When handling and pouring molten metals such as babbitt, aluminum, or bronze, wear a face shield and gloves. Do not pour molten metals where there

is a concrete floor unless it is covered with sand. When molten metal spills onto a concrete floor, it will cause a piece of the concrete to explode upward along with hot metal. If the molten metal falls onto a water puddle or even on wet sand, it will explode upward possibly causing injury to anyone in the vicinity.

When **heat treating,** always wear a face shield and heavy gloves (Figure A1.9). There is a definite hazard to the face and eyes when cooling tool steel by oil **quenching,** that is, submerging it in oil. The oil, hot from the steel, tends to fly upward, so you should stand to one side of the oil tank.

Certain metals, when divided as finely as a powder or even as coarse machining chips, can ignite with a spark or just by the heat of machining. Magnesium and zirconium are two such metals. The fire, once started, is difficult to extinguish, and if water or a water-based fire extinguisher is used, the fire will only increase in intensity. Chloride-based power fire extinguishers are commercially available. These are effective for such fires as they prevent water absorption and form an air-excluding crust over the burning metal. Sand is also used to smother fires in magnesium.

Hazardous Fumes

Some metals such as zinc give off **toxic fumes** when heated above their boiling point. Some of these fumes when inhaled cause temporary sickness, but other fumes can be severe or even fatal. The fumes of mercury and lead are especially dangerous, as their effect is cumulative in your body and can cause irreversible damage. Cadmium and beryllium com-

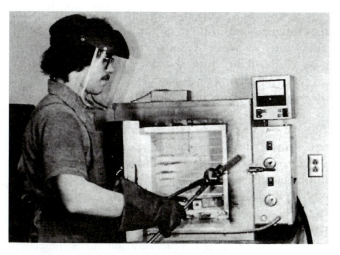

FIGURE A1.9 Face shield and gloves are worn for protection while heat treating and grinding.

pounds are also very poisonous. Therefore, when welding, burning, or heat treating metals, adequate ventilation is an absolute necessity. This is also true when parts are being carburized with compounds containing potassium cyanide. These *cyanogen compounds* are deadly poisonous and every precaution should be taken when using them. Kasenite, a trade name for a carburizing compound that is not toxic, is often found in school shops and in machine shops. Uranium salts are toxic and all **radioactive materials** are extremely dangerous.

PERSONAL SAFETY

Clothing, Hair, and Jewelry

Wear a short sleeve shirt or roll up long sleeves above the elbow. Keep your shirt tucked in and remove your necktie. It is recommended that you wear a shop apron. If you do, keep it tied behind you. If apron strings become entangled in the machine, you may be reeled in as well. A shop coat may be worn as long as you roll up long sleeves. Do not wear fuzzy sweaters around machine tools.

If you have long hair, keep it secured properly. In industry, you may be required to wear a hair net so that your hair cannot become entangled in a moving machine. The result of this can be disastrous.

Remove your wristwatch and rings before operating any machine tool. These can cause serious injury if they should be caught in a moving machine part.

Hand Protection

There is really no device that will totally protect your hands from injury. Next to your eyes, your hands are the most important tools that you have. It is up to you to keep them out of danger. Use a brush to remove chips from a machine (Figure A1.10). Do not use your hands. Chips are not only razor sharp; they are often extremely hot. Resist the temptation to grab chips as they come from a cut. Long chips are extremely dangerous. These can often be eliminated by properly sharpening your cutting tools. Chips should *not* be removed with a rag. The metal particles become inbedded in the cloth and they may cut you. Furthermore, the rag may be caught in a moving machine. Gloves must not be worn around most machine tools, although they are acceptable when working with a band saw blade or when cleaning chip pans while the machine is stopped. If a glove should be caught in a moving part, it will be pulled in, along with the hand inside it.

FIGURE A1.10   Use a brush to clear chips.

Various cutting oils, coolants, and solvents may affect your skin. The result may be a rash or possible infection. Avoid direct contact with these products as much as possible and wash your hands as soon as possible after contact.

Grinding Dust

Grinding dust is produced by **abrasive** wheels and consists of extremely fine metal particles and abrasive wheel particles. These should not be inhaled. In the machine shop, most grinding machines have a vacuum dust collector (Figure A1.11). Grinding may be done with coolants that aid in dust control. A machinist may be involved in portable grinding operations. This is common in such industries as shipbuilding. You should wear an approved respirator if you are exposed to grinding dust. Change the respirator filter at regular intervals. Grinding dust can pre-

FIGURE A1.11   Vacuum dust collector on grinders.

sent a great danger to health. Examples include the dust of such metals as beryllium or the presence of radioactivity in nuclear systems. In these situations, the spread of grinding dust must be carefully controlled.

Lifting

Improper lifting can result in a permanent back injury that can limit or even end your career. Back injury can be avoided if you lift properly at all times. If you must lift a large or heavy object, get some help or make use of a hoist or forklift. Don't try to be a "superman" and lift something that you know is too heavy. It is not worth the risk.

Objects within your lifting capability can be lifted safely by the following procedure (Figures A1.12*a* and A1.12*b*):

1. Keep your back straight.
2. Squat down, bending your knees.
3. Lift smoothly; let the muscles in your legs do the work. Keep your back straight. Bending over the load puts an excessive stress on your spine.
4. Position the load so that it is comfortable to carry. Watch where you are walking when carrying a load.
5. If you are placing the load back to floor level, lower it in the same manner you picked it up.

Scuffling and Horseplay

The machine shop is no place for scuffling and horseplay. This activity can result in a serious injury to you, a fellow student, or worker. Practical joking is also very hazardous. What might appear to be a comical situation to you could result in a disastrous accident to someone else. In industry, horseplay and practical joking are often grounds for dismissal of an employee.

Injuries

If you should be injured, report it immediately to your instructor.

IDENTIFYING SHOP HAZARDS

A machine shop is a potentially dangerous place. One of the best ways to be safe is to be able to identify shop **hazards** before they involve you in an accident. By being aware of potential danger, you can better make safety part of your work in the machine shop. Although industrial hazards exist, they are controlled by safety programs and employee-employer cooperation. The machine shop is a relatively safe place, but safety rules must be observed.

FIGURE A1.12*a* Proper way of lifting.

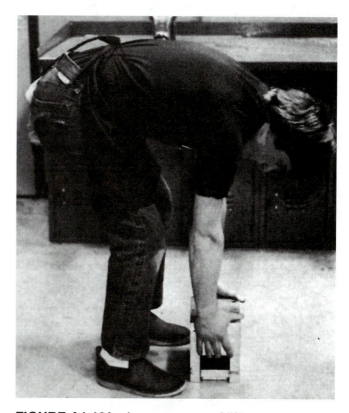

FIGURE A1.12*b* Improper way of lifting.

Compressed Air

Most machine shops have compressed air. This is needed to operate certain machine tools. Often flexible air hoses are hanging about the shop. Few people realize the large amount of energy that can be stored in a compressed gas such as air. When this energy is released, extreme danger may be present. You may be tempted to blow chips from a machine tool using compressed air. This is not good practice. The air will propel metal particles at high velocity. They can injure you or someone on the other side of the shop. Use a brush to clean chips from the machine. Do not blow compressed air on your clothing or skin. The air can be dirty and the force can implant dirt and germs into your skin. Air can be a hazard to ears as well. An eardrum can be ruptured.

Should an air hose break, or the nozzle on the end come unscrewed, the hose will whip wildly. This can result in an injury if you happen to be standing nearby. When an air hose is not in use, it is good practice to shut off the supply valve. The air trapped in the hose should be vented. When removing an air hose, even one that has a quick disconnect, from its supply valve, be sure that the supply is turned off and the hose has been vented. Removing a charged air hose will result in a sudden venting of air. This can surprise you and an accident might result if you are not careful.

Housekeeping

Keep the floor and aisles clear of stock and tools. This will insure that all exits are clear if the building should have to be evacuated. Material on the floor, especially round bars, can cause falls. Clean oils or coolants that may have spilled on the floor. Several preparations designed to absorb oil are available. These may be used from time to time in the shop. Keep oily rags in an approved safety can (Figure A1.13). This will prevent possible fire from spontaneous combustion. Rag containers should be emptied every night.

Electricity

Electricity is another potential danger in a machine shop. Your exposure to electrical hazard will be minimal unless you become involved with machine maintenance. A machinist is mainly concerned with the on and off switch on a machine tool. However, if you are adjusting the machine or accomplishing maintenance, you should unplug it from the electri-

FIGURE A1.13 Store oil-soaked rags in an approved safety can.

cal service. If it is permanently wired, the circuit breaker may be switched off and tagged with an appropriate warning.

MACHINE HAZARDS

There are many machine hazards. Each section of this book will discuss the specific hazards applicable to that type of machine tool. Remember that a machine has no intelligence of its own. It cannot distinguish between cutting metal and cutting fingers. Do not think that you are strong enough to stop a machine should you become tangled in moving parts. You are not. When operating a machine, think about what you are going to do before you do it. Go over a checklist.

Safety Checklist

1. Do I know how to operate this machine?
2. What are the potential hazards involved?
3. Are all guards in place?
4. Are my procedures safe?

5. Am I doing something that I probably should not do?
6. Have I made all the proper adjustments and tightened all locking bolts and clamps?
7. Is the workpiece secured properly?
8. Do I have proper safety equipment?
9. Do I know where the stop switch is?
10. Do I think about safety in everything that I do?

SAFETY IN MATERIAL HANDLING

Machinists were once expected to lift pieces of steel weighing a hundred pounds or more into awkward positions. This was a dangerous practice that resulted in too many injuries. Hoists and cranes are used to lift all but the smaller parts. Steel weighs about 487 pounds per cubic foot; water weighs 62.5 pounds per cubic foot, thus steel is a very heavy material for its size. You can easily be misled into thinking that a small piece of steel does not weigh much. Follow these two rules in all lifting that you do: Don't lift more than you can *easily* handle, and bend your knees and keep your back straight. If a material is too heavy or awkward for you to position it on a machine such as a lathe, use a hoist. Once the workpiece has been hoisted to the required level, it can hang in that position until the clamps or chuck jaws on the machine have been secured.

Hoisting

When lifting heavy metal parts with a mechanical or electric hoist, always stand in a safe position no matter how secure the slings and hooks seem to be. They don't often break, but it can and does

happen, and if a careless foot is under the edge, a painful or crippling experience is sure to follow. Slings should not have less than a 30-degree angle with the load (Figure A1.14). It should be noted that when a sling, as shown in Figure A1.14, is at 30 degrees, it will support only one-half the load that the cable or chain will in a vertical lifting position. At 45 degrees, a sling will support about two-thirds, and at 60 degrees about three-fourths of a vertical rigging arrangement. When hoisting long bars or shafts, a spreader bar (Figure A1.15) should be used so the slings cannot slide together and unbalance the load. When operating a crane, make sure that no one is standing in the way of the load or hook.

Fire Extinguishers

It is an important safety consideration to know the correct fire extinguisher to use for a particular fire. For example, if you should use a water-based extinguisher on an electrical fire, you could receive a severe or fatal electrical shock. Fires are classed according to types as given in Table A1.2.

There are four basic types of fire extinguishers used other than tap water:

1. The dry chemical type is effective on Classes B and C fires.
2. The pressurized water and loaded stream types are safe only on Class A fires. This type may actually spread an oil or gasoline fire.
3. The dry chemical multipurpose extinguisher may be safely used on Classes A, B, and C fires.
4. Pressurized carbon dioxide (CO_2) can be used on Classes B and C fires.

You should always make yourself aware of the locations of fire extinguishers in your working area. Take time to look at them closely and note their types and capabilities. This way, if there should ever be an oil-based fire or electrical fire in your area, you will know how to safely put it out.

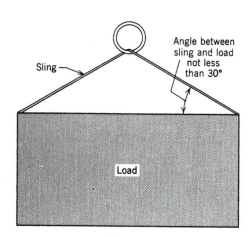

FIGURE A1.14 Load sling.

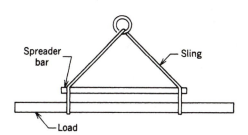

FIGURE A1.15 Sling for lifting long bars.

TABLE A1.2 Types of extinguishers used on the classes of fire (Brodhead-Garrett Company)

	Pressurized Water	Loaded Stream	CO_2	Regular Dry Chemical	All Use Dry Chemical
Class A Fires Paper, wood, cloth, etc. where quenching by water or insulating by general purpose dry chemical is effective.	Yes— Excellent	Yes— Excellent	Small surface fires only	Small surface fires only	Yes— Excellent Forms smothering film, prevents reflash
Class B Fires Burning liquids (gasoline, oils, cooking fats, etc.) where smothering action is required.	No— Water will spread fire	Yes— Has limited capability	Yes— Carbon dioxide has no residual effects on food or equipment	Yes— Excellent Chemical smothers fire	Yes— Excellent Smothers fire, prevents reflash
Class C Fires Fire in live electrical equipment (motors, switches, appliances, etc.) where a nonconductive extinguishing agent is required.	No— Water is a conductor of electricity	No— Water is a conductor of electricity	Yes— Excellent CO_2 is a nonconductor, leaves no residue	Yes— Excellent Nonconducting smothering film. Screens operator from heat	Yes— Excellent Nonconducting smothering film. Screens operator from heat

SOURCE: By permission, Brodhead-Garrett Company

SELF-TEST

1. What is the primary piece of safety equipment in the machine shop?
2. What can you do if you wear prescription glasses?
3. Describe proper dress for the machine shop.
4. What can be done to control grinding dust?
5. What hazards exist from coolants, oils, and solvents?
6. Describe proper lifting procedure.
7. Describe at least two compressed air hazards.
8. Describe good housekeeping procedures.
9. How should long pieces of material be carried?
10. List at least five points from the safety checklist for a machine tool.
11. Should you lift a 100-pound workpiece by hand to put it into a lathe chuck? Explain.
12. a. Locate and identify all of the fire extinguishers in your area.
    b. Name the types of extinguishers that can be safely used on electrical fires.

Selection and Identification of Metals

When the village smithy plied his trade, there were only wrought iron and carbon steel for making tools, implements, and horseshoes, so the task of separating metals was relatively simple. As industry grew, more alloy steels and special metals were needed and thus gradually developed. Today many hundreds of these metals are used. Without some means of reference to or identification of metals, work in the machine shop would be chaotic. Therefore, this unit introduces you to several systems used for marking steels, known as **ferrous** metals, and some ways to choose among them. You will also learn how to identify many nonferrous metals, metals other than iron and steel.

OBJECTIVE

After completing this unit, you should be able to:

Identify different types of metals by various means of shop testing.

STEEL IDENTIFICATION SYSTEMS

Color coding is one means of identifying a particular type of steel. Its main disadvantage is that there is no universal color coding system. Manufacturers and local shops all have different codes.

The most common systems in the United States used to classify steels by chemical composition were developed by the Society of Automotive Engineers (SAE) and the American Iron and Steel Institute (AISI). The SAE and AISI systems use a four- or five-digit number (Table A2.1). The first number indicates the type of steel. Carbon, for instance, is denoted by the number 1, 2 is a nickel steel, 3 is a nickel-chromium steel, and so on. The second digit indicates the approximate percentage of the predominant alloying element. The third and fourth digits, represented by x, always denote the percentage of carbon in hundredths of one percent. For example, SAE 1040 denotes plain carbon steel with .40 percent carbon; SAE 4140 denotes a chromium-molybdenum steel containing .40 percent carbon and about 1.0 percent of the major alloy (molybdenum). Plain carbon steels, alloys, and tool steels can contain anywhere from .06 to 1.70 percent carbon. Steels having over 1 percent carbon require a five-digit number. Certain corrosion- and heat-resisting alloys are classified with a five-digit number in order to identify the approximate alloy composition of the steel.

The AISI numerical system is basically the same as the SAE system with certain capital-letter prefixes. These prefixes designate the process used to make the steel. The lowercase letters from a to i as a suffix denote special conditions in the steel.

AISI prefixes:

B—Acid Bessemer, carbon steel

C—Basic open-hearth carbon steel

CB—Either acid Bessemer or basic open-hearth carbon steel at the option of the manufacturer

D—Acid open-hearth carbon steel

E—Electric furnace alloy steel

STAINLESS STEEL

It is the element chromium (Cr) that makes stainless steels stainless. Steel must contain a minimum of about 11 percent chromium in order to gain resistance to atmospheric corrosion. Higher percentages of chromium make steel even more resistant to corrosion and high temperatures. Nickel is added to improve **ductility,** corrosion resistance, and other properties.

TABLE A2.1 SAE-AISI numerical designation of alloy steels (*x* represents percentage of carbon in hundredths)

Carbon steels	
Plain carbon	10xx
Free-cutting, resulfurized	11xx
Manganese steels	13xx
Nickel steels	
.50% nickel	20xx
1.50% nickel	21xx
3.50% nickel	23xx
5.00% nickel	25xx
Nickel-Chromium steels	
1.25% nickel, .65% chromium	31xx
1.75% nickel, 1.00% chromium	32xx
3.50% nickel, 1.57% chromium	33xx
3.00% nickel, .80% chromium	34xx
Corrosion and heat-resisting steels	303xx
Molybdenum steels	
Chromium	41xx
Chromium-nickel	43xx
Nickel	46xx and 48xx
Chromium steels	
Low-chromium	50xx
Medium-chromium	51xx
High-chromium	52xx
Chromium-Vanadium steels	6xxx
Tungsten steels	7xxx
Triple alloy steels	8xxx
Silicon-Manganese steels	9xxx
Leaded steels	11Lxx (example)

SOURCE: John E. Neely, *Practical Metallurgy and Materials of Industry*, John Wiley and Sons, Inc., Copyright © 1979, New York.

Excluding the precipitation hardening types that harden over a period of time after solution heat treatment, there are three basic types of stainless steels: the martensitic and ferritic types of the 400 series, and the austenitic types of the 300 series.

The martensitic, hardenable type has carbon content up to 1 percent or more, so it can be hardened by heating to a high temperature and then **quenching** (cooling) in oil or air. The cutlery grades of stainless are to be found in this group. The ferritic type contains little or no carbon. It is essentially soft iron that has 11 percent or more chromium content. It is the least expensive of the stainless steels and is used for such things as building trim, pots, and pans. Both ferritic and martensitic types are **magnetic.**

Austenitic stainless steel contains chromium and nickel, little or no carbon, and cannot be hardened by quenching, but it readily work hardens while retaining much of its ductility. For this reason it can be work hardened until it is almost as hard as a hardened martensitic steel. Austenitic stainless steel is somewhat magnetic in its work-hardened condition but nonmagnetic when **annealed** or soft. See the Glossary for definitions of these metallurgical terms.

Table A2.2 illustrates the method of classifying the stainless steels. Only a very few of the basic types are given here. You should consult a manufacturer's catalog for further information.

TOOL STEELS

Special carbon and alloy steels called *tool steels* have their own classification. There are six major tool steels for which one or more letter symbols have been assigned:

1. Water-hardening tool steels
   W—High-carbon steels
2. Shock-resisting tool steels
   S—Medium carbon, low alloy
3. Cold-work tool steels
   O—Oil-hardening types
   A—Medium-alloy air-hardening types
   D—High-carbon, high-chromium types
4. Hot-work tool steels
   H—H1 to H19, chromium-base types; H20 to H39, tungsten-base types; H40 to H59, molybdenum-base types
5. High-speed tool steels
   T—Tungsten-base types
   M—Molybdenum-base types
6. Special-purpose tool steels
   L—Low-alloy types
   F—Carbon-tungsten types
   P—Mold steels P1 to P19, low-carbon types; P20 to P39, other types

Several metals can be classified under each group, so that an individual type of tool steel will also have a suffix number that follows the letter symbol of its alloy group. The carbon content is given only in those cases where it is considered an identifying element of that steel.

Type of Steel	Examples
Water hardening:	
straight carbon tool steel	W1, W2, W4
Manganese, chromium, tungsten:	
oil-hardening tool steel	O1, O2, O6
Chromium (5.0%):	
air-hardening die steel	A2, A5, A10
Silicon, manganese, molybdenum:	
punch steel	S1, S5
High-speed tool steel	M2, M3, M30
	T1, T5, T15

TABLE A2.2 Classification of stainless steels

Class	Alloy Content	Metallurgical Structure	Ability to be Heat Treated
I	Chromium	Martensitic	Hardenable (types 410, 414, 416, 405, 403, 420, 440, 431)
II	Chromium	Ferritic	Nonhardenable (types 430, 442, 409, 400, 446, Armco 18 SR)
III	Chromium-nickel	Austenitic	Nonhardenable (types 302, Armco 18-9 LW, 301, 303, 304, 305, 308, 309, 310, 347, 316, 321, 348, 317)
	Chromium-nickel-manganese	Austenitic	Nonhardenable (Armco Nitronic 32, Armco Nitronic 33, Armco Nitronic 40, Armco Nitronic 50, Armco Nitronic 60)
IV	Chromium-nickel (copper added)	Martensitic	Precipitation-hardening (Armco 17-4 PH, Armco 15-5 PH, Armco 13-8 Mo)
	Chromium-nickel	Semiaustenitic	Precipitation-hardening (Armco 17-7 PH, Armco PH 15-7 Mo, Armco PH 14-8 Mo)
	Chromium-nickel	Austenitic	Precipitation-hardening (Armco 17-10 P)

SOURCE: Armco Inc., Middletown, Ohio, *Development of the Stainless Steels,* Copyright © 1977, Armco.

NOTE: The precipitation-hardening stainless steels (for example Armco 17-4 PH) are normally purchased in the solution-treated condition and somewhat martensitic, but machinable (Rc34 to Rc38). Solution treatment consists of heating to 1875 to 1950°F (1024 to 1065°C) for one-half hour and oil quenching. Further hardening is done by reheating 900 to 1150°F (482 to 621°C) and air cooling. This produces a Rockwell hardness up to Rc44.

SHOP TESTS FOR IDENTIFYING STEELS

One of the disadvantages of steel identification systems is that the marking is often lost. The end of a shaft is usually marked. If the marking is obliterated or cut off and the piece is separated from its proper storage rack, it is very difficult to determine its carbon content and alloy group. This shows the necessity of returning stock material to its proper rack. It is also good practice to leave the identifying mark on one end of the stock material and always to cut off the other end.

Unfortunately, in the shop there are always some short ends and otherwise useful pieces that have become unidentified. In addition, when repairing or replacing parts for old or nonstandard machinery, there is usually no record available for material selection. There are many shop methods a worker may use to identify the basic type of steel in an unknown sample. By the process of elimination, the worker can then determine which of the several steels of that type in the shop is most comparable to his sample. The following are several methods of shop testing that you can use.

Visual

Some metals can be identified by visual observation of their surface finishes (Figure A2.1). Heat scale or black mill scale is found on all **hot-rolled** (HR) steels. These can be either low carbon (.05 to .30 percent), medium carbon (.30 to .60 percent), high carbon (.60 to 1.70 percent), or alloy steels. Other surface coatings that might be detected are the sherardized, plated, case-hardened, or nitrided

FIGURE A2.1 The contrast of surface textures of hot-rolled (HR) and cold-rolled (CR) steels can be seen in this photograph. HR is on the left and CR on the right.

surfaces (Figure A2.2). Sherardizing is a process in which zinc vapor is inoculated into the surface of iron or steel.

Cold-rolled (CR) steel usually has a metallic luster. Not all cold-worked steel is rolled; some bars are cold-drawn through a die. **Cold finish** (CF) is a term used to designate the finish on all types of **cold-worked** steel. Ground and polished (G and P) steel (for example, drill rod) has a bright, shiny finish with closer dimensional tolerances than CF (Figure A2.2). Also, cold-drawn **ebonized**, or black, finishes are sometimes found on alloy and resulfurized shafting.

Chromium-nickel stainless steel, which is austenitic and nonmagnetic, usually has a white appearance. Straight 12 to 13 percent chromium is ferritic and magnetic with a bluish-white color. Manganese steel is blue when polished, but copper colored when oxidized. White cast-iron **fractures** will appear silvery or white. Gray cast-iron fractures appear dark gray and will smear a finger with a gray **graphite** smudge when touched.

Magnet Test

All **ferrous** metals such as iron and steel are magnetic; that is, they are attracted to a magnet. With the exception of nickel, cobalt, and a few rare earth metals, nonferrous metals are generally not attracted to a magnet. United States "nickel" coins contain about 25 percent nickel and 75 percent copper and so they do not respond to the magnet test, but Canadian "nickel" coins are attracted to a magnet because they contain a much larger proportion of nickel. Ferritic and martensitic (400 series) stainless steels are also attracted to a magnet and so cannot be separated from other steels by

this method. Austenitic (300 series) stainless steel is not attracted to a magnet unless it is work hardened (Figure A2.3).

Hardness Test

Wrought iron is very soft since it contains almost no carbon or any other alloying element. Generally speaking, the more carbon (up to 2 percent) and other elements that steel contains, the harder, stronger, and less ductile it becomes, even if in an annealed state. Thus, the hardness of a sample can help us to separate low-carbon steel from an alloy steel or a high-carbon steel.

The **Rockwell Hardness Tester** (Figure A2.4) and the **Brinell Hardness Tester** are the most commonly used types of hardness testers for industrial and metallurgical purposes. Heat treaters, inspectors, and many others in industry often use these machines. These instruments provide a quick, accurate method of determining hardness by the use of a given weight and a penetrator. The Brinell test gives a reading in BHN (Brinell hardness number). The Rockwell test has several scales of which the B and C scales are most used: the B scale for soft metals such as aluminum, and the C scale for hard metals such as steel. Mild steel can be in the range of 10 to 30 Rc (Rockwell), 180 to 290 BHN. Spring steel is about 45 Rc, 430 BHN, and file hard is about 62 Rc, 685 BHN. See Table 1 in Appendix I.

Not all shops have hardness testers available, in which case the scratch test or the file test may be used.

FIGURE A2.2 Round bars having various surface finishes. Left to right: aluminum, ground and polished, cold-finished steel, hot-rolled steel, sherardized surface, and zinc dip surface.

FIGURE A2.3 The ferritic stainless steel is attracted to a magnet and the austenitic is not.

FIGURE A2.4 The Rockwell Hardness Tester.

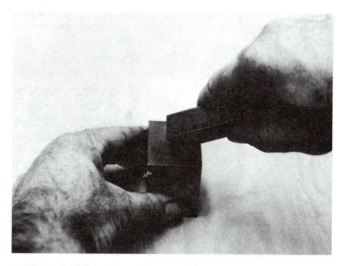

FIGURE A2.5a A piece of keystock (mild steel) is scratched across an unknown sample. Since the sample is not scratched, it is harder than the keystock and probably is an alloy or tool steel.

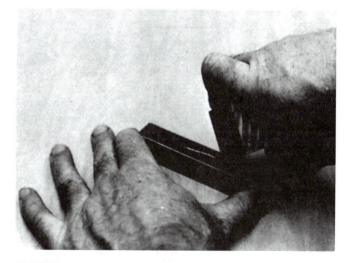

FIGURE A2.5b The sample is now scratched against the keystock as a further test and it does scratch the keystock.

Scratch Test

Geologists and "rock hounds" scratch rocks against items of known hardness for identification purposes. The same method can be used to check metals for relative hardness. Simply scratch one sample with another and the softer sample will be marked. Be sure all scale or other surface impurities have been removed before scratch testing (Figures A2.5a and A2.5b). A variation of this method is to strike two similar edges of two samples together. The one receiving the deepest indentation is the softer of the two.

File Test

Files can be used to establish the relative hardness between two samples, as in the scratch test, or they can determine an approximate hardness of a piece on a scale of many steels. Table A2.3 gives the Rockwell and Brinell hardness numbers for this file test when using new files. This method can only be as accurate, however, as the skill that the user has acquired through practice.

Care must be taken not to damage the file, since filing on hard materials may ruin the file. Testing should be done on the tip or near the edge of the file.

Spark Test

Spark testing is a useful way to test for carbon content in many steels. The metal tested, when held against a grinding wheel, will display a particular spark pattern depending on its content. Spark testing provides a convenient means of distinguishing between tool steel (of medium or high carbon) and

TABLE A2.3 File test and hardness table

Type Steel	Rockwell B	C	Brinell	File Reaction
Mild steel	65		100	File bites easily into metal. (Machines well but makes built-up edge on tool.)
Medium-carbon steel		16	212	File bites into metal with pressure. (Easily machined with high-speed tools.)
High-alloy steel		30	290	File does not bite into metal except with difficulty.
High-carbon steel		31	294	(Readily machinable with carbide tools.)
Tool steel		42	390	Metal can only be filed with extreme pressure. (Difficult to machine even with carbide tools.)
Hardened tool steel		50	481	File will mark metal but metal is nearly as hard as the file and machining is impractical; should be ground.
Case-hardened parts and hardened tool steel		64	739	Metal is as hard as the file; should be ground.

SOURCE: John E. Neely, *Practical Metallurgy and Materials of Industry,* John Wiley and Sons, Inc., Copyright © 1979, New York.

NOTE: Rockwell and Brinell hardness numbers are only approximations since file testing is not an accurate method of hardness testing.

low-carbon steel. High-carbon steel (Figure A2.6) shows many more bursts than low-carbon steel (Figure A2.7).

Almost all tool steel contains some alloying elements besides the carbon, which affects the carbon burst. Chromium, molybdenum, silicon, aluminum, and tungsten suppress the carbon burst. For this reason, spark testing is not very useful in determining the content of an unknown sample of steel. It is useful, however, as a comparison test. Comparing the spark of a known sample to that of an unknown sample can be an effective method of identification for the trained observer. Cast iron may be distinguished from steel by the characteristic spark stream (Figure A2.8). High-speed steel can also be readily identified by spark testing (Figure A2.9).

When spark testing, always wear safety glasses or a face shield. Adjust the wheel guard so the spark will fly outward and downward and away from you. A coarse grit wheel that has been freshly dressed to remove contaminants should be used.

Machinability

Machinability will be studied further in another unit, but as a simple comparison test, it can be useful to determine a specific type of steel. For example, two unknown samples identical in appearance

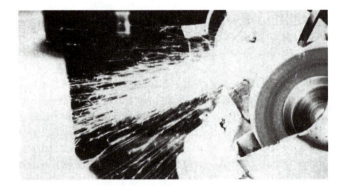

FIGURE A2.6 High-carbon steel. Short, very white or light-yellow carrier lines with considerable forking, having many starlike bursts. Many of the sparks follow around the wheel.

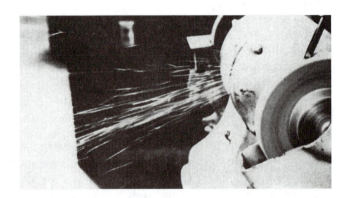

FIGURE A2.7 Low-carbon steel. Straight carrier lines having yellowish color with very small amount of branching and very few carbon bursts.

FIGURE A2.8 Cast iron. Short carrier lines with many bursts, which are red near the grinder and orange-yellow farther out. Considerable pressure is required on cast iron to produce sparks.

FIGURE A2.9 High-speed steel. Carrier lines are orange, ending in pear-shaped globules with very little branching or carbon sparks. High-speed steel requires moderate pressure to produce sparks.

and size can be test cut in a machine tool, using the same speed and feed for both. The ease of cutting should be compared and chips observed for heating color and curl (Figure A2.10).

SELECTION OF STEEL

Several properties should be considered when selecting a piece of steel for a job: strength, machinability, hardenability, weldability, fatigue resistance, and corrosion resistance. Manufacturer's catalogs and reference books are available for selection of standard structural shapes, bars, and other steel products. Others are available for the stainless steels, tool steels, and finished carbon steel and alloy shafting. Many of these steels are known by a trade name.

FIGURE A2.10 Machinability test. When one of two samples having the same feed and speed shows a darker color (blue) on the chip, it can be assumed that it is a harder, stronger metal.

A worker is often called upon to select a shaft material from which to machine finish a part. Shafting is manufactured with two kinds of surface finish: cold finished (CF), found on low-carbon steel, and ground and polished (G and P), found mostly on alloy steel shafts. Tolerances are kept much closer on ground and polished shafts. The following are some common alloy steels:

1. SAE 4140 is a chromium–molybdenum alloy with 0.40 percent carbon. It lends itself readily to heat treating, forging, and welding. It provides a high resistance to torsional and reversing stresses such as drive shafts.
2. SAE 1140 is a resulfurized, drawn, free-machining bar stock. This material has good resistance to bending stresses because of its **fibrous** qualities and good **tensile strength,** but it has low resistance to high torque values. SAE 1140 is also useful where stiffness is required. It should not be heat treated or welded.
3. Leaded steels have all the free-machining qualities and finishes of resulfurized steels. Leaded alloy steels such as SAE41L40 have the superior strength of 4140 but are much easier to machine.
4. SAE 1040 is a medium carbon steel that has a normalized tensile strength of about 85,000 **PSI** (pounds per square inch). It can be heat treated, but large sections will be hardened only on the surface and the core will still be in a normalized or relatively soft condition.

5. SAE 1020 is a low-carbon steel that has good machining characteristics. It normally comes as CF shafting and is very commonly used for shafting in industrial applications. It has a lower tensile strength than the alloy steels or higher-carbon steels.

NONFERROUS METALS

Nonferrous metals such as gold, silver, copper, and tin were in use hundreds of years before the smelting of iron, and yet some nonferrous metals have appeared relatively recently in common industrial use. For example, aluminum was first commercially extracted from ore in 1886 by the Hall-Heroult process; titanium, a space-age metal, was produced in commercial quantities only after World War II.

In general, nonferrous metals are more costly than ferrous metals. It is not always easy to distinguish a nonferrous metal from a ferrous metal, nor to separate one from another, although they often vary widely in color and density or pounds per cubic inch (lb/in^3).

Aluminum

Aluminum is white or white-gray in color and can have any surface finish from dull to shiny when polished. An **anodized** surface is frequently found on aluminum products. Aluminum weighs 168.5 pounds per cubic foot (lb/ft^3), as compared to 487 pounds per cubic foot for steel, and has a melting point of 1220°F (660°C) when pure. It is readily machinable and can be manufactured into almost any shape or form (Figure A2.11).

FIGURE A2.11 Structural aluminum shapes used for building trim provide a pleasing appearance.

TABLE A2.4   Aluminum and aluminum alloys

Code Number	Major Alloying Element
1xxx	None
2xxx	Copper
3xxx	Manganese
4xxx	Silicon
5xxx	Magnesium
6xxx	Magnesium and silicon
7xxx	Zinc
8xxx	Other elements
9xxx	Unused (not yet assigned)

Magnesium is also a much lighter metal than steel, as it weighs 108.6 pounds per cubic foot, and looks much like aluminum. In order to distinguish between the two metals, it is sometimes necessary to make a chemical test. A zinc chloride solution in water, or copper sulfate, will blacken magnesium immediately but will not change aluminum.

There are several numerical systems used to identify aluminums, such as federal specifications, military specifications, and the American Society for Testing Materials (ASTM) and SAE specifications. The system most used by manufacturers, however, is one adopted by the Aluminum Association in 1954.

From Table A2.4, you can see that the first digit of a number in the aluminum alloy series indicates the alloy type. The second digit, represented by an x in the table, indicates any modifications that were made to the original alloy. The last two digits indicate the numbers of similar aluminum alloys of an older marking system, except in the 1100 series, where the last two digits indicate the amount of pure aluminum above 99 percent contained in the metal.

EXAMPLES

An aluminum alloy numbered 5056 is an aluminum–magnesium alloy, where the first 5 represents the alloy magnesium, the second digit represents modifications to the alloy, and 56 stands for a similar aluminum alloy of an older marking system. An aluminum numbered 1120 contains no major alloy and has 0.20 percent pure aluminum above 99 percent.

Aluminum and its alloys are produced as castings or as **wrought** (cold-worked) shapes such as sheets, bars, and tubing. They can be either cold worked by rolling, drawing, or extruding, or hot worked by **forging**. Cast aluminum is not generally as strong as wrought aluminum. Aluminum alloys are harder than pure aluminum and will scratch the softer

(1100 series) aluminum. A 10 percent solution of sodium hydroxide (caustic soda) will not stain pure aluminum but will leave a dark stain on the aluminum alloy.

Pure aluminum and some of its alloys cannot be hardened by heat treating, but they can be annealed to soften them by heat treatment. One method of hardening aluminum is by cold working (strain hardening). These temper (hardness) designations are indicated by a letter that follows the four-digit alloy series number:

—F As fabricated. No special control over strain hardening or temper designation is noted.
—O Annealed, recrystallized wrought products only. Softest temper.
—H Strain-hardened, wrought products only. Strength is increased by work hardening.

This letter —H is always followed by two or more digits. The first digit—1, 2, or 3—denotes the final degree of strain hardening.

—H1 Strain hardened only
—H2 Strain hardened and partially annealed
—H3 Strain hardened and stabilized

The second digit denotes higher strength **tempers** obtained by heat treatment:

2 $\frac{1}{4}$ hard
4 $\frac{1}{2}$ hard
6 $\frac{3}{4}$ hard
8 Full hard
9 Extra hard

EXAMPLE

5056-H18 is an aluminum-magnesium alloy, strain hardened to a full hard temper.

Another method of hardening some aluminum alloys is a process called **solution heat treatment** and **precipitation hardening** or **aging.** This process involves heating the aluminum alloy to a temperature where the alloying element is dissolved into a solid solution. The aluminum alloy is then quenched in water and allowed to age or is artificially aged by heating slightly. The aging produces an internal strain that hardens and strengthens the aluminum. Some other nonferrous metals are also hardened by this process. For these aluminum alloys the letter —T follows the four-digit series number. The numbers 2 to 10 follow this letter to indicate the sequence of treatment.

TABLE A2.5 Cast aluminum alloy designations

Code Number	Major Alloy Element
1xx.x	None, 99 percent aluminum
2xx.x	Copper
3xx.x	Silicon with Cu and/or Mg
4xx.x	Silicon
5xx.x	Magnesium
6xx.x	Zinc
7xx.x	Tin
8xx.x	Unused series
9xx.x	Other major alloys

—T2   Annealed (cast products only)
—T3   Solution heat treated and cold worked
—T4   Solution heat treated, but naturally aged
—T6   Solution heat treated and artificially aged
—T8   Solution heat treated, cold worked, and artificially aged
—T9   Solution heat treated, artificially aged, and cold worked
—T10 Artificially aged and then cold worked

EXAMPLE

2024-T6 Aluminum-copper alloy, solution heat treated and artificially aged.

Cast aluminum alloys generally have lower tensile strength than wrought alloys. Sand castings, permanent mold, and die casting alloys are of this group. They owe their mechanical properties to solution heat treatment and precipitation or to the addition of alloys. A classification system similar to that of wrought aluminum alloys is used (Table A2.5).

The cast aluminum 108F, for example, has an ultimate tensile strength of 24,000 PSI in the as-fabricated condition and contains no alloy. The 220.T4 copper aluminum alloy has a tensile strength of 48,000 PSI.

MACHINABILITY OF ALUMINUM AND MAGNESIUM ALLOYS

Aluminum alloys are relatively easy to machine using tools with positive back rakes (making the cutting edge keener). Negative back rakes (making the cutting edge less sharp) on tools will produce a built-up edge (BUE), causing tearing of the metal and a rough finish. Pure aluminum, the 1100 series, is the most difficult to machine because it is very soft.

Cadmium

Cadmium has a blue-white color and is commonly used as a protective plating on parts such as screws, bolts, and washers. It is also used as an alloying element to make metal alloys that melt at low temperature, such as bearing metals, solder, type casting metals, and storage batteries. Cadmium compounds such as cadmium oxide are toxic and can cause illness when breathed. These toxic fumes can be produced by welding, cutting, or machining on cadmium-plated parts. Breathing the fumes should be avoided by using adequate ventilation systems. The melting point of cadmium is 610°F (321°C). Its weight is 539.6 lb/ft^3.

Copper and Copper Alloys

Copper is a soft, heavy metal that has a reddish color. It has high electrical and thermal **conductivity** when pure but loses these properties to a certain extent when alloyed. It must be strain hardened when used for electric wire. Copper is very ductile and can be easily drawn into wire or tubular products. It is so soft that it is difficult to machine and it has a tendency to adhere to tools. Copper can be work hardened or hardened by solution heat treatment when alloyed with beryllium. The melting point of copper is 1981°F (1083°C). Its weight is 554.7 lb/ft^3.

BERYLLIUM COPPER Beryllium copper is an alloy of copper and beryllium that can be hardened by heat treating for making nonsparking tools and other products (Figure A2.12). Machining of this metal should be done after solution heat treatment and aging and not when it is in the annealed state. Ma-

chining or welding beryllium copper can be very hazardous if safety precautions are not followed. Machining dust or welding fumes should be removed by a heavy coolant flow or by a vacuum exhaust system, and a respirator type of face mask should be worn. The melting point of beryllium is 2345°F (1285°C), and its weight is 115 lb/ft^3.

BRASS Brass is an alloy of zinc and copper. Brass colors usually range from white to yellow or, in some alloys, red to yellow. Brasses range from gilding metal used for jewelry (95 percent copper, 5 percent zinc) to Muntz metal (60 percent copper, 40 percent zinc) used for bronzing rod and sheet stock. Brasses are easily machined. Brass is usually tougher than bronze and produces a stringy chip when machined. The melting point of brasses ranges from 1616 to 1820°F (880 to 993°C), and their weights range from 512 to 536 lb/ft^3.

BRONZE Bronze is found in many combinations of copper and other metals, but copper and tin are the original elements combined to make bronze. Bronze colors usually range from red to yellow. Phosphor bronze contains 92 percent copper, 0.05 percent phosphorus, and 8 percent zinc. Aluminum bronze is often used in the shop for making bushings or bearings that support heavy loads (Figure A2.13). (Brass is not normally used for making antifriction bushings.) The melting point of bronze is about 1841°F (1005°C) and its weight is about 548 lb/ft^3. Bronzes are usually harder than brasses but are easily machined with sharp tools. The chip produced is often granular. Some bronze alloys are used as brazing rods. Tools for brass and bronze machining should have a zero rake (straight, neither positive nor negative).

FIGURE A2.12 Beryllium copper chisel being used to cut a chip in mild steel.

FIGURE A2.13 Flanged bronze bushing.

Chromium

Chromium is a slightly gray metal that can take a high polish. It has a high resistance to corrosion by most reagents; exceptions are dilute hydrochloric and sulfuric acids. Chromium is widely used as a decorative plating on automobile parts and other products.

Chromium is not a very ductile or malleable metal and its brittleness limits its use as an unalloyed metal. It is commonly alloyed with steel to increase hardness and corrosion resistance. Chrome-nickel and chrome-molybdenum are two very common chromium alloys. Chromium is also used in electrical heating elements such as chromel or nichrome wire. The melting point of chromium is 2939°F (1615°C) and its weight is 432.4 lb/ft^3.

Die Cast Metals

Finished castings are produced with various metal alloys by the process of die casting. **Die casting** is a method of casting molten metal by forcing it into a mold. After the metal has solidified, the mold opens and the casting is ejected. Carburetors, door handles, and many small precision parts are manufactured using this process (Figure A2.14). Die cast alloys, often called **"pot metals,"** are classified in six groups:

1. Tin-base alloys
2. Lead-base alloys
3. Zinc-base alloys
4. Aluminum-base alloys
5. Copper, bronze, or brass alloys
6. Magnesium-base alloys

The specific content of the alloying elements in each of the many die cast alloys may be found in handbooks or other references on die casting.

Lead and Lead Alloys

Lead is a heavy metal that is silvery when newly cut and gray when oxidized. It has a high density, low tensile strength, low ductility (cannot be easily drawn into wire), and high malleability (can be easily compressed into a thin sheet).

Lead has a high corrosion resistance and is alloyed with antimony and tin for various uses. It is used as shielding material for nuclear and X-ray radiation, for cable sheathing, and battery plates. Lead is added to steels, brasses, and bronzes to improve machinability. Lead compounds are very toxic and they are also cumulative in the body. Small amounts ingested over a period of time can be fatal. The melting point of lead is 621°F (327°C); its weight is 707.7 lb/ft^3.

A **babbitt metal** is a soft, antifriction alloy metal often used for bearings and is usually tin or lead based (Figure A2.15). Tin babbitts usually contain from 65 to 90 percent tin with antimony, lead, and a small percentage of copper added. These are the higher grade and generally the more expensive of the two types. Lead babbitts contain up to 75 percent lead with antimony, tin, and some arsenic making up the difference.

Cadmium-base babbitts resist higher temperatures than other tin- and lead-base types. These alloys contain from 1 to 15 percent nickel or a small percentage of copper and up to 2 percent silver. The melting point of babbitt is about 480°F (249°C).

Magnesium

When pure, magnesium is a soft, silver-white metal that closely resembles aluminum but is lighter in density. In contrast to aluminum, magnesium will readily

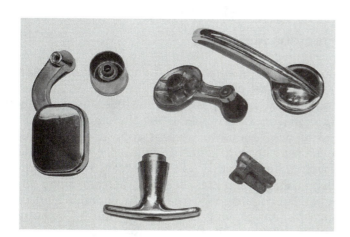

FIGURE A2.14 Die cast parts.

FIGURE A2.15 Babbitted pillow block bearings.

burn with a brilliant white light. Cast and wrought magnesium alloys are designated by SAE and ASTM numbers, which may be found in metals reference handbooks such as *Machinery's Handbook*.

Magnesium, which is similar to aluminum in density and appearance, presents some quite different machining problems because magnesium chips can burn in air, and applying water will only cause the chips to burn more fiercely. Sand or special compounds should be used to extinguish these fires and should always be on hand when machining magnesium. Thus, when working with magnesium, a water-based coolant should never be used. Magnesium can be machined dry when light cuts are taken and the heat is dissipated. Compressed air is sometimes used as a coolant. **Anhydrous** (containing no water) oils having a high flash point and low viscosity are used in most production work. Magnesium is machined with very high surface speeds and with tool angles similar to those used for aluminum. The melting point of magnesium is 1204°F (651°C), and its weight is 108.6 lb/ft³.

Molybdenum

As a pure metal, molybdenum is used for high-temperature applications and, when machined, it chips like gray cast iron. It is used as an alloying element in steel to promote deep hardening and to increase its tensile strength and toughness. Pure molybdenum is used for filament supports in lamps and in electron tubes. The melting point of molybdenum is 4748°F (2620°C). Its weight is 636.5 lb/ft³.

Nickel

Nickel is noted for its resistance to corrosion and oxidation. It is a whitish metal used for electroplating and as an alloying element in steel and other metals to increase ductility and corrosion resistance. It resembles pure iron in some ways but has a greater corrosion resistance. **Electroplating** is the coating or covering of another material with a thin layer of metal, using electricity to deposit the layer.

When spark tested, nickel throws short orange carrier lines with no sparks or sprigs (Figure A2.16). Nickel is attracted to a magnet but becomes nonmagnetic near 680°F (360°C). The melting point of nickel is 2646°F (1452°C), and its weight is 549.1 lb/ft³.

Nickel-Base Alloys

Monel is an alloy of 67 percent nickel and 28 percent copper, plus impurities such as iron, cobalt, and manganese. It is a tough, but machinable, duc-

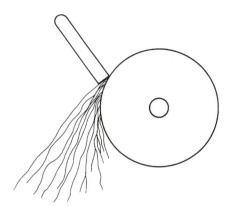

FIGURE A2.16 Nickel. Short carrier lines with no forks or sprigs. Average stream length is 10 inches, having an orange color.

tile, and corrosion-resistant alloy. Its tensile strength (resistance of a metal to a force tending to tear it apart) is 70,000 to 85,000 lb/in². Monel metal is used to make marine equipment such as pumps, steam valves, and turbine blades. On a spark test, monel shoots orange-colored, straight sparks about 10 inches long, similar to those of nickel. K-monel contains 3 to 5 percent aluminum and can be hardened by heat treatment.

Chromel and nichrome are two nickel-chromium-iron alloys used as resistance wire for electric heaters and toasters. Nickel-silver contains nickel and copper in similar proportions to monel but also contains 17 percent zinc. Other nickel alloys such as inconel are used for parts that are exposed to high temperatures for extended periods.

Inconel, a high-temperature- and corrosion-resistant metal consisting of nickel, iron, and chromium, is often used for aircraft exhaust manifolds because of its resistance to high-temperature oxidation (scaling). The nickel alloys' melting points range from 2425 to 2950°F (1329 to 1621°C).

Precious Metals

Gold has a limited industrial value and is used in dentistry, the electronic and chemical industries, and jewelry. In the past, gold has been used mostly for coinage. Gold coinage is usually hardened by alloying with about 10 percent copper. Silver is alloyed with 8 to 10 percent copper for coinage and jewelry. Sterling silver is 92.5 percent silver in English coinage and has been 90 percent silver for American coinage. Silver has many commercial uses such as an alloying element for mirrors, photographic compounds, and electrical equipment. It has a very high electrical conductivity. Silver is used in silver solders that are stronger and have a higher melting point than lead-tin solders.

Platinum, palladium, and iridium, as well as other rare metals are even more rare than gold. These metals are used commercially because of their special properties such as extremely high resistance to corrosion, high melting points, and high hardness. The melting points of some precious metals are: gold 1945°F (1063°C), iridium 4430°F (2443°C), platinum 3224°F (1773°C), and silver 1761°F (961°C). Gold has a weight of 1204.3 lb/ft³. The weight of silver is about 654 lb/ft³. Platinum is one of the heaviest of metals with a weight of 1333.5 lb/ft³. Iridium is also a heavy metal, weighing 1397 lb/ft³.

Tantalum

Tantalum is a bluish-gray metal that is difficult to machine because it is quite soft and ductile and the chip clings to the tool. It is immune to attack from all corrosive acids except hydrofluoric and fuming sulfuric acids. It is used for high-temperature operations above 2000°F (1093°C). It is also used for surgical implants and in electronics. Tantalum carbides are combined with tungsten carbides for cutting tools that have extreme wear resistance. The melting point of tantalum is 5162°F (2850°C). Its weight is 1035.8 lb/ft³.

Tin

Tin has a white color with a slightly bluish tinge. It is whiter than silver or zinc. Since tin has a good corrosion resistance, it is used to plate steel, especially for the food processing industry. Tin is used as an alloying element for solder, babbitt, and pewter. A popular solder is an alloy of 50 percent tin and 50 percent lead. Tin is alloyed with copper to make bronze. The melting point of tin is 449°F (232°C). Its weight is 454.9 lb/ft³.

Titanium

The strength and light weight of this silver-gray metal make it very useful in the aerospace industries for jet engine components, heat shrouds, and rocket parts. Pure titanium burns in air at 1156°F (610°C) and has a tensile strength of 60,000 to 110,000 PSI, similar to that of steel; by alloying titanium, its tensile strength can be increased considerably. Titanium weighs about half as much as steel and, like stainless steel, is a relatively difficult metal to machine. Machining can be accomplished with rigid setups, sharp tools, slower surface speed, and

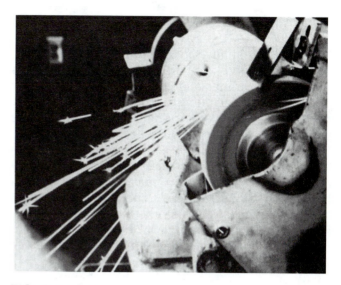

FIGURE A2.17 Titanium spark testing.

the use of proper coolants. When spark tested, titanium throws a brilliant white spark with a single burst on the end of each carrier (Figure A2.17). The melting point of titanium is 3272°F (1800°C), and its weight is 280.1 lb/ft³.

Tungsten

Typically, tungsten has been used for incandescent light filaments. It has the highest known melting point (6098°F or 3370°C) of any metal but is not resistant to oxidation at high temperatures. Tungsten is used for rocket engine nozzles and welding electrodes and as an alloying element with other metals. Machining pure tungsten is very difficult with single-point tools, and grinding is preferred for finishing operations. Tungsten carbide compounds are used to make extremely hard and heat-resistant lathe tools and milling cutters by compressing the tungsten carbide powder into a briquette and **sintering** it in a furnace. Tungsten weighs about 1180 lb/ft³.

Zinc

The familiar galvanized steel is actually steel plated with zinc and is used mainly for its high corrosion resistance. The fumes of galvanized steel should be avoided when welding. Zinc alloys are widely used as die casting metals. Zinc and zinc-based die cast metals conduct heat much more slowly than aluminum. The rate of heat transfer on similar shapes of aluminum and zinc is a means of distinguishing between them. The melting point of zinc is 787°F (419°C), and it weighs about 440 lb/ft³.

Zirconium

Zirconium is similar to titanium in both appearance and physical properties. It was once used as an explosive primer and as a flashlight powder for photography. Machining zirconium, like titanium, requires rigid setups and slow surface speeds. Pure zirconium will burn in air when heated to 958°F (500°C). Zirconium has an extremely high resistance to corrosion from acids and seawater. Zirconium alloys are used in nuclear reactors, flash bulbs, and surgical implants such as screws, pegs, and skull plates. When spark tested, it produces a spark that is similar to that of titanium. The melting point of zirconium is 3182°F (1750°C). Its weight is 399 lb/ft^3.

SELF-TEST

1. By what universal coding system is carbon steel and alloy steel designated?
2. What are three basic types of stainless steels and what is the number series assigned to them? What are their basic differences?
3. If your shop stocked the following steel shafting, how would you determine the content of an unmarked piece of each, using shop tests as given in this unit?
   a. AISI C1020 CF
   b. AISI B1140 (G and P)
   c. AISI C4140 (G and P)
   d. AISI 8620 HR
   e. AISI B1140 (ebony)
   f. AISI C1040
4. A small part has obviously been made by a casting process. How can you determine whether it is a ferrous or a nonferrous metal, or if it is steel, or white or gray cast iron?
5. What is the meaning of the symbols O1 and W1 when applied to tool steels?
6. When checking the hardness of a piece of steel with the file test, the file slides over the surface without cutting.
   a. Is the steel piece readily machinable?
   b. What type of steel is it most likely to be?
7. Steel that is nonmagnetic is called _____.
8. What nonferrous metal is magnetic?
9. List at least four properties of steel that should be kept in mind when you select the material for a job.
10. What advantage do aluminum and its alloys have over steel alloys? What disadvantages?
11. Describe the meaning of the letter *H* when it follows the four-digit number that designates an aluminum alloy. The meaning of the letter *T*.
12. Name two ways in which magnesium differs from aluminum.
13. What is the major use of copper? How can copper be hardened?
14. What is the basic difference between brass and bronze?
15. Name two uses for nickel.
16. Lead and tin have one useful property in common. What is it?
17. Molybdenum and tungsten are both used in _____ steels.
18. Babbitt metals, used for bearings, are made in what major basic types?
19. What type of metal can be injected under pressure into a permanent mold?
20. Which is stronger, cast or wrought (worked) aluminum?
21. What can be done to avoid building up an edge on the tool bit when machining aluminum?
22. Should a water-based coolant be used when machining magnesium? Explain.
23. Should the rake angles on tools for brasses and bronzes be zero, positive, or negative? Explain.
24. How is tungsten used for cutting tools?

Cutoff Machines

Sawing machines constitute some of the most important machine tools found in the machine shop. Common types of cutoff machines include reciprocating saws, horizontal band saws, universal tilt frame band saws, and abrasive saws. This unit will introduce you to safety practices and some of the many uses for these machines.

OBJECTIVES

After completing this unit, you should be able to:

1. Identify several types of metal cutting saws used in the shop and name their parts.
2. Explain the uses and advantages of cutoff saws and vertical band machines.
3. Describe correct safety habits when using metal cutting saws.

TYPES OF CUTOFF MACHINES

Reciprocating Saws

The first machine tool that you will probably encounter in the metals shop is a cutoff machine. In the shop, cutoff machines are generally found near the stock supply area. The primary function of the cutoff machine is to reduce mill lengths of bar stock material into lengths suitable for holding in other machine tools. In a large production machine shop where stock is being supplied to many machine tools, the cutoff saw will be constantly busy working on many materials. Material can also be cut with a shear; however, much of the stock used is in the form of a round bar. Mill lengths of round bar are almost always cut by sawing. Shearing is applied to materials in the form of sheets and flat bar; it produces a distorted and rough cut that is not suitable for machine shop use.

The reciprocating saw is often called a *power hacksaw.* The reciprocating saw is used in many machine shops; however, it is giving way to the band machine. The reciprocating saw is built much like the metal-cutting hand hacksaw. Basically, the machine consists of a frame that holds a blade.

Figure A3.1 shows the basic parts of a reciprocating saw. The blade is tightened with a tensioner nut or screw and the workpiece is held securely in a vise. The saw frame is mounted on a heavy base that usually holds coolant, which is pumped to the saw blade. Reciprocating hacksaw blades are wider and thicker than those used in the hand hacksaw. The reciprocating motion is provided by hydraulics or a crankshaft mechanism.

Reciprocating saws are either the hinge type, as in Figure A3.1, or the column type. The saw frame on the hinge type pivots around a single point at the rear of the machine. On the column type, both ends of the frame rise vertically. The size of a reciprocating saw is determined by the largest piece of square material that can be cut. Sizes range from about 5 by 5 in. to 24 by 24 in. Large-capacity reciprocating saws are often of the column design.

Horizontal Band Cutoff Machine

One disadvantage of the reciprocating saw is that it only cuts in one direction of the stroke. The band machine uses a steel band blade with the teeth on one edge. The horizontal band machine has a high cutting efficiency because the band is cutting at all times with no wasted motion. Band saws are the mainstay of production stock cutoff in the shop (Figure A3.2).

The major parts of the horizontal band cutoff machine are shown in Figure A3.3. Some of the parts are similar to those in the reciprocating saw: The blade is

FIGURE A3.1 Reciprocating cutoff saw. (Kasto-Racine, Inc.)

here endless—that is, it is a band—and the saw frame pivots on a base. There is a vise to hold the material being sawed and a stock length adjustment for cutting off many pieces of the same length. The blade has a tensioner and the blade guides support it near the material being sawed for maximum rigidity.

A modern band saw may be equipped with a variable speed drive. This permits the most efficient cutting speed to be selected for the material being cut. The feed rate through the material may also be varied. The size of the horizontal band machine is determined by the largest piece of square material that the machine can cut. Large-capacity horizontal band saws are designed to handle large-dimension workpieces that can weigh as much as 10 tons. With a wide variety of band types available, plus many special workholding devices, the band saw is an extremely valuable and versatile machine tool.

Universal Tilt Frame Cutoff

The universal tilt frame band saw is much like its horizontal counterpart. This machine has a vertical band blade, and the frame can be tilted from side to side (Figure A3.4). The tilt frame machine is particularly useful for making angle cuts on large structural shapes such as I-beams or pipe.

FIGURE A3.3 Horizontal band cutoff machine. (The DoAll Company)

FIGURE A3.2 This large horizontal band cutoff saw is used for production cutting. (The DoAll Company)

FIGURE A3.4 Tilt frame band saw. (The DoAll Company)

Abrasive Cutoff Machine

The abrasive cutoff machine (Figure A3.5) uses a thin, circular abrasive wheel for cutting. Abrasive saws are very fast cutting. They can be used to cut a number of nonmetallic materials such as glass, brick, and stone. The major advantages of the abrasive cutoff machine are speed and the ability to cut nonmetals. Each particle of abrasive acts as a small tooth and actually cuts a small bit of material. Abrasive saws are operated at very high speeds. Blade speed can be as high as 10,000 to 15,000 surface feet per minute.

CUTOFF MACHINE SAFETY

Reciprocating Saws

When you are preparing to use the cutoff saw, be sure that all guards around moving parts are in place before starting the machine. The saw blade must be properly installed with the teeth pointing in the right direction. Check for correct blade tension. Be sure that the width of the workpiece is less than the distance of the saw stroke. The frame will be broken if it hits the workpiece during the stroke. Be sure that the cutting speed and the rate of feed are correct for the material being cut.

When operating a saw with coolant, see that the coolant does not run on the floor during the cutting operation. This can cause an extremely dangerous slippery area around the machine tool.

Horizontal Band Saws

Recent, new regulations require that the blade of the horizontal band machine be fully guarded except at the point of cut (Figure A3.6). Make sure that the blade tensions are correct on reciprocating and band saws. Check band tensions, especially after installing a new band. New bands may stretch and loosen during their run-in period. Band teeth are sharp. When installing a new band, it should be handled with gloves. This is one of the few places that gloves may be worn around the shop. They must not be worn when operating any machine tool.

Band blades are often stored in double coils. Be careful when unwinding them, as they are under tension. The coils may spring apart and could cause an injury.

Check to see that the band is tracking properly on the wheels and in the blade guides. If a band should break, it could be ejected from the machine and cause an injury.

Make sure that the material being cut is properly secured in the workholding device. If this is a vise, be sure that it is tight. If you are cutting off short pieces of material, the vise jaw must be supported at both ends (Figure A3.7). It is not good practice to attempt to cut pieces of material that are quite short. The stock cannot be secured properly and may be pulled from the vise by the pressure of the cut (Figure A3.8). This can cause damage to the machine as well as possible injury to the operator. Stock should extend at least halfway through the vise at all times.

Many cutoff machines have a rollcase that supports long bars of material while they are being cut. The stock should be brought to the saw on a rollcase

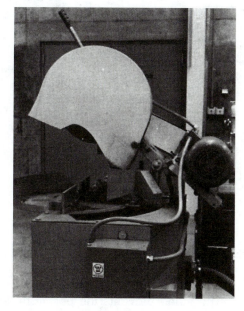

FIGURE A3.5 Typical abrasive cutoff machine.

FIGURE A3.6 The horizontal band blade is guarded except in the immediate area of the cut.

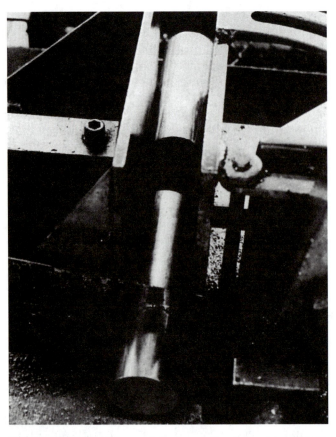

FIGURE A3.7 Support both ends of the vise when cutting short material.

FIGURE A3.8 Result of cutting stock that is too short.

(Figure A3.9) or a simple rollstand. The pieces being cut off can sometimes be several feet long and should be similarly supported. Sharp burrs left from the cutting should be removed immediately with a file. You can acquire a nasty cut by sliding your hand over one of these burrs.

Be careful around a rollcase, since bars of stock can roll, pinching fingers and hands (Figure A3.10). Also, be careful that heavy pieces of stock do not fall off the stock table or saw and injure feet or toes. Get help when lifting heavy bars of material. This will save your back and possibly your career.

On an abrasive saw, inspect the cutting edge of the blade for cracks and chips (Figure A3.11). Replace the blade if it is damaged. Always operate an abrasive saw blade at the proper **RPM** (Figure A3.12). Overspeeding the blade can cause it to fly apart. If an abrasive saw blade should fail at high speed, pieces of the blade can be thrown out of the machine at extreme velocities. A very serious injury can result if you happen to be in the path of these bulletlike projectiles.

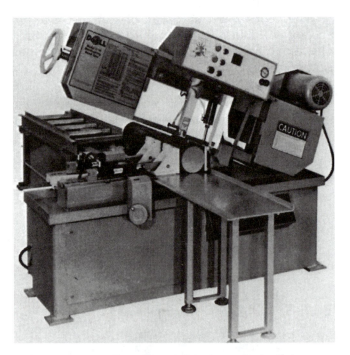

FIGURE A3.9 The material is brought into the saw on the rollcase (opposite side) and when pieces are cut off, they are supported by the stand (this side of the saw). The stand prevents the part from falling to the floor. (The DoAll Company)

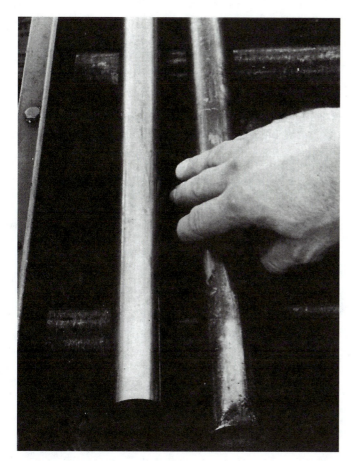

FIGURE A3.10 Stock on a roller table can pinch fingers and hands.

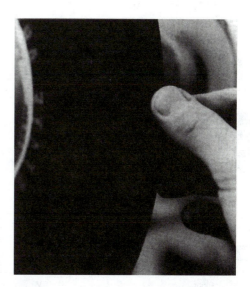

FIGURE A3.11 Inspecting the abrasive wheel for chips and cracks.

FIGURE A3.12 Abrasive wheel must be operated at the correct RPM.

Chips produced by any sawing operation should be brushed away with a suitable brush. Do not use your hands and do not use a rag as it may be caught in a moving part of the machine. A machine cannot distinguish between rags and clothing. Roll up your sleeves, keep your shirt tucked in, and if you have long hair, keep it properly secured and out of the way. Remove wristwatches and rings before operating any machine tool.

Safety checklist for sawing machines:

1. Remove wristwatches and rings and wear eye protection.
2. Properly secure the workpiece.
3. See that all guards are in place.
4. Select correct blade and inspect for chips and cracks.
5. Set blade tension correctly.
6. Set feeds and speeds accurately.
7. Adjust blade guides as close as possible to the work.
8. Remember to remove the material stop before making the cut.

MAKING THE CUT

Select the appropriate strokes-per-minute speed rate for the material being cut. Be sure to secure the workpiece properly. If you are cutting material with a sharp corner, begin the cut on a flat side if possible (Figure A3.13). When it is absolutely necessary to start a cut on a sharp corner, as on an angle, hold up on the saw frame and gently feed the saw into the work. When a flat cut is made that is wider than three saw teeth, allow the saw to continue in a normal manner. Before making the cut, go over the safety checklist. Make sure that the length of the workpiece does not exceed the

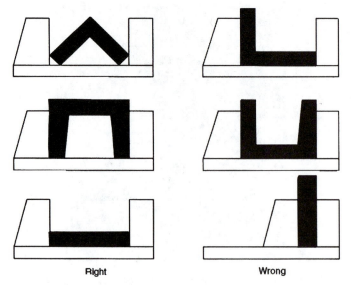

Right   Wrong

FIGURE A3.13 Cutting workpieces with sharp corners.

capacity of the stroke. This can break the frame if it should hit the workpiece. Bring the saw gently down until the blade has a chance to start cutting.

Apply the proper feed. On reciprocating saws, feed is regulated with a sliding weight or feeding mechanism. If chips produced in the cut are blue, too much feed is being used. The blade will be damaged rapidly. Very fine powderlike chips indicate too little pressure. This will dull the blade. If the blade is replaced after starting a cut, turn the workpiece over and begin a new cut. Do not attempt to saw through the old cut. This will damage the new blade. After a new blade has been used for a short time, recheck the tension and adjust if necessary.

SELF-TEST

1. List three types of cutoff machines.
2. What type uses a back-and-forth motion when cutting?
3. List three safety considerations for reciprocating and band saws.
4. List three safety considerations for abrasive saws.
5. What hazards exist around a roller stock table?
6. What can be the result of overspeeding an abrasive saw blade?

SECTION B
Dimensional Measurement

The ability to make accurate measurements on machined parts is a vital necessity for both manual and CNC machinists. The systems and measuring standards must be understood and the ability to correctly use precision measuring instruments must be learned.

Machined parts are often required to be within tolerances of plus or minus a few ten thousandths of an inch or within a hundredth of a millimeter.

Unit 1   Systems of Measurement
Unit 2   Using Micrometer Instruments and Calipers
Unit 3   Using Comparison and Angular Measuring Instruments
Unit 4   Tolerances, Fits, Geometric Dimensions, and Statistical Process Control (SPC)
Unit 5   Reading Shop Drawings

UNIT 1   Systems of Measurement

Throughout history there have been many systems of measurement. Prior to the era of national and international industrial operations, individual craftsmen were often responsible for the manufacture of a complete product. Since they made all the necessary parts and did the required assembly, they needed to conform only to their particular system of measurement. However, as machines replaced people and diversified mass production was established on a national and international basis, the need for standardization of measurement became readily apparent. Total standardization of measurement throughout the world still has not been fully realized. Most measurement in the modern world does, however, conform to either the English (inch-pound-second) or the metric (meter-kilogram-second) system. Metric measurement is now predominant in most of the industrialized nations of the world. The inch system is still used to a great extent in the United States. However, because of the interdependence of the world's industrial community, even the United States is turning more and more toward the use of metric measurement. This unit will introduce you to these systems of measurement and some methods of conversion from one system to the other.

OBJECTIVES

After completing this unit, you should be able to:

1. Identify common methods of measurement conversion.
2. Convert inch dimensions to metric equivalents and convert metric dimensions to inch equivalents.

THE ENGLISH SYSTEM OF MEASUREMENT

The English system of measurement uses the units of inches, pounds, and seconds to represent the measurement of length, mass, and time. Since we are primarily concerned with the measurement of length in the machine shop, we will simply refer to the English system as the *inch* system.

Subdivisions and Multiples of the Inch

The following table shows the common subdivisions and multiples of the inch that are used by the machinist:

Common Subdivisions	
.000001	one-millionth
.00001	one hundred-thousandth
.0001	one ten-thousandth
.001	one-thousandth
.01	one-hundredth
.1	one-tenth
1.00	one *unit inch*

In the machine shop, most common inch dimensions on a blueprint are given in thousandths or ten-thousandths. For example, $\frac{3}{8}$ in. would be expressed as .375 in.—375-thousandths of an inch. However, if the measurement were two ten-thousandths in. larger, it would be expressed as .3752 in.—3752 ten-thousandths of an inch.

Common Multiples
12.00 = 1 foot
36.00 = 1 yard

Other Common Subdivisions of the Inch are:	
$\frac{1}{128}$	.007810 (decimal equivalent)
$\frac{1}{64}$	.015625
$\frac{1}{32}$	.031250
$\frac{1}{20}$	.050000
$\frac{1}{16}$	.062500
$\frac{1}{8}$	.125000
$\frac{1}{4}$	.250000
$\frac{1}{2}$	.500000

Multiples of Feet	Multiples of Yards
3 feet = 1 yard	1760 yards = 1 mile
5280 feet = 1 mile	

See Section D, Unit 2, Table D2.1 for fractional-decimal conversion. When measuring closer than $\frac{1}{64}$ in., thousandths of an inch is the standard subdivision in the metalworking trades.

THE METRIC SYSTEM AND THE INTERNATIONAL SYSTEM OF UNITS—SI

The basic unit of length in the metric system is the meter. Originally the length of the meter was defined by a natural standard, specifically, a portion of the earth's circumference. Later, more convenient metal standards were constructed. In 1886, the metric system was legalized in the United States, but its use was not mandatory. Since 1893 the yard has been defined in terms of the metric meter by the ratio

$$1 \text{ yard} = \frac{3600}{3937} \text{ meter}$$

Although the metric system had been in use for many years in many different countries, it still lacked complete standardization among its users. Therefore, an attempt was made to modernize and standardize the metric system. From this effort has come the *Système Internationale d'Unités,* known as *SI* or the *International Metric System.*

The basic unit of length in SI is the meter or metre (in the common international spelling). The SI meter is defined by a physical standard that can be reproduced anywhere with unvarying accuracy.

1 meter = 1,650,763.73 wavelengths in a
vacuum of the orange-red light
spectrum of the Krypton-86 atom

Probably the primary advantage of the metric system is that of convenience in computation. All subdivisions and multiples use 10 as a divisor or multiplier. This can be seen in the following table:

.000001	(one-millionth meter or micrometer)
.001	(one-thousandth meter or millimeter)
.01	(one-hundredth meter or centimeter)
.1	(one-tenth meter or decimeter)
1.00	*unit meter*
10	(ten meters or one dekameter)
100	(100 meters or one hectometer)
1000	(1000 meters or one kilometer)
1,000,000	(one million meters or one megameter)

METRIC SYSTEM EXAMPLES

1. One meter (m) = _____ millimeters (mm).
   Since a mm is $\frac{1}{1,000}$ part of a m, there are 1000 mm in a meter.
2. 50 mm = _____ centimeters (cm).
   Since 1 cm = 10 mm, $\frac{50}{10}$ = 5 cm in 50 mm.
3. Four kilometers (km) = _____ m.
   Since 1 km = 1000 m then 4 km = 4000 m.
4. 582 mm = _____ cm.
   Since 10 mm = 1 cm, $\frac{582}{10}$ = 58.2 cm.

Millimeters and hundredths of a millimeter are the standard metric measure in the metalworking trades.

CONVERSION BETWEEN SYSTEMS

Much of the difficulty with working in a two-system environment is experienced in converting from one system to the other. This can be of particular concern to the machinist as he must exercise due caution in making conversions. Arithmetic errors can be easily made. Therefore the use of a calculator and metric measuring tools for metric work is recommended.

Conversion Factors and Mathematical Conversion

Since the historical evolution of the inch and metric systems is quite different, there are no obvious relationships between length units of the two systems. You simply have to memorize the basic conversion factors. We know from the preceding discussion that the yard has been defined in terms of the meter. Knowing this relationship, you can derive mathematically any length unit in either system. However, the conversion factor

$$1 \text{ yard} = \frac{3600}{3937} \text{ meter}$$

is a less common factor for the machinist. A more common factor can be determined by the following:

$$1 \text{ yard} = .91440 \text{ meter}$$
$$\left(\frac{3600}{3937} \text{ expressed in decimal form}\right)$$

Then

$$1 \text{ inch} = \frac{1}{36} \text{ of } .91440 \text{ meter}$$

So

$$\frac{.91440}{36} = .025400 \text{ meter}$$

We know that

$$1 \text{ m} = 1000 \text{ millimeters}$$

Therefore

$$1 \text{ inch} = .025400 \times 1000$$

Or

$$1 \text{ in.} = 25.4000 \text{ mm}$$

The conversion factor 1 in. = 25.4 mm is very common and should be memorized. From the example shown it should be clear that in order to find inches knowing millimeters, you must divide millimeters by 25.4.

$$1000 \text{ mm} = \underline{\hspace{1cm}} \text{ in.}$$

$$\frac{1000}{25.4} = 39.37 \text{ in.}$$

In order to simplify the arithmetic, any conversion can always take the form of a multiplication problem.

EXAMPLE

Instead of $\frac{1000}{25.4}$, multiply by the reciprocal of 25.4, which is $\frac{1}{25.4}$ or .03937. Therefore, $1000 \times .03937 = 39.37$ in.

EXAMPLES OF CONVERSIONS (INCH TO METRIC)

1. 17 in. = _____ cm.
   Knowing inches, to find centimeters multiply inches by 2.54: 2.54 × 17 in. = 43.18 cm.
2. .807 in. = _____ mm.
   Knowing inches, to find millimeters multiply inches by 25.4: 25.4 × .807 in. = 20.49 mm.

EXAMPLES OF CONVERSIONS (METRIC TO INCH)

1. .05 mm = _____ in.
   Knowing millimeters, to find inches multiply millimeters by .03937: .05 × .03937 = .00196 in.
2. 1.63 m = _____ in.
   Knowing meters, to find inches, multiply meters by 39.37: 1.63 × 39.37 m = 64.173 in.

Conversion Factors to Memorize

$$1 \text{ in.} = 25.4 \text{ mm or } 2.54 \text{ cm}$$

$$1 \text{ mm} = .03937 \text{ in.}$$

Other Methods of Conversion

The conversion chart (Figure B1.1) is a popular device for making conversions between systems. Conversion charts are readily available from many manufacturers. However, most conversion charts give equivalents for whole millimeters or standard fractional inches. If you must find an equivalent for a factor that does not appear on the chart, you must interpolate. In this instance, knowing the common conversion factors and determining the equivalent mathematically is more efficient.

Several electronic calculators designed to convert directly from system to system are available. Of course, any calculator can and should be used to do a conversion problem. The direct converting calculator does not require you to remember any conversion constant. These constants are permanently programmed into the calculator memory.

MILLIMETERS TO INCHES
(Basis: 1 inch = 25.4 millimeters)

Millimeters	Inches	Millimeters	Inches	Millimeters	Inches	Millimeters	Inches
1	0.039370	26	1.023622	51	2.007874	76	2.992126
2	.078740	27	1.062992	52	2.047244	77	3.031496
3	.118110	28	1.102362	53	2.086614	78	3.070866
4	.157480	29	1.141732	54	2.125984	79	3.110236
5	.196850	30	1.181102	55	2.165354	80	3.149606
6	.236220	31	1.220472	56	2.204724	81	3.188976
7	.275591	32	1.259843	57	2.244094	82	3.228346
8	.314961	33	1.299213	58	2.283465	83	3.267717
9	.354331	34	1.338583	59	2.322835	84	3.307087
10	.393701	35	1.377953	60	2.362205	85	3.346457
11	.433071	36	1.417323	61	2.401575	86	3.385827
12	.472441	37	1.456693	62	2.440945	87	3.425197
13	.511811	38	1.496063	63	2.480315	88	3.464567
14	.551181	39	1.535433	64	2.519685	89	3.503937
15	.590551	40	1.574803	65	2.559055	90	3.543307
16	.629921	41	1.614173	66	2.598425	91	3.582677
17	.669291	42	1.653543	67	2.637795	92	3.622047
18	.708661	43	1.692913	68	2.677165	93	3.661417
19	.748031	44	1.732283	69	2.716535	94	3.700787
20	.787402	45	1.771654	70	2.755906	95	3.740157
21	.826772	46	1.811024	71	2.795276	96	3.779528
22	.866142	47	1.850394	72	2.834646	97	3.818898
23	.905512	48	1.889764	73	2.874016	98	3.858268
24	.944882	49	1.929134	74	2.913386	99	3.897638
25	.984252	50	1.968504	75	2.952756	100	3.937008

Note: The above table is approximate: 1/25.4 0.039370078740

FIGURE B1.1 Metric conversion table. (MTI Corporation)

FIGURE B1.2 Inch/metric conversion dials for machine tools. (The Monarch Machine Tool Company, Sidney, Ohio)

Converting Machine Tools

With the increase in metric measurement in industry, which predominantly uses the inch system, several devices have been developed that permit a machine tool to function in either system. These conversion devices eliminate the need to convert all dimensions prior to beginning a job.

Conversion equipment includes conversion dials (Figure B1.2) that can be attached to lathe cross slide screws as well as milling machine saddle and table screws. The dials are equipped with gear ratios that permit a direct metric reading to appear on the dial.

Metric mechanical and electronic travel indicators can also be used. The mechanical dial travel indicator (Figure B1.3) uses a roller that contacts a moving part of a machine tool. Travel of the machine component is indicated on the dial. This type of travel indicator discriminates to .01 millimeter. Whole millimeters are counted on the 1-millimeter counting wheel. Mechanical dial travel indicators are used in many applications such as reading the travel of a milling machine saddle and table (Figure B1.4).

The electronic travel indicator uses a sensor that is attached to the machine tool. Machine tool component travel is indicated on an electronic digital display. The equipment can be switched to read travel in inch or metric dimensions.

FIGURE B1.3 Metric mechanical dial travel dial indicator. (Southwestern Industries, Inc. Trav-A-Dial® is a registered trademark of Southwestern Industries, Inc., Los Angeles, California)

FIGURE B1.4 Metric mechanical dial travel indicators reading milling machine saddle and table movement. (Southwestern Industries, Inc. Trav-A-Dial® is a registered trademark of Southwestern Industries, Inc., Los Angeles, California)

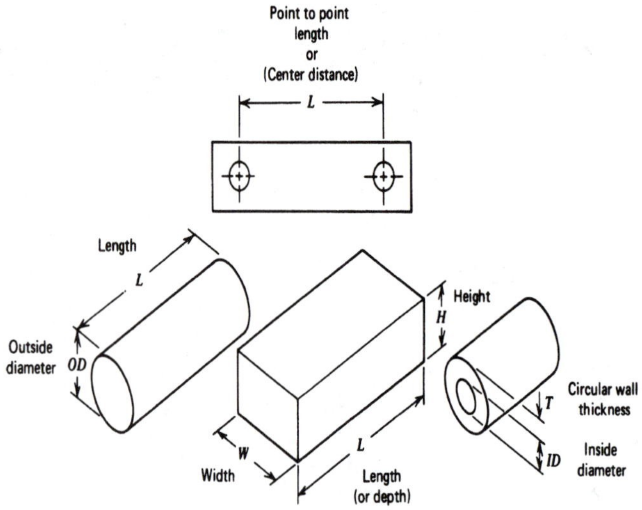

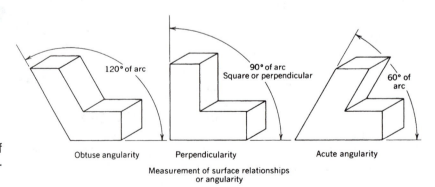

FIGURE B1.5 The measurement of length may appear under several different names.

Metric conversion devices can be fitted to existing machine tools for a moderate expense. Many new machine tools, especially those built abroad, have dual system capability built into them.

HOW THINGS ARE MEASURED

The measurement of length is the distance along a line between two points (Figure B1.5). It is also length that defines the longer or longest dimension of an object. Depth is often called length when the object is long. Width is the dimension taken at right angles to the length. Height is the distance from the bottom to the top of an object standing upright. Depth is the direct linear measurement from the point of viewing, usually from the front to back of an object or the perpendicular measurement downward from a surface.

Length is measured in such basic linear units as inches, millimeters, and in advanced metrology, wavelengths of light. In addition, measurements are sometimes made to measure the relationship of one surface to another, which is commonly called **angularity** (Figure B1.6). **Squareness,** which is closely related to angularity, is the measure of deviation from true perpendicularity. A craftsman will measure angularity in the basic units of angular measure, **degrees, minutes,** and **seconds of arc.**

In addition to the measure of length and angularity, you also need to measure such things as surface finish (Figure B1.7), concentricity, straightness, and flatness. Occasionally, you will come in contact with measurements that involve circularity, sphericity, and alignment (Figure B1.8). However, many of these more specialized measurement techniques are in the realm of the inspector or laboratory metrologist and appear infrequently in general shop work. To be reliable, measurements must be taken in line with the axis of measurement (Figure B1.9). This is

FIGURE B1.6 Measurement of surface relationships or angularity.

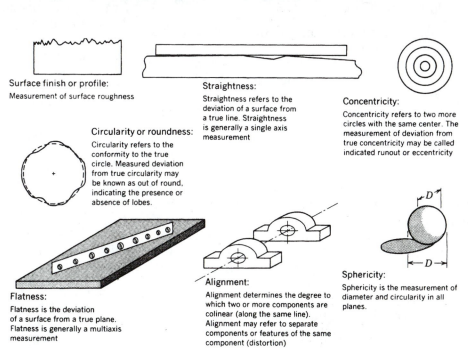

FIGURE B1.7 Visual surface finish comparator gage. (The DoAll Company)

FIGURE B1.8 Other measurements encountered by the machinist.

Surface finish or profile:
Measurement of surface roughness

Straightness:
Straightness refers to the deviation of a surface from a true line. Straightness is generally a single axis measurement

Concentricity:
Concentricity refers to two more circles with the same center. The measurement of deviation from true concentricity may be called indicated runout or eccentricity

Circularity or roundness:
Circularity refers to the conformity to the true circle. Measured deviation from true circularity may be known as out of round, indicating the presence or absence of lobes.

Flatness:
Flatness is the deviation of a surface from a true plane. Flatness is generally a multiaxis measurement

Alignment:
Alignment determines the degree to which two or more components are colinear (along the same line). Alignment may refer to separate components or features of the same component (distortion)

Sphericity:
Sphericity is the measurement of diameter and circularity in all planes.

FIGURE B1.9 The axis of a linear measuring instrument must be in line with the axis of measurement.

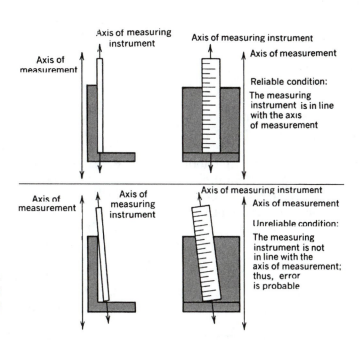

Axis of measuring instrument

Axis of measurement

Reliable condition:
The measuring instrument is in line with the axis of measurement

Axis of measuring instrument

Axis of measurement

Unreliable condition:
The measuring instrument is not in line with the axis of measurement; thus, error is probable

true for all types of measuring tools, rules, calipers, micrometers, and transfer measuring tools.

CONSIDERATIONS IN MEASUREMENT

The units in this section discuss most of the common measuring tools available to a worker. The capabilities, discrimination, and reliability as well as procedures for use are examined. It is, of course, the responsibility of a worker to select the proper measuring instrument for the job at hand. When faced with a need to measure, the following questions should be considered:

1. What degree of accuracy and precision must this measurement meet?
2. What degree of measuring tool discrimination does this required accuracy and precision demand?
3. What is the most reliable tool for this application?

Calibration

Accurate and reliable measurement places a considerable amount of responsibility on the worker, who must keep the measuring tools being used to an acceptable standard. This is the process known as **calibration.** Most of the common measuring instruments provide a factory standard. Even though calibration cannot be carried out under laboratory conditions, the instruments should at least be checked periodically against available standards.

Variables in Measuring

You should be further aware that any measurement is relative to the conditions under which it is taken. A common expression that is often used around the machine shop is: "The measurement is right on." There is, of course, little probability of obtaining a measurement that is truly exact. Each measurement has a certain degree of deviation from the theoretical exact size. This degree of error is dependent on many variables, including the measuring tool selected, the procedure used, the temperature of the part, the temperature of the room, the cleanliness of the air in the room, and the cleanliness of the part at the time of measurement. Of course, the **discrimination** of the measuring tool must be taken into account. Discrimination refers to the extent to which a unit of length has been divided. If a micrometer is divided to read thousandths of an inch, then it has a .001 in. discrimination.

The deviation of a measurement from exact size is taken into consideration by the designer. Every measurement has a **tolerance,** meaning that the measurement is acceptable within a specific range. Tolerance can be quite small depending on design requirements. When this condition exists, reliable measurement becomes more difficult because it is more heavily influenced by the many variables present. Therefore, before you make any measurement, you should stop for a moment and consider the possible variables involved. You should then consider what might be done to control as many of these variables as possible.

In practical terms, in most machine shops measurements are made to three or four decimal places in inch measure since micrometers and vernier calipers will measure to .001 in. and, if the micrometer has a vernier calibration, to .0001 in. It is common shop practice in order to avoid confusion over decimal numbers to refer to thousandths only, and add the ten thousandths. Thus, the decimal equivalent of $\frac{7}{16}$ in. = .4375 would be read "four hundred thirty-seven and five-tenths," instead of "four thousand three hundred seventy-five ten-thousandths." This would be easily confused with $\frac{3}{8}$ = .375 in.

Metric micrometers measure to .01 mm; therefore, only two decimal places are necessary when rounding off a calculation. Most metric measurements in machine shops are made in millimeters. Ten-thousandths inch measurements are somewhat finer than one-hundredth millimeter measurements; .01 mm = .0003937 in., almost four ten-thousandths of an inch, while .0001 in. = .00254 mm, a dimension smaller than a standard metric micrometer can measure.

SELF-TEST

1. Convert 35 mm into in.
2. Convert .125 in. into mm.
3. Convert 6.273 in. into cm.
4. The greatest dimension of a rectangular object is called its _____.
5. Angularity is expressed in _____, _____, and _____.
6. What is meant by calibration of a measuring instrument?
7. Express the tolerance ±.050 in. in metric terms to the nearest $\frac{1}{100}$ mm.
8. To find cm knowing mm, do you multiply or divide by 10?
9. Express the tolerance ±.02 mm in terms of inches to the nearest $\frac{1}{10,000}$ in.
10. Can an inch machine tool be converted to work in metric units?

Using Micrometer Instruments and Calipers

Micrometer measuring instruments are the most commonly used precision measuring tools found in industry. Correct use of them is essential to anyone engaged in making, inspecting, or assembling machined parts.

OBJECTIVES

After completing this unit and with the use of appropriate measuring kits, you should be able to:

1. Measure and record the dimensions of 10 objects, using outside micrometers, to an accuracy of plus or minus .001 of an inch.
2. Measure and record the diameters of five holes in test objects to an accuracy of plus or minus .001 in., using an inside micrometer.
3. Measure and record five depth measurements on a test object using a depth micrometer to an accuracy of plus or minus .001 in.
4. Measure and record the dimensions of 10 objects, using a metric micrometer, to an accuracy of plus or minus .01 mm.
5. Measure and record the dimensions of five objects, using a vernier micrometer, to an accuracy of plus or minus .0001 in. (assuming proper measuring conditions).

OUTSIDE MICROMETERS AND THEIR CARE

You should be familiar with the names of the major parts of the typical outside micrometer (Figure B2.1). The micrometer uses the movement of a precisely threaded rod turning in a nut for precision measurements. The accuracy of micrometer measurements is dependent on the quality of its construction, the care the tool receives, and the skill of the user. Consider some of the important factors in the care of the micrometer: A micrometer should be wiped clean of dust and oil before and after it is used. A micrometer should not be opened or closed by holding it by the thimble and spinning the frame around the axis of the spindle. Make sure that the micrometer is not dropped. Even a fall of a short distance can spring the frame; this will cause misalignment between the anvil and spindle faces and destroy the accuracy of this precision tool. A mi-

crometer should be kept away from chips on a machine tool. The instrument should be placed on a clean tool board (Figure B2.2) or on a clean shop towel close to where it is needed.

Always remember that the machinist is responsible for any measurements that he may make. To excuse an inaccurate measurement on the grounds that a micrometer was not properly adjusted or cared for would be less than professional. When a micrometer is stored after use, make sure that the spindle face does not touch the anvil. Perspiration, moisture from the air, or even oils promote corrosion between the measuring faces with a corresponding reduction in accuracy.

Prior to using a micrometer, clean the measuring faces. The measuring faces of many newer micrometers are made from an extremely hard metal called tungsten carbide. These instruments are often known as carbide-tipped micrometers. If you examine the measuring faces of a carbide-tipped micrometer, you will see where the carbide has

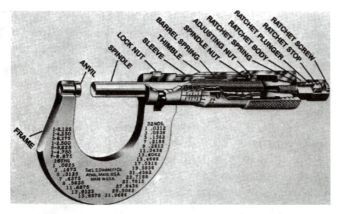

FIGURE B2.1 Parts of the outside micrometer. (The L.S. Starrett Co.)

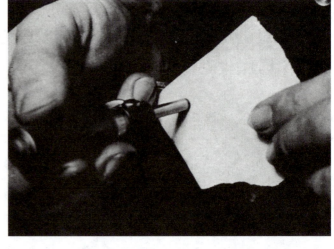

FIGURE B2.3 Cleaning the measuring faces.

been attached to the face of the anvil and spindle. Carbide-tipped micrometers have very durable and long-wearing measuring faces. Screw the spindle down lightly against a piece of paper held between it and the anvil (Figure B2.3). Slide the paper out from between the measuring faces and blow away any fuzz that clings to the spindle or anvil. At this time, you should test the zero reading of the micrometer by bringing the spindle slowly into contact with the anvil (Figure B2.4). Use the ratchet stop or friction thimble to perform this operation. The ratchet stop or friction thimble found on most micrometers is designed to equalize the gaging force. When the spindle and anvil contact the workpiece, the ratchet stop or friction thimble will slip as a predetermined amount of torque is ap-

plied to the micrometer thimble. If the micrometer does not have a ratchet device, use your thumb and index finger to provide a slip clutch effect on the thimble. Never use more pressure when checking the zero reading than when making actual measurements on the workpiece. If there is a small error, it may be corrected by adjusting the index line to the **zero point** (Figure B2.5). The manufacturer's instructions provided with the micrometer should be followed when making this adjustment. Also, follow the manufacturer's instructions for correcting a loose thimble-to-spindle connection or incorrect friction thimble or ratchet stop action. One drop of instrument oil applied to the micrometer thread at monthly intervals will help it to provide many years of reliable service.

FIGURE B2.2 Micrometers should always be kept on a tool board when used near a machine tool.

FIGURE B2.4 Checking the zero reading.

READING INCH MICROMETERS

Dimensions requiring the use of micrometers will generally be expressed in decimal form to three decimal places. In the case of the inch instrument, this would be the thousandths place. You should think in terms of thousandths whenever reading decimal fractions. For example, the decimal .156 of an inch would be read as "one hundred and fifty-six thousandths" of an inch. Likewise, .062 would be read as "sixty-two thousandths."

On the sleeve of the micrometer is a **graduated** scale with 10 numbered divisions, each one being $\frac{1}{10}$ of one inch or .100 (one hundred-thousandths) apart. Each of these major divisions is further subdivided into four equal parts, which makes the distance between these graduations $\frac{1}{4}$ of .100 or .025 (twenty-five thousandths) (Figure B2.6). The spindle screw of a micrometer has 40 threads per inch. When the spindle is turned one complete revolution, it has moved $\frac{1}{40}$ of one inch or, expressed as a decimal, .025 (twenty-five thousandths).

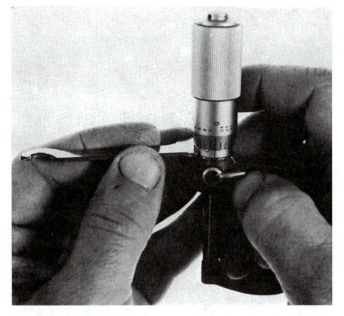

FIGURE B2.5   Adjusting the index line to zero.

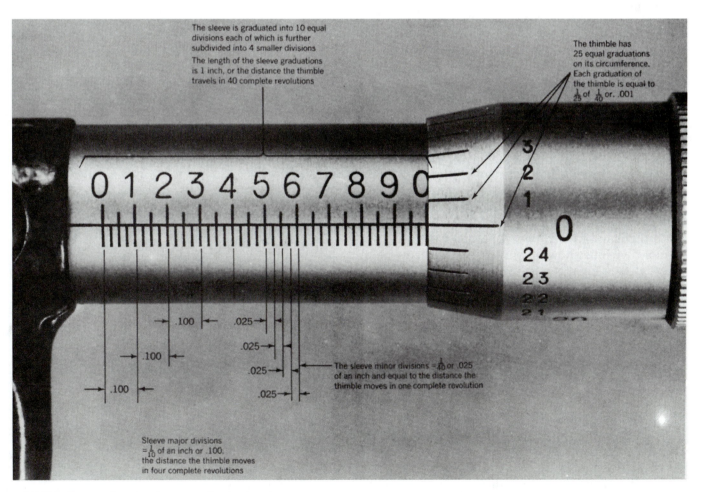

The sleeve is graduated into 10 equal divisions each of which is further subdivided into 4 smaller divisions

The length of the sleeve graduations is 1 inch, or the distance the thimble travels in 40 complete revolutions

The thimble has 25 equal graduations on its circumference. Each graduation of the thimble is equal to $\frac{1}{25}$ of $\frac{1}{40}$ or .001

The sleeve minor divisions = $\frac{1}{40}$ or .025 of an inch and equal to the distance the thimble moves in one complete revolution

Sleeve major divisions = $\frac{1}{10}$ of an inch or .100, the distance the thimble moves in four complete revolutions

FIGURE B2.6   Graduations on the inch micrometer.

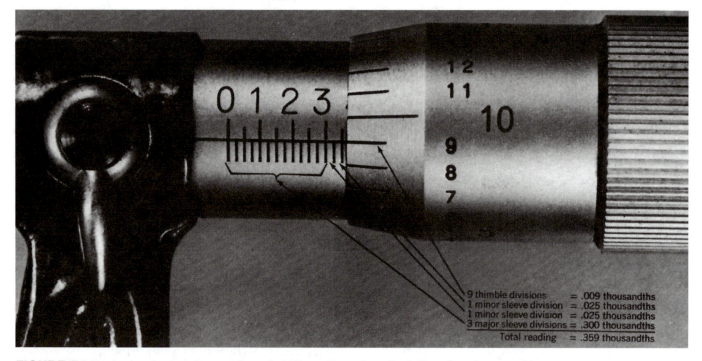

9 thimble divisions	= .009 thousandths
1 minor sleeve division	= .025 thousandths
1 minor sleeve division	= .025 thousandths
3 major sleeve divisions	= .300 thousandths
Total reading	= .359 thousandths

FIGURE B2.7 Inch micrometer reading of .359, or three hundred fifty-nine thousandths.

When we examine the thimble, we find 25 evenly spaced divisions around its circumference (Figure B2.6). Because each complete revolution of the thimble causes it to move a distance of .025 in., each thimble graduation must be equal to $\frac{1}{25}$ of .025, or .001 in. (one-thousandth). On most micrometers, each thimble graduation is numbered to facilitate reading the instrument. On older micrometers only every fifth line may be numbered.

When reading the micrometer (Figure B2.7), first determine the value indicated by the lines exposed on the sleeve. The edge of the thimble exposes three major divisions. This represents .300 in. (three hundred-thousandths). However, there are also two minor divisions showing on the sleeve. The value of these is .025, for a total of .050 in. (fifty-thousandths). The reading on the thimble is 9, which indicates .009 in. (nine-thousandths). The final micrometer reading is determined by adding the total of the sleeve and thimble readings. In the example shown (Figure B2.7), the sleeve shows a total of .350 in. Adding this to the thimble, the final reading becomes .350 in. + .009 in., or .359 in.

USING THE MICROMETER

When used to measure small parts held in the hand, the micrometer should be gripped by the frame (Figure B2.8) leaving the thumb and forefinger free to operate the thimble. When possible, take micrometer readings while the instrument is in contact with the workpiece (Figure B2.9). Use only enough pressure on the spindle and anvil to yield a reliable result. This is what is referred to as *feel*. The proper feel of a micrometer will come only from experience. Obviously, excessive pressure will not only result in an inaccurate measurement, it will also distort the frame of the micrometer and possibly damage it permanently. You should also

FIGURE B2.8 Proper way to hold a micrometer.

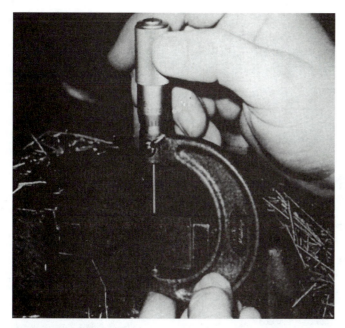

FIGURE B2.9 Read a micrometer while still in contact with the workpiece.

remember that too light a pressure on the part by the measuring faces can yield an unreliable result.

When used to measure parts in a machine, the micrometer should be held in both hands whenever possible. This is especially true when measuring cylindrical workpieces (Figure B2.10). Holding the

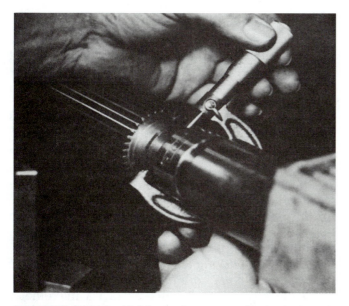

FIGURE B2.10 Hold a micrometer in both hands when measuring a round part.

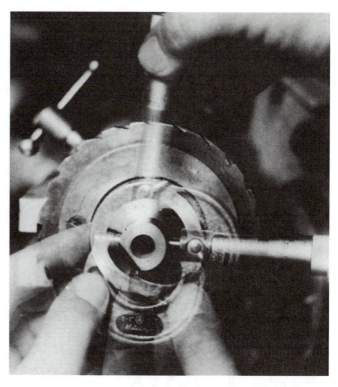

FIGURE B2.11 When measuring round parts, take two readings 90 degrees apart.

instrument in one hand does not permit sufficient control for reliable readings. Furthermore, cylindrical workpieces should be checked at least twice with measurements made 90 degrees apart. This is to check for an out-of-round condition (Figure B2.11). When critical dimensions are measured, that is, any dimension where a very small amount of tolerance is acceptable, make at least two consecutive measurements. Both readings should indicate identical results. If two identical readings cannot be determined, then the actual size of the part cannot be stated reliably. All critical measurements should be made at a temperature of 68°F (20°C). A workpiece warmer than this temperature will be larger because of heat expansion.

Outside English-measure micrometers usually have a measuring range of one inch. They are identified by size as to the largest dimensions they measure. A two-inch micrometer will measure from one to two inches. A three-inch micrometer will measure from two to three inches. The capacity of the tool is increased by increasing the size of the frame. Typical outside micrometers range in capacity from 0 to 168 in. It requires a great deal more skill to get consistent measurements with large-capacity micrometers.

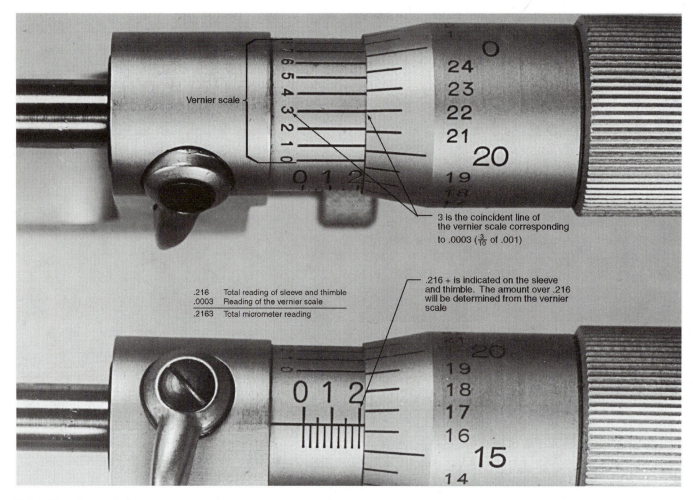

.216	Total reading of sleeve and thimble
.0003	Reading of the vernier scale
.2163	Total micrometer reading

3 is the coincident line of the vernier scale corresponding to .0003 ($\frac{3}{10}$ of .001)

.216 + is indicated on the sleeve and thimble. The amount over .216 will be determined from the vernier scale

FIGURE B2.12 Inch vernier micrometer reading of .2163 in.

READING VERNIER INCH MICROMETERS

When measurements must be made to a discrimination greater than .001 in., a standard micrometer is not sufficient. With a **vernier micrometer,** readings can be made to a ten-thousandth part of an inch (.0001 in.). This kind of micrometer is commonly known as a *tenth mike.* A vernier scale is part of the sleeve graduations. The vernier scale consists of 10 lines parallel to the index line and located above it (Figure B2.12).

If the 10 spaces on the vernier scale were compared to the spacing of the thimble graduations, the 10 vernier spacings would correspond to 9 spacings on the thimble. Therefore, the vernier scale spacing must be smaller than the thimble spacing. That is, in fact, precisely the case. Since 10 vernier spacings compare to 9 thimble spacings, the vernier spacing

is $\frac{1}{10}$ smaller than the thimble space. We know that the thimble graduations correspond to .001 in. (one-thousandth). Each vernier spacing must then be equal to $\frac{1}{10}$ of .001 in., or .0001 in. (one ten-thousandth). Thus, according to the principle of the vernier, each thousandth of the thimble is subdivided into 10 parts. This permits the vernier micrometer to discriminate to .0001 in.

To read a vernier micrometer, first read to the nearest thousandth as on a standard micrometer. Then, find the line on the vernier scale that coincides with a graduation on the thimble. The value of this coincident vernier scale line is the value in ten-thousandths, which must be added to the thousandths reading thus making up the total reading. *Remember to add the value of the vernier scale line and not the number of the matching thimble line.*

In the lower view (Figure B2.12), a micrometer reading of slightly more than .216 in. is indicated.

In the top view, on the vernier scale, the line numbered 3 is in alignment with the line on the thimble. This indicates that .0003 (three ten-thousandths) must be added to the .216 in. for a total reading of .2163. This number is read "two hundred sixteen-thousandths and three-tenths."

RELIABILITY AND EXPECTATION OF ACCURACY IN MICROMETER INSTRUMENTS

The standard inch micrometer will discriminate to .001 ($\frac{1}{1000}$) of an inch. In its vernier form, the discrimination is increased to .0001 ($\frac{1}{10,000}$) of an inch. The common metric micrometer discriminates to .01 ($\frac{1}{100}$) of a millimeter. The same rules apply to micrometers as apply to all measuring instruments. The tool should not be used beyond its discrimination. A standard micrometer with .001 discrimination should not be used in an attempt to ascertain measurements beyond that point. In order to measure to a discrimination of .0001 with the vernier micrometer, certain special conditions of accuracy must be met.

You must exercise cautious judgment when attempting to measure to a tenth of a thousandth using a vernier micrometer. The tenth measure should be carried out under controlled conditions if truly reliable results are to be obtained. The finish of the workpiece must be extremely smooth. Contact pressure of the measuring faces must be very consistent. The workpiece and instrument must be temperature stabilized. Furthermore, the micrometer must be carefully calibrated against a known standard. Only under these conditions can true reliability be realized.

The micrometer has increased reliability over the vernier. One reason for this is readability of the instruments. The .001 graduations that dictate the maximum discrimination of the micrometer are placed on the circumference of the thimble. The distance between the marks is therefore increased, making them easier to see.

The micrometer will yield very reliable results to .001 discrimination if the instrument is properly cared for, properly calibrated, and correct procedure for use is followed. **Calibration** is the process by which any measuring instrument is compared to a known standard. If the tool deviates from the standard, it may then be adjusted to conformity. This is an additional advantage of the micrometer over the vernier. The micrometer must be periodically calibrated if reliable results are to be obtained.

Can a micrometer measure reliably to within .001? The answer is "no" for the standard micrometer since it cannot discriminate to smaller units than .001 in. You cannot be sure if it is "right on" when the thimble lines do not quite match up with the index line, for example. The answer is "yes" for the vernier micrometer, but only under controlled conditions. What then, is an acceptable expectation of accuracy that will yield maximum reliability? This is dependent to some degree on the tolerance specified and can be summarized in the following table:

Tolerance Specified	Acceptability of the Standard Micrometer	Acceptability of the Vernier Micrometer
±.0001	No	Yes (under controlled conditions)
±.001	Yes	Yes (vernier will not be required)

For a specified tolerance within .001 in., the vernier micrometer should be used. Plus or minus .001 in. is a total range of .002 in. or within the capability of the standard micrometers.

The micrometer is indeed a marvelous example of precision manufacturing. These rugged tools are produced in quantity with each one conforming to equally high standards. Micrometer instruments, in all their many forms, constitute one of the fundamental measuring instruments for the machinist.

READING METRIC MICROMETERS

The metric **micrometer** (Figure B2.13) has a spindle thread with a .5 mm lead. This means that the spindle will move .5 mm when the thimble is turned one complete revolution. Two revolutions of the thimble will advance the spindle 1 millimeter. In precision machining, metric dimensions are usually expressed in terms of .01 ($\frac{1}{100}$) of a millimeter. On the metric micrometer the thimble is graduated into 50 equal divisions with every fifth division numbered (Figure B2.14). If one revolution of the thimble is .5 mm, then each division on the thimble is equal to .5 mm divided by 50 or .01 mm. The sleeve of the metric micrometer is divided into 25 main divisions above the index line with every fifth division numbered. These are whole millimeter graduations. Below the index

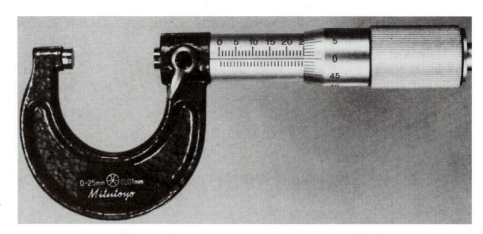

FIGURE B2.13 Metric micrometer.

line are graduations which fall halfway between the divisions above the line. The lower graduations represent half or .5 mm values. The thimble edge (Figure B2.15) leaves the 12 mm line exposed with no .5 mm line showing. The thimble reading is 32, which is .32 mm. Adding the two figures results in a total of 12.32 mm.

The 15 mm mark (Figure B2.16) is exposed on the sleeve plus a .5 mm graduation below the index line. The thimble reads 20 or .20 mm. Adding these three values, 15.00 + .50 + .20, results in a total of 15.70 mm.

Any metric micrometer should receive the same care discussed for inch micrometers.

COMBINATION METRIC/INCH OR INCH/METRIC MICROMETER

The combination micrometer is designed for dual system use in metric and inch measurement. The tool has a digital reading scale for one system while the other system is read from the sleeve and thimble.

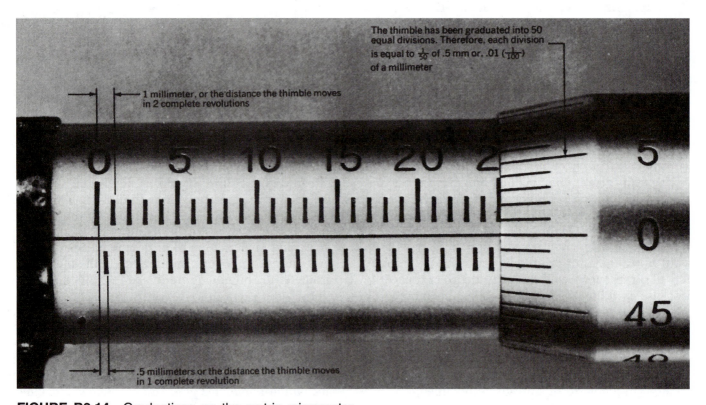

FIGURE B2.14 Graduations on the metric micrometer.

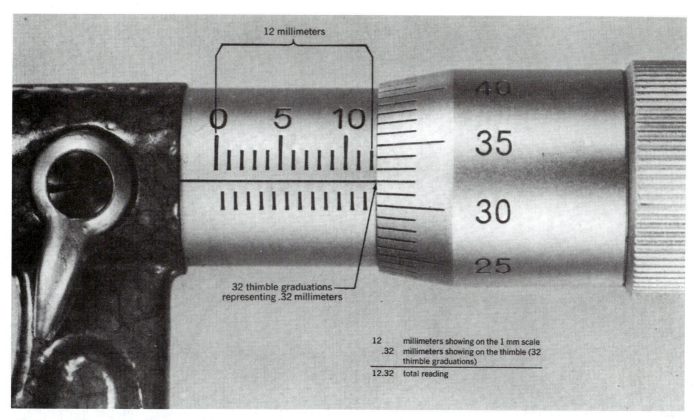

12	millimeters showing on the 1 mm scale
.32	millimeters showing on the thimble (32 thimble graduations)
12.32	total reading

FIGURE B2.15 Metric micrometer reading of 12.32 millimeters.

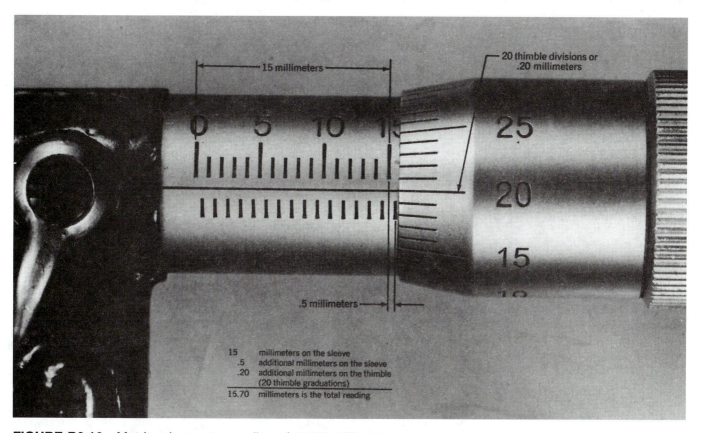

15	millimeters on the sleeve
.5	additional millimeters on the sleeve
.20	additional millimeters on the thimble (20 thimble graduations)
15.70	millimeters is the total reading

FIGURE B2.16 Metric micrometer reading of 15.70 millimeters.

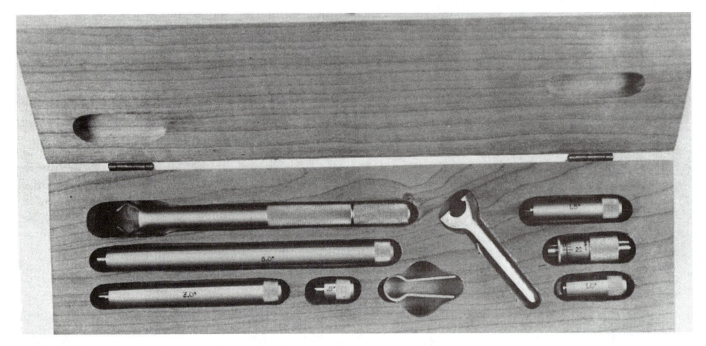

FIGURE B2.17 Tubular type inside micrometer set.

DIRECT-READING MICROMETER

The direct-reading micrometer, which may also be known as a high-precision micrometer, reads directly to .0001 $\left(\frac{1}{10,000}\right)$ of an inch.

USING INSIDE MICROMETERS

Inside micrometers are equipped with the same graduations as outside micrometers. Inside inch micrometers discriminate to .001 in. and have a measuring capacity ranging from 1.5 to 20 in. or more. A typical tubular type inside micrometer set (Figure B2.17) consists of the micrometer head with detachable hardened anvils and several tubular measuring rods with hardened contact tips. The lengths of these rods differ in increments of .5 in. to match the measuring capacity of the micrometer head, which in this case is .5 in. A handle is provided to hold the instrument into places where holding the instrument directly would be difficult. Another common type of inside micrometer comes equipped with relatively small-diameter solid rods that differ in inch increments, even though the head movement is .5 in. In this case, a .5 in. spacing collar is provided. This can be slipped over the base of the rod before it is inserted into the measuring head.

Inside micrometer heads have a range of .250, .500, 1.000, or 2.000 in. depending on the total capacity of the set. For example, an inside micrometer set with a head range of .500 in. will be able to measure from 1.500 to 12.500 in.

The measuring range of the inside micrometer is changed by attaching the extension rods. Extension rods may be solid or tubular. Tubular rods are lighter in weight and are often found in large-range inside micrometer sets. Tubular rods are also more rigid. It is very important that all parts be extremely clean when changing extension rods (Figure B2.18). Even small dust particles can affect the accuracy of the instrument.

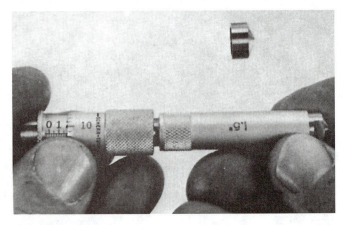

FIGURE B2.18 Attaching 1.5-in. extension rod to inside micrometer head.

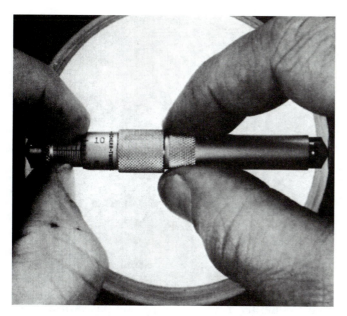

FIGURE B2.19 Placing the inside micrometer in the bore to be measured.

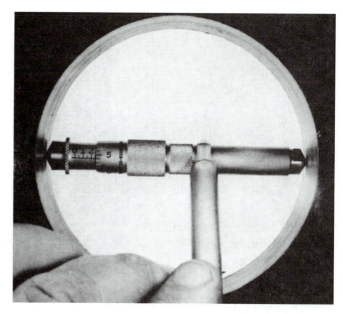

FIGURE B2.20 The inside micrometer head used with a handle.

When making internal measurements, set one end of the inside micrometer against one side of the hole to be measured (Figure B2.19). An inside micrometer should not be held in the hands for extended periods, as the resultant heat may affect the accuracy of the instrument. A handle is usually provided, which eliminates the need to hold the instrument and also facilitates insertion of the micrometer into a bore or hole (Figure B2.20). One end of the micrometer will become the center of the arcing movement used when finding the centerline of the hole to be measured. The micrometer should then be adjusted to the size of the hole. When the correct hole size is reached, there should be a very light drag between the measuring tip and the work when the tip is moved through the centerline of the hole. The size of the hole is determined by adding the reading of the micrometer head, the length of the

extension rod, and the length of the spacing collar, if one was used. Read the micrometer *while it is still in place if possible.* If the instrument must be removed to be read, the correct range can be determined by checking with a rule (Figure B2.21). A skilled craftsman will usually use an accurate outside micrometer to verify a reading taken with an inside micrometer. In this case, the inside micrometer becomes an easily adjustable transfer measuring tool (Figure B2.22). Take at least two readings 90 degrees apart to obtain the size of a hole or bore. The readings should be identical unless the bore is out-of-round. Inside micrometers do not have a spindle lock. Therefore, to prevent the spindle from turning while establishing the correct feel, the adjusting nut should be maintained slightly tighter than normal.

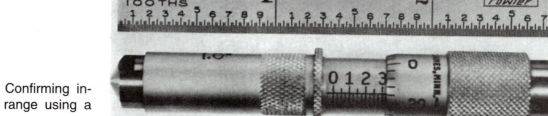

FIGURE B2.21 Confirming inside micrometer range using a rule.

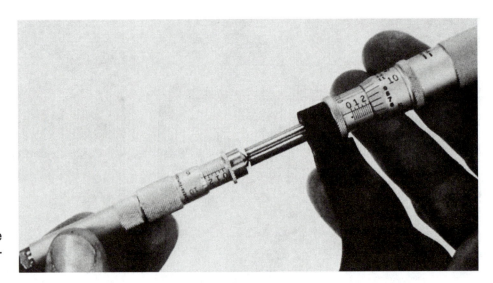

FIGURE B2.22 Checking the inside micrometer with an outside micrometer.

USING DEPTH MICROMETERS

A depth micrometer is a tool that is used to precisely measure depths of holes, grooves, shoulders, and recesses. As other micrometer instruments, it will discriminate to .001 in. Depth micrometers usually come as a set with interchangeable rods to accommodate different depth measurements (Figure B2.23). The basic parts of the depth micrometer are the base, sleeve, thimble, extension rod, thimble cap,

and frequently a ratchet stop. The bases of a depth micrometer can be of various widths. Generally, the wider bases are more stable, but in many instances space limitations dictate the use of narrower bases. Some depth micrometers are made with only a half base for measurements in confined spaces.

The extension rods are installed or removed by holding the thimble and unscrewing the thimble cap. Make sure that the seat between the thimble cap and rod adjusting nuts is clean before reassembling the micrometer. Do not overtighten when replacing the thimble cap. Furthermore, do not attempt to adjust the rod length by turning the adjusting nuts. These rods are factory adjusted and matched as a set. The measuring rods from a specific depth micrometer set should always be kept with that set. Since these rods are factory adjusted and matched to a specific instrument, transposing measuring rods from set to set will usually result in incorrect measurements.

When making depth measurements, it is very important that the micrometer base has a smooth and flat surface on which to rest. Furthermore, sufficient pressure must be applied to keep the base in contact with the reference surface. When a depth micrometer is used without a ratchet, a slip clutch effect can be produced by letting the thimble slip while turning it between the thumb and index finger (Figure B2.24).

READING INCH-DEPTH MICROMETERS

When a comparison is made between the sleeve of an outside micrometer and the sleeve of a depth micrometer, note that the graduations are numbered in the opposite direction (Figure B2.25). When reading a depth micrometer, the distance to

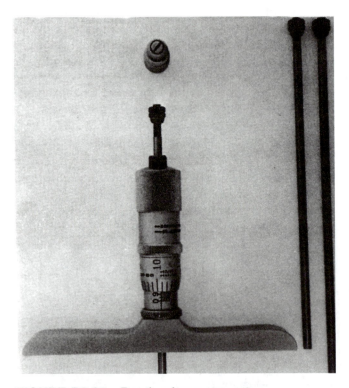

FIGURE B2.23 Depth micrometer set.

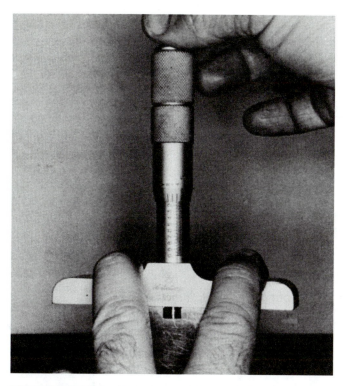

FIGURE B2.24 Proper way to hold the depth micrometer.

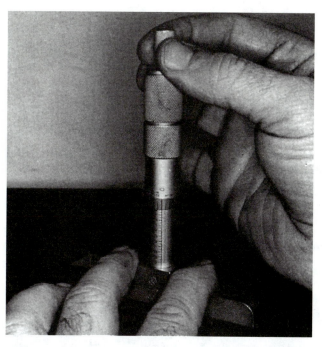

FIGURE B2.26 Checking a depth micrometer for zero adjustment using the surface plate as a reference surface.

be measured is the value covered by the thimble. Consider the reading shown (Figure B2.25). The thimble edge is between the numbers 5 and 6. This indicates a value of at least .500 in. on the sleeve

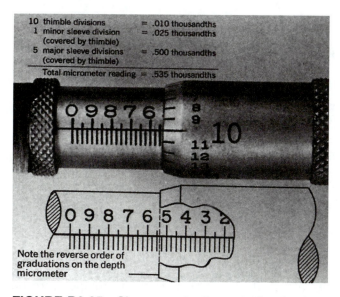

10 thimble divisions	= .010 thousandths
1 minor sleeve division (covered by thimble)	= .025 thousandths
5 major sleeve divisions (covered by thimble)	= .500 thousandths
Total micrometer reading	= .535 thousandths

Note the reverse order of graduations on the depth micrometer

FIGURE B2.25 Sleeve graduations on the depth micrometer are numbered in the opposite direction as compared to the outside micrometer.

major divisions. The thimble also covers the first minor division on the sleeve. This has a value of .025 in. The value on the thimble circumference indicates .010 in. Adding these three values results in a total of .535 in., or the amount of extension of the rod from the base.

A depth micrometer should be tested for accuracy before it is used. When the 0- to 1-in. rod is used, retract the measuring rod into the base. Clean the base and contact surface of the rod. Hold the micrometer base firmly against a flat, highly finished surface, such as a surface plate, and advance the rod until it contacts the reference surface (Figure B2.26). If the micrometer is properly adjusted, it should read zero. When testing for accuracy with the 1-in. extension rod, set the base of the micrometer on a 1-in. gage block and measure to the reference surface (Figure B2.27). Other extension rods can be tested in a like manner.

DIAL CALIPERS

The dial caliper (Figure B2.28) is a very popular measuring instrument. This instrument has a discrimination of .001 in. and can be used for outside, inside, and depth measurement.

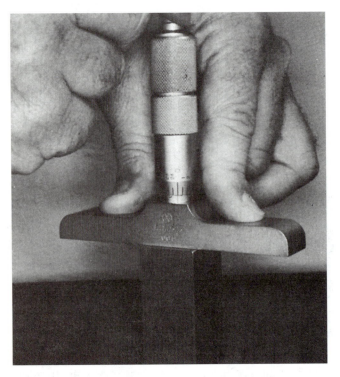

FIGURE B2.27 Checking the depth micrometer calibration at the 1.000-in. position using the 0–1-in. rod and a 1-in. square or "Hoke type" gage block.

FIGURE B2.28 Dial caliper.

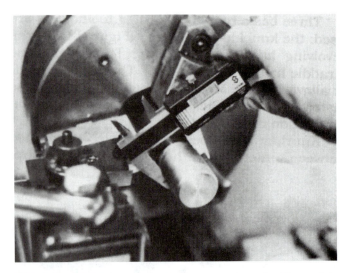

FIGURE B2.29 Groove being measured with a sliding caliper with digital readout.

Though a useful tool, the dial caliper is somewhat delicate and cannot withstand rough usage. It is generally being replaced by sliding digital calipers (Figure B2.29). These tools are very versatile in that they can measure narrow grooves and make both inside and outside measurements in increments of .001 in. They can also measure depth at the end of the beam. Both of these measuring tools have replaced vernier calipers, which are difficult to read.

SELF-TEST

1. Why should a micrometer be kept clean and protected?
2. Why should a micrometer be stored with the spindle out of contact with the anvil?
3. Why are the measuring faces of the micrometer cleaned before measuring?
4. How precise is the standard inch micrometer?
5. What affects the accuracy of a micrometer?
6. What is the difference between the sleeve and thimble?
7. Why should a micrometer be read while it is still in contact with the object to be measured?
8. How often should an object be measured to verify its actual size?
9. What effect has an increase in temperature on the size of a part?
10. What is the purpose of the friction thimble or ratchet stop on the micrometer?

Exercise

Read and record the five outside micrometer readings in Figures B2.30a to B2.30e.

Figure B2.30a _____   Figure B2.30d _____
Figure B2.30b _____   Figure B2.30e _____
Figure B2.30c _____

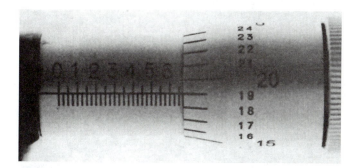

FIGURE B2.30a

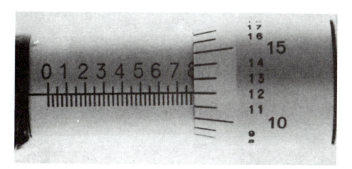

FIGURE B2.30b

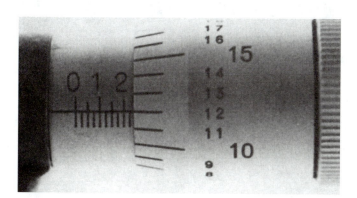

FIGURE B2.30c

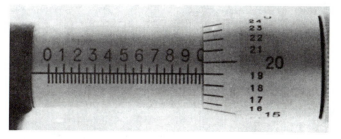

FIGURE B2.30d

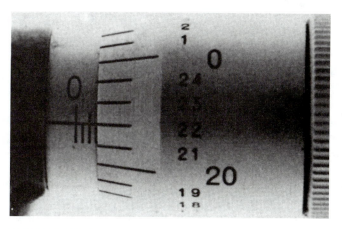

FIGURE B2.30e

SELF-TEST

Read and record the five inside micrometer readings (Figures B2.31a to B2.31e). Micrometer head is 1.500 in. when zeroed.

Figure B2.31a _____

Figure B2.31b _____

Figure B2.31c _____ (.5-in. extension)

Figure B2.31d _____ (1.0-in. extension)

Figure B2.31e _____ (1.0-in. extension)

Exercise

Obtain an inside micrometer set from your instructor and practice using the instrument on objects around your laboratory. Measure examples such as lathe spindle holes, bushings, bores of roller bearings, hydraulic cylinders, and tubing.

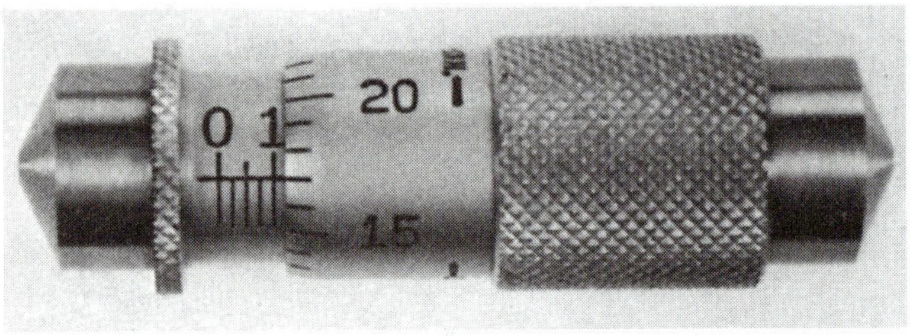

FIGURE B2.31c   .5-in. extension.

FIGURE B2.31a

FIGURE B2.31d   1.0-in. extension.

FIGURE B2.31b

FIGURE B2.31e   1.0-in. extension.

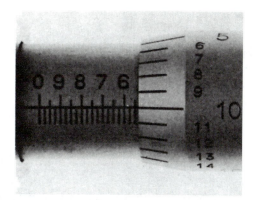

FIGURE B2.32a

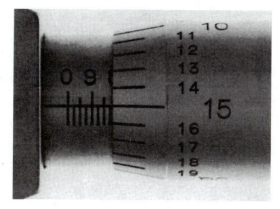

FIGURE B2.32b

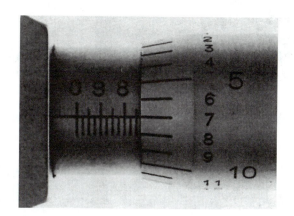

FIGURE B2.32c

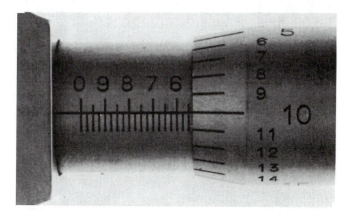

FIGURE B2.32d

SELF-TEST

Read and record the five depth micrometer readings in Figures B2.32a to B2.32e.

Figure B2.32a _____

Figure B2.32b _____

Figure B2.32c _____

Figure B2.32d _____

Figure B2.32e _____

Exercise

Obtain a depth micrometer from your instructor and practice measuring objects in your shop.

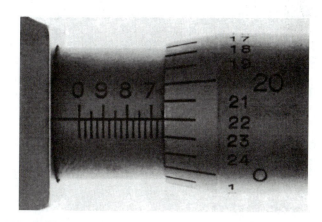

FIGURE B2.32e

Using Comparison and Angular Measuring Instruments

As a machinist or CNC machine operator you will often use a number of measuring instruments that have no capacity within themselves to show a measurement. These tools will be used in comparison measurement applications where they are compared to a known standard or used in conjunction with an instrument that has the capability to show a measurement. In this unit you are introduced to the principles of comparison measurement, the common tools of comparison measurement and their applications, and the use of some mechanical dial and angular measuring instruments.

OBJECTIVES

After completing this unit, you should be able to:

1. Define comparison measurement.
2. Identify common comparison measuring tools.
3. Given a measuring situation, select the proper comparison tool for the measuring requirement.
4. Identify common angular measuring tools.

MEASUREMENT BY COMPARISON

All of us, at some time, were probably involved in constructing something in which we used no measuring instruments of any kind. For example, suppose that you had to build some wooden shelves. You have the required lumber available with all boards longer than the shelf spaces. You hold a board to the shelf space and mark the required length for cutting. By this procedure, you have *compared* the length of the board (*the unknown length*) to the shelf space (*the known length or standard*). After cutting the first board to the marked length, it is then used to determine the lengths of the remaining shelves. The board, in itself, has no capacity to show a measurement. However, in this case, it became a measuring instrument.

A great deal of comparison measurement often involves the following steps:

1. A device that has no capacity to show measurement is used to establish and represent an unknown distance.

2. This representation of the unknown is then *transferred* to an instrument that has the capability to show a measurement.

This is commonly known as *transfer measurement*. In the example of cutting shelf boards, the shelf space was transferred to the first board, and then the length of the first board was transferred to the remaining boards.

Transfer of measurements may involve some reduction in reliability. This factor must be kept in mind when using comparison tools requiring that a transfer be made. Remember that an instrument with the capability to show measurement directly is always best.

GAGES

Telescoping Gages

Telescoping gages are widely used in the machine shop, and they can accomplish a variety of measuring requirements.

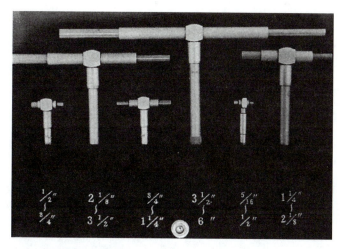

FIGURE B3.1 Set of telescoping gages.

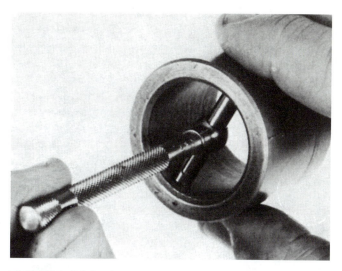

FIGURE B3.2 Inserting the telescoping gage into the bore.

Telescoping gages generally come in a set of six (Figure B3.1). The range of the set is usually $\frac{5}{16}$ to 6 in. (8 to 150 mm). The gage consists of two telescoping plungers (some types have only one) with a handle and locking screw. The gage is inserted into a bore or slot, and the plungers are permitted to extend, thus conforming to the size of the feature. The gage is then removed and transferred to a micrometer where the reading is determined. The telescoping gage can be a reliable and versatile tool if proper procedure is used in its application.

Procedure for Using the Telescoping Gage

Step 1. Select the proper gage for the desired measurement range.

Step 2. Insert the gage into the bore to be measured and release the handle lock screw (Figure B3.2). Rock the gage sideways to insure that you are measuring at the full diameter (Figure B3.3). This is especially important in large diameter bores.

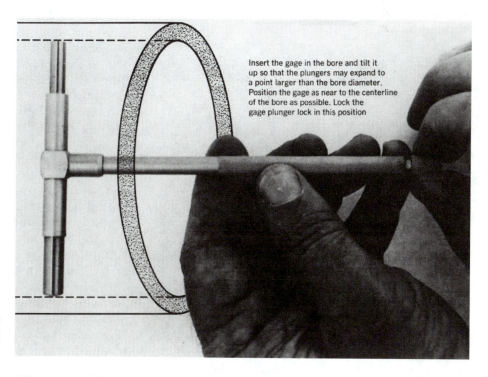

Insert the gage in the bore and tilt it up so that the plungers may expand to a point larger than the bore diameter. Position the gage as near to the centerline of the bore as possible. Lock the gage plunger lock in this position

FIGURE B3.3 Release the lock and let the plungers expand larger than the bore.

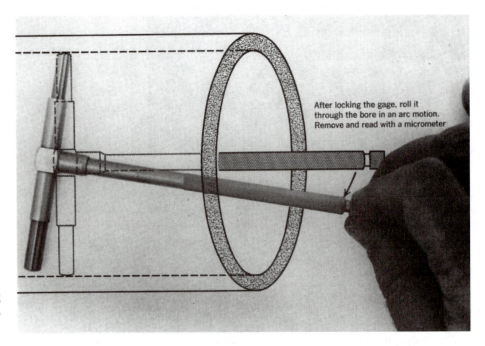

After locking the gage, roll it through the bore in an arc motion. Remove and read with a micrometer

FIGURE B3.4 Tighten the lock and roll the gage through the bore.

Step 3. Lightly tighten the locking screw.

Step 4. Using a downward or upward motion, move the gage through the diameter of the bore. The plungers will be pushed in, thus conforming to the bore diameter (Figure B3.4). Tighten the locking screw firmly.

Step 5. Remove the gage and measure with an outside micrometer (Figure B3.5). Place the gage between the micrometer spindle and anvil. Try to deter-

mine the same feel on the gage with the micrometer as you felt while the gage was in the bore. Excessive pressure with the micrometer will depress the gage plungers and cause an incorrect reading.

Step 6. Take at least two readings or more with the telescoping gage in order to verify reliability. If the readings do not agree, repeat Steps 2 to 6.

Small Hole Gages

Small hole gages, like telescoping gages come in sets with a range of $\frac{1}{8}$ to $\frac{1}{2}$ in. (4 to 12 mm). One type of small hole gage consists of a split ball that is connected to a handle (Figure B3.6). A tapered rod is drawn between the split ball halves causing them to extend and contact the surface to be measured (Figure B3.7). The split ball small hole gage has a flattened end so that a shallow hole or slot may be measured. After the gage has been expanded in the feature to be measured, it should be moved back and forth and rotated to determine the proper feel (Figures B3.8 and B3.9). The gage is then removed and measured with an outside micrometer (Figure B3.10).

A second type of small hole gage consists of two small balls that can be moved out to contact the surface to be measured. This type of gage is available in a set ranging from $\frac{1}{16}$ to $\frac{1}{2}$ in. (1.5 to 12 mm) (Figure B3.11). Once again, the proper feel must be obtained when using this type of small hole gage

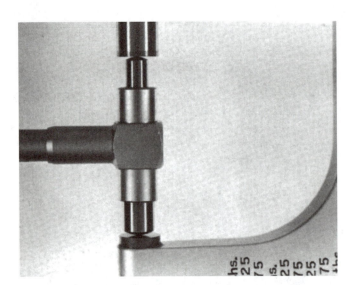

FIGURE B3.5 Checking the telescoping gage with an outside micrometer.

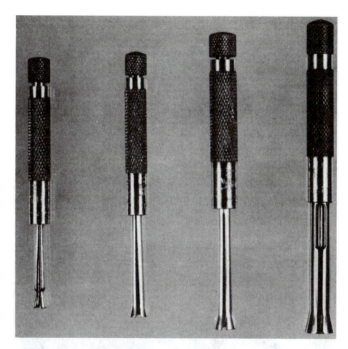

FIGURE B3.6 Set of small hole gages.

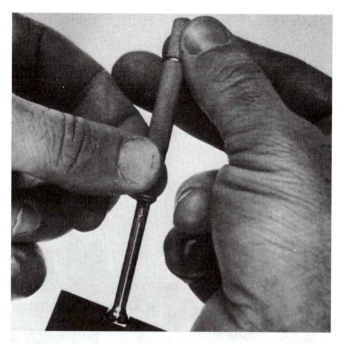

FIGURE B3.8 Rock the gage to one side to determine the proper feel.

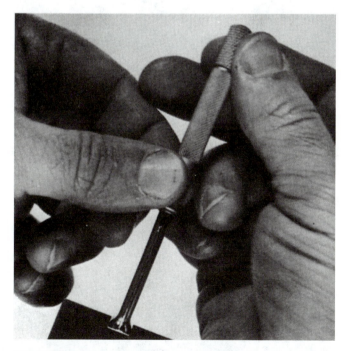

FIGURE B3.7 Insert the small hole gage in the slot to be measured. Rotate to locate large diameter of gage.

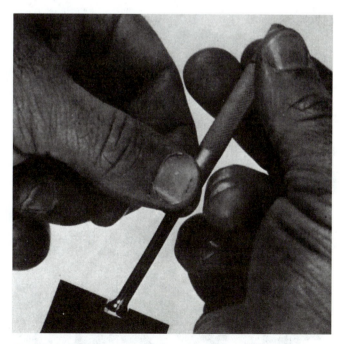

FIGURE B3.9 Rock the gage to the other side to determine proper feel.

FIGURE B3.10 Withdraw the gage and measure with an outside micrometer. Rotate to locate large diameter of gage.

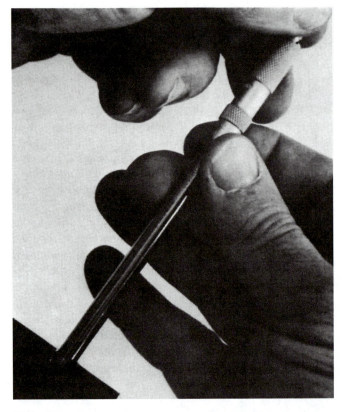

FIGURE B3.12 Using the twin ball small hole gage.

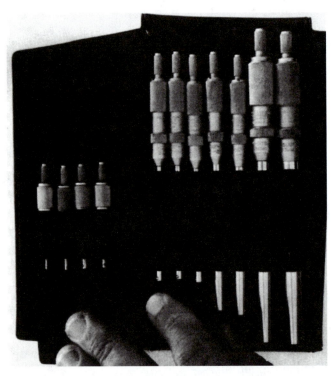

FIGURE B3.11 Twin ball small hole gage set.

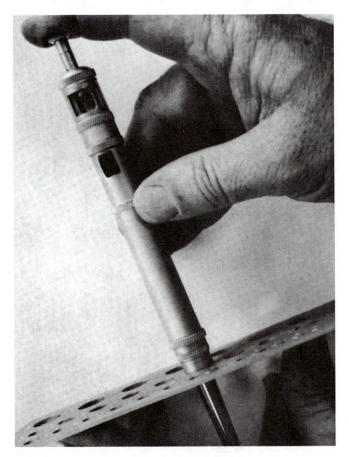

FIGURE B3.13 Direct reading small hole gage.

(Figure B3.12). After the gage is set, it is removed and measured with a micrometer.

Another type of small hole gage is the direct reading type (Figure B3.13). This is not a comparison instrument as it has the capability to display the measurement directly. The gage will discriminate to .001 of an inch and is read from the dial in the handle (Figure B3.14).

Adjustable Parallels

For the purpose of measuring slots, grooves, and keyways, the adjustable parallel may be used. Adjustable parallels are available in sets ranging from about $\frac{3}{8}$ to $1\frac{1}{2}$ in. (10 to 38 mm). They are precision ground for accuracy.

The typical adjustable parallel consists of two parts that slide together on an angle. Adjusting screws are provided so that clearance in the slide may be adjusted or the parallel locked after setting for a measurement. As the halves of the parallel slide, the width increases or decreases depending on direction. The parallel is placed in the groove or slot to be measured and expanded until the parallel edges conform to the width to be measured. The parallel is then locked with a small screwdriver and measured with a micrometer (Figure B3.15). If possible, an adjustable parallel should be left in place while being measured.

Radius Gages

The typical radius gage set ranges in size from $\frac{1}{32}$ to $\frac{1}{2}$ in. (.8 to 12 mm). Larger radius gages are also available. The gage can be used to check the radii of grooves and external or internal fillets (rounded corners). Radius gages may be separate (Figure B3.16) or the full set may be contained in a convenient holder (Figure B3.17).

Thickness Gages

The thickness gage (Figure B3.18) is often called a **feeler gage.** It is probably best known for its various automotive applications. However, a machinist

FIGURE B3.14   Reading the small hole gage.

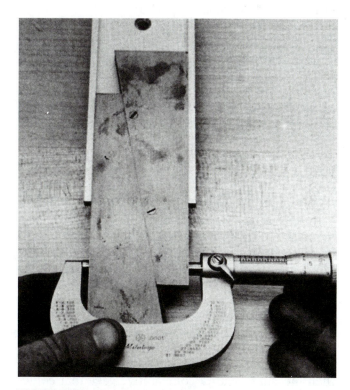

FIGURE B3.15   Using adjustable parallels.

FIGURE B3.16 Using an individual radius gage.

FIGURE B3.17 Radius gage set.

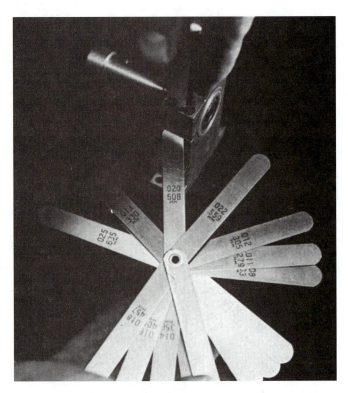

FIGURE B3.18 Using a feeler or thickness gage.

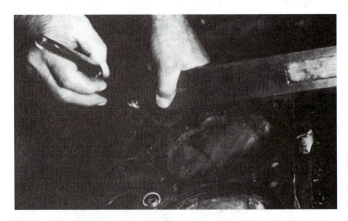

FIGURE B3.19 Straightedge and feeler gage being used to check for flatness on an engine block.

may use a thickness gage for such measurements as the thickness of a shim, setting a grinding wheel above a workpiece, or determining the height difference of two parts. The thickness gage is not a true comparison measuring instrument, as each leaf is marked for size. However, it is good practice to check a thickness gage with a micrometer, especially when a number of leaves are stacked together. The feeler gage is often used together with a precision **straightedge** to determine flatness. A common example is in checking an engine head or block for warpage (Figure B3.19). Another thickness gage, sometimes called a plasti-gage, is used by automotive mechanics to check for crankshaft bearing clearance. The gage is used only once and is placed in the bearing cap, which is then tightened in place. When it is removed, the gage has compressed to the amount of clearance and it is then measured with a comparative gage that comes with the plasti-gage.

SQUARES

The square is an important and useful tool for the machinist. A square is a comparative measuring instrument in that it compares its own degree of perpendicularity (or squareness) with an unknown degree of perpendicularity on the workpiece.

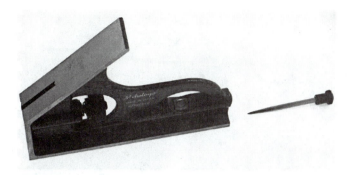

FIGURE B3.20 Combination square head with scriber.

Machinist's Combination Square

The combination square (Figure B3.20) is part of the combination set (Figure B3.21). The combination set consists of a graduated rule, square head, bevel protractor, and center head. The square head slides on the graduated rule and can be locked at any position (Figure B3.22). This feature makes the tool useful for layout because the square head can be set according to the rule graduations. The combination square head also has a 45-degree angle along with a spirit level and layout **scriber.** The

combination set is one of the most versatile tools of the machinist.

INDICATORS

The many types of dial indicators are some of the most valuable and useful tools for the machinist or machine operator. There are two general types of indicators in general use. These are *dial indicators* (see Figure B3.23) and *dial test indicators* (see Figure B3.30 on page 67). Both types generally take the form of a spring-loaded spindle that, when depressed, actuates the hand of an indicating dial. At the initial examination of a dial or test indicator, you will note that the dial face is usually graduated in thousandths of an inch or subdivisions of thousandths. This might lead you to the conclusion that the indicator spindle movement corresponds directly to the amount shown on the indicator face. However, this conclusion is to be arrived at only with the most cautious judgment. *Dial test indicators should not be used to make direct linear measurements.* Reasons for this will be developed in the following sections. Dial indicators can be used to make linear measurements, but only if they are specifically designed to do so and under proper conditions.

FIGURE B3.21 Machinist's combination set.

FIGURE B3.22 Using the combination square.

Dial Indicators

Dial indicators have discriminations that typically range from .0001 to .001 in. In metric dial indicators, the discriminations typically range from .002 to .01 mm. Indicator ranges (the total reading capacity) of the instrument may commonly range from .003 to 2.000 in., or .2 to 50 mm for metric instruments. On the "balanced" indicator (Figure B3.23), the face

numbering goes both clockwise and counterclockwise from zero. This is convenient for comparator applications where readings above and below zero need to be indicated. The indicator shown has a lever-actuated stem. This permits the stem to be retracted away from the workpiece if desired.

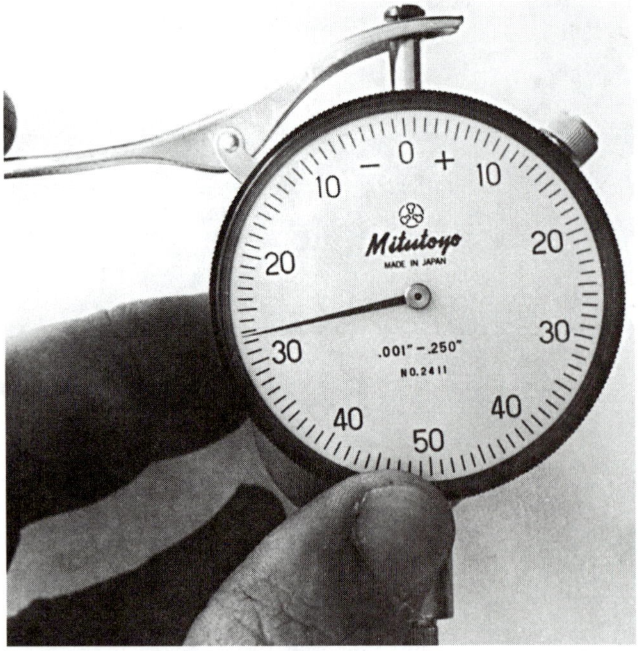

The continuous reading indicator (Figure B3.24) is numbered from zero in one direction. This indicator has a discrimination of .0005 and a total range of 1 inch. The small center hand counts revolutions of the large hand. Note that the center dial counts each .100 in. of spindle travel. This indicator is also equipped with tolerance hands that can be set to mark a desired limit. Many dial indicators are designed for high discrimination and short range (Figure B3.25). This indicator has a .0001 in. discrimination and a range of .025 in.

The "back plunger" indicator (Figure B3.26) has the spindle in the back or at right angles to the face. This type of indicator usually has a range of about .200 in. with .001 in. discrimination. It is a very popular model for use on a machine tool. The indicator usually comes with a number of mounting accessories.

Indicators are equipped with a rotating face or **bezel.** This feature permits the instrument to be set to zero at any desired place. Many indicators also have a bezel lock. Dial indicators may have removable spindle tips, thus permitting use of different shaped tips as required by the specific application (Figure B3.27).

FIGURE B3.23 Balanced dial indicator.

FIGURE B3.24 Dial indicator with 1-inch travel.

FIGURE B3.25 Dial indicator with .025 in. range and .0001 in. discrimination.

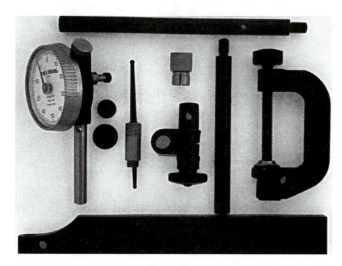

FIGURE B3.26 Back plunger indicator with mounting accessories.

FIGURE B3.27 Dial indicator tips with holder. (Rank Scherr-Tumico Inc.)

Care and Use of Indicators

Dial indicators are precision instruments and should be treated accordingly. They *must not be dropped* and should *not be exposed to severe shocks.* Dropping an indicator may bend the spindle and render the instrument useless. Shocks, such as hammering on a workpiece while an indicator is still in contact, may damage the delicate operating mechanism. The spindle should be kept free from dirt and grit, which can cause binding that results in damage and false readings. It is important to *check* indicators *for free travel* before using. When an indicator is not in use, it should be stored carefully with a protective device around the spindle.

One of the problems encountered by indicator users is *indicator mounting.* All indicators must be *mounted solidly* if they are to be reliable. Indicators must be clamped or mounted securely when used on a machine tool. A number of mounting devices are in common use. Some of these have magnetic bases that permit an indicator to be attached at any convenient place on a machine tool. The permanent magnet indicator base (Figure B3.28) is a useful accessory. This type of indicator base is equipped with an adjusting screw that can be used to set the instrument to zero. Another useful magnetic base has a provision for turning off the magnet by mechanical means (Figure B3.29). This feature makes for easy location of the base prior to turning on the magnet. A number of bases making use of flexible link indicator holding arms are also in general use. Often they are not adequately rigid for reliability. In addition to attaching an indicator to a magnetic base, it may be clamped to a machine setup by the use of any suitable clamps.

Dial Test Indicators

Dial test indicators frequently have a discrimination of .0005 in. and a range of about .030 in. The test indicator is frequently quite small (Figure B3.30) so

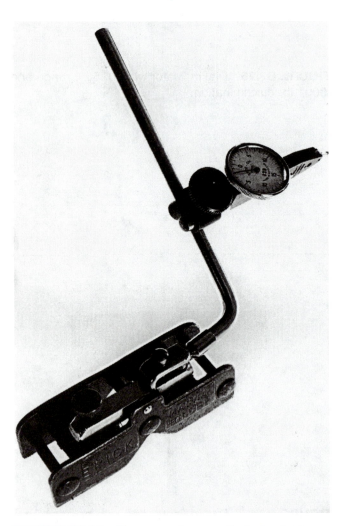

FIGURE B3.28 Permanent magnetic indicator base.

FIGURE B3.29 Magnetic base indicator holder with on/off magnet. (The L.S. Starrett Company)

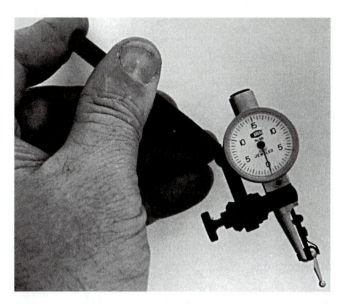

FIGURE B3.30 Dial test indicator.

that it can be used to indicate in locations inaccessible to other indicators. The spindle or tip of the test indicator can be swiveled to any desired position. Test indicators are usually equipped with a movement reversing lever. This means that the indicator can be actuated by pressure from either side of the tip. The instrument need not be turned around. Test indicators, like dial indicators, have a rotating bezel for zero setting. Dial faces are generally of the balanced design. The same care given to dial indicators should be extended to test indicators.

POTENTIAL FOR ERROR IN USING DIAL INDICATORS Indicators must be used with appropriate caution if reliable results are to be obtained. The spindle of a dial indicator usually consists of a gear rack that engages a pinion and other gears that drive the indicating hand. In any mechanical device, there is always some clearance between the moving parts. There are also minute errors in the machining of the indicator parts. Because of these, small errors may creep into an indicator reading. This is especially true in long-travel indicators. For example, if a 1-inch travel indicator with .001 in. discrimination had plus or minus 1 percent error at full travel, the following error could occur if the instrument were to be used for a direct measurement.

Say you wish to determine if a certain part is within the tolerance of .750 ± .003 in. The 1-inch travel indicator has the capacity for this; but remember, it is only accurate to plus or minus 1 percent of full travel. Therefore, .01 × 1.000 in. is equal to ± .010 in. or the total possible error. To calculate the error per thousandth of indicator travel, divide .010 in. by 1000. This is equal to .00001 in., which is the average error per thousandth of indicator travel. This means that at a travel amount of .750 in., the indicator error could be as much as .00001 in. × 750, or ± .0075 in. In a direct measurement of the part, the indicator could read anywhere from .7425 to

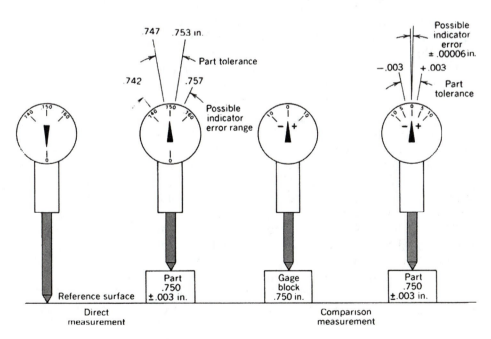

FIGURE B3.31 Potential for errors in indicator travel.

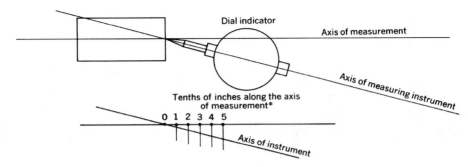

FIGURE B3.32 Misalignment of the dial indicator will cause a measuring error.

.7575. As you can see, this is well outside the part tolerance and would hardly be reliable.

The indicator should be used as a comparison measuring instrument according to the following procedure (Figure B3.31). The indicator is set to zero on a .750-in. gage block. The part to be measured is then placed under the indicator spindle. In this case, the error caused by a large amount of indicator travel is greatly reduced, because the travel is never greater than the greatest deviation of a part from the basic size. The total part tolerance is .006 in. (± .003 in.). Therefore, 6 × .00001 in. error per thousandth is equal to only ± .00006 in. This is well within the part tolerance and, in fact, cannot even be read on a .001-in. discrimination indicator.

Of course, you will not know what the error amounts to on any specific indicator. This can only be determined by a calibration procedure. Furthermore, you would probably not use a long-travel indicator in this particular application. A moderate- to short-travel indicator would be more appropriate. Keep in mind that any indicator may contain some travel error and that by using only a fraction of that travel, this error can be reduced considerably.

The axis of a linear measurement instrument must be in line with the axis of measurement. If a dial indicator is misaligned with the axis of measurement, you can see that an error in measurement will result (Figure B3.32). Even if a dial indicator is only being used to check for eccentricity (runout) when adjusting a part in a lathe chuck and the indicator is misaligned to the axis of measurement, a faulty reading will result and correct adjustments will be extremely difficult.

When using dial test indicators, watch for arc versus chord length errors (Figure B3.33) The tip of the test indicator moves through an arc. This distance may be considerably greater than the chord distance of the measurement axis. Dial test indicators should *not* be used to make direct measure-

ments. They should only be applied in comparison applications.

ANGULAR MEASURE

In the metals shop, you will find the need to measure **acute angles, right angles,** and **obtuse angles** (Figure B3.34). Acute angles are less than 90 degrees. Obtuse angles are more than 90 degrees but less than 180 degrees. Ninety-degree or right angles are generally measured with squares. However, the amount of angular deviation from perpendicularity may have to be determined. This would require that an angular measuring instrument be used. Straight

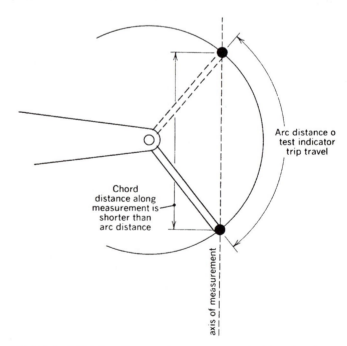

FIGURE B3.33 Potential for error in dial test indicator tip movement.

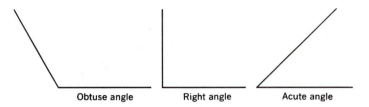

Obtuse angle   Right angle   Acute angle

FIGURE B3.34 Obtuse, right, and acute angles.

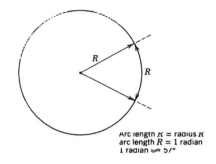

Arc length R = radius R
arc length R = 1 radian
1 radian ≈ 57°

FIGURE B3.35 Radian measure.

angles, or those containing 180 degrees, generally fall into the category of straightness or flatness and are measured by other types of instruments.

Units of Angular Measure

In the inch system, the unit of angular measure is the **degree.**

> Full circle = 360 degrees
> 1 degree   = 60 minutes of arc (1° = 60′)
> 1 minute   = 60 seconds of arc (1′ = 60″)

In the metric system, the unit of angular measure is the **radian.** A radian is the length of an arc on the circle **circumference** that is equal in length to the radius of the circle (Figure B3.35). Since the circumference of a circle is equal to 2 pi r (radius), there are 2 pi radians in a circle. Converting radians to degrees gives the equivalent:

$$1 \text{ radian} = \frac{360}{2 \text{ pi } r}$$

Assuming a radius of 1 unit:

$$1 \text{ radian} = \frac{360}{2 \text{ pi}}$$
$$= 57° \ 17′ \ 44″ \text{ (approximately)}$$

It is unlikely that you will come in contact with much radian measure. All of the common comparison measuring tools you will use read in degrees and fractions of degrees. Metric angles expressed in radian measure can be converted to degrees by the equivalent shown.

Reviewing Angle Arithmetic

You may find it necessary to perform angle arithmetic. Use your calculator, if you have one available.

ADDING ANGLES Angles are added just like any other quantity. One degree contains 60 minutes. One minute contains 60 seconds. Any minute total of 60 or larger must be converted to degrees. Any second total of 60 or larger must be converted to minutes.

EXAMPLE

3° 15′ + 7° 49′ = 10° 64′
Since 64′ = 1° 4′, the final result is 11° 4′

EXAMPLE

265° 15′ 52″ + 10° 55′ 17″ + 275° 70′ 69″
Since 69″ = 1′ 9″ and 70′ = 1° 10′, the final result is 276° 11′ 9″

SUBTRACTING ANGLES When subtracting angles, where borrowing is necessary, degrees must be converted to minutes and minutes must be converted to seconds.

EXAMPLE

15° − 8° = 7°

EXAMPLE

15° 3′ − 6° 8′ becomes 14° 63′ − 6° 8′ = 8° 55′

EXAMPLE

39° 18′ 13″ − 17° 27′ 52″ becomes 38° 77′ 73″ − 17° 27′ 52″ = 21° 50′ 51″

ANGULAR MEASURING INSTRUMENTS

Plate Protractors

Plate protractors have a discrimination of 1 degree and are useful in such applications as layout and checking the point angle of a drill (Figure B3.36).

Bevel Protractors

The bevel protractor is part of the machinist's combination set. This protractor can be moved along the rule and locked in any position. The protractor has a flat base permitting it to rest squarely on the workpiece (Figure B3.37). The combination set protractor has a discrimination of 1 degree.

COMPARATORS

Comparators are exactly what their name implies. They are instruments that are used to compare the size or shape of the workpiece to a known standard. Types include dial indicators and optical, electrical, and electronic comparators. Comparators are used

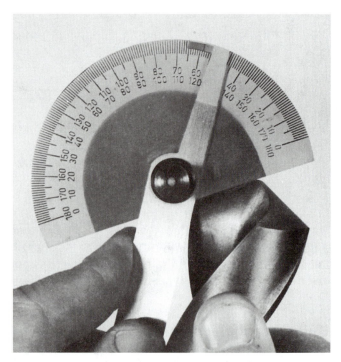

FIGURE B3.36 Plate protractor measuring a drill point angle.

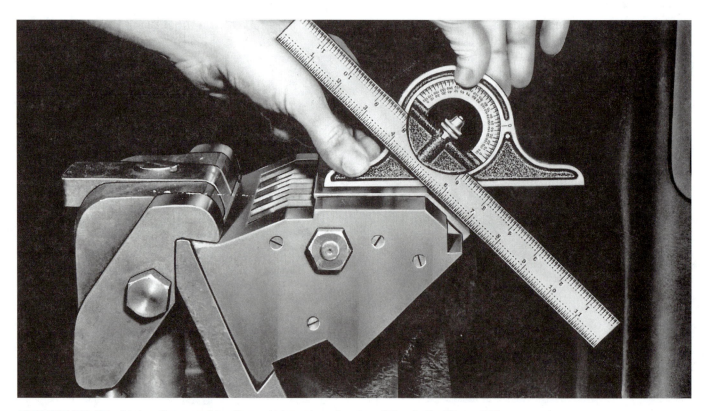

FIGURE B3.37 Using the combination set bevel protractor. (The L.S. Starrett Company)

where parts must be checked to determine acceptable tolerance. They may also be used to check the geometry of such things as threads, gears, and formed machine tool cutters. The electronic comparator may be found in the inspection area, toolroom, or gage laboratory and used in routine inspection and calibration of measuring tools and gages.

Dial Indicator Comparators

The dial indicator comparator is no more than a dial indicator attached to a rigid stand. These are dial indicator instruments such as the ones previously discussed. However, in their application as comparator instruments, the fixed design of the instrument's components eliminates as many errors as possible. The indicator is set to zero at the desired dimension by use of gage blocks (Figure B3.38). When using a dial indicator comparator, keep in mind the *potential for error* in indicator travel and instrument alignment along the axis of measurement. Once the indicator has been set to zero, parts can be checked for acceptable tolerance (Figure B3.39). A particularly useful comparator indicator for this is one equipped with tolerance hands (Figure B3.40). The tolerance hands can be set to establish an upper and lower limit for part size. On this type of comparator indicator, the spindle can be lifted clear of the workpiece by using the cable mechanism. This permits the indicator to always travel downward as it comes into contact with the work. This is an ad-

FIGURE B3.39 Using the dial indicator comparator.

ditional compensation for any mechanical error in the indicator mechanism.

Electronic Digital Indicator Comparators

Microelectronic technology has been adapted to indicator comparators as well as many other instruments. With digital readouts, these instruments are easy to read and calibrate, and they demonstrate

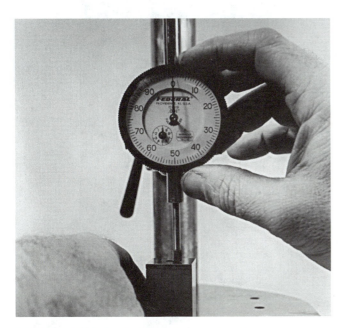

FIGURE B3.38 Setting the dial indicator to zero using gage blocks.

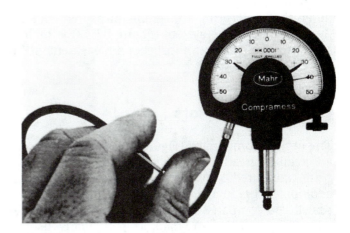

FIGURE B3.40 Dial comparator indicator with cable lift. (Harry Smith & Associates)

FIGURE B3.41 Digital electronic comparator. (MTI Corporation)

FIGURE B3.42 Digital electronic comparator with computer and printer. (MTI Corporation)

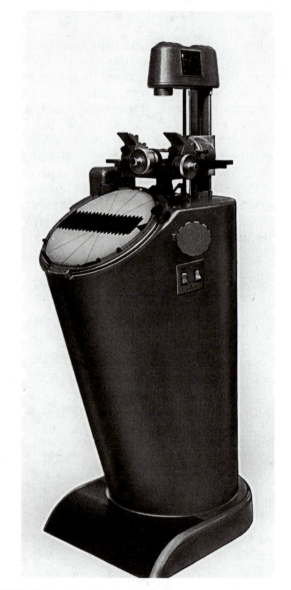

FIGURE B3.43 Optical comparator checking a screw thread. (Rank Scherr-Tumico, Inc.)

high reliability and high discrimination. One example is the digital indicator comparator (Figure B3.41).

This instrument may be coupled to a microcomputer that will print and graph measurements as they are taken. The system (Figure B3.42) is highly suited to quality control inspection measurement in the machine shop.

Optical Comparators

The optical comparator (Figure B3.43) projects onto a screen a greatly magnified profile of the object being measured. Various templates or patterns in addition to graduated scales can be placed on the screen and compared to the projected shadow of the part. The optical comparator is particularly useful for inspecting the geometry of screw threads, gears, and formed cutting tools.

Electronic and digital readouts also appear on the optical comparator. These features increase this instrument's reliability, ease of operation, metric/inch selection, high discrimination, and high sensitivity (Figure B3.44).

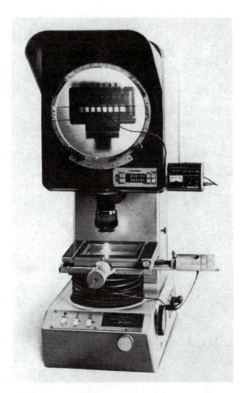

FIGURE B3.44 Optical comparator with digital electronic readouts in inch and metric units. (MTI Corporation)

SELF-TEST

1. Define comparison measurement.
2. What can be said of most comparison measuring instruments?

Match the following measuring situations with the list of comparison measuring tools. Answers may be used more than once.

3. A milled slot 2 in. wide with a tolerance of ±.002 in.
4. A height transfer measurement.
5. The shape of a form lathe cutter.
6. Checking a combination square to determine its accuracy.
7. The diameter of a $1\frac{1}{2}$-in. hole.
8. Measuring a shim under a piece of machinery.
   a. Spring caliper
   b. Telescope gage
   c. Adjustable parallel
   d. Radius gage
   e. Thickness gage
   f. Combination square
   g. Micrometer square
   h. Dial indicator
   i. Dial test indicator
   j. Dial indicator comparator
   k. Optical comparator
   l. Electronic comparator

UNIT 4

Tolerances, Fits, Geometric Dimensions, and Statistical Process Control (SPC)

Almost no product today is totally manufactured by a single maker. Although you might think that a complex product like an automobile or aircraft is made by a single manufacturer, if you look behind the scenes you would discover that the manufacturer uses many suppliers that make components for the final assembly.

In order to make all the component parts fit together, or **interface,** to form a complex assembly, you have already learned that a standardized system of measurement is essential. You have further learned that the measurement instruments must be compared to known standards in the process of **calibration** in order to maintain their accuracy.

Although standardized measurement is essential to modern industry, perhaps even more important are the design specifications that indicate the **dimensions** of a part. These dimensions control the size of a part, its features, and/or their location on the part or relative to other parts. Through this, parts are able to be interchanged and mated to each other to form complete assemblies.

Although the size of part features is important, dimensions that pertain to part geometry may be equally important. For example, a part with a specific pattern of holes may require that the location of the holes be as exact as the size of the holes themselves. In this case, the geometric dimensions and tolerances pertaining to true position of part features apply and must be met in the machining process as well as in the inspection of the final product.

The methods and tools of statistical process control (SPC) are used to track the dimensions of machined parts, especially those produced by computer-controlled machining processes (CNC). SPC can determine whether the machining process is producing in-tolerance parts and can measure and record many other factors about the parts including tolerance, average size, deviation from nominal dimensions, and distributions of part dimensions about the nominal dimensional specifications. SPC is very useful in determining the exact point where a manufacturing process is failing to meet specifications.

The purpose of this unit is to introduce the basic terminology and concepts of tolerances, fits, geometric dimensions, and statistical process control.

OBJECTIVES

After completing this unit, you should be able to:

1. Describe basic reasons for tolerance specifications.
2. Recognize common geometric dimensions and tolerances.
3. Describe the reasons for press fits and know where to find press-fit allowance information.
4. Describe in general terms the purpose of SPC.

LIMIT AND TOLERANCE

Since it is impossible to machine a part to an exact size, a designer must specify an acceptable range of sizes that will still permit the part to fit and function as intended. The maximum and minimum sizes in part dimensions that are acceptable are limits between which the actual part dimension must fall. The difference between the maximum and minimum limits is **tolerance,** or the total amount by

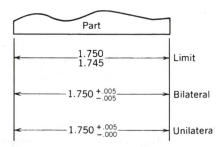

FIGURE B4.1 Tolerance notations.

which a part dimension may vary. Tolerances on drawings are often indicated by specifying a limit, or by plus and minus notations (Figure B4.1). With plus and minus tolerancing, when the tolerance is both above and below the nominal (true theoretical) size, it is said to be *bilateral* (two sides). When the tolerance is indicated all on one side of nominal, it is said to be *unilateral* (one sided).

Tolerance Expression: Decimal Inch

UNILATERAL TOLERANCE

$$.750 + .005 \qquad .750 + .000$$
$$\qquad - .000 \qquad \qquad - .005$$

The zero tolerance must have the same number of decimal places as the numeric tolerance and must include a plus or minus sign.

BILATERAL TOLERANCE

$$.750 \pm .005 \quad \text{not} \quad .75 \pm .005$$

When bilateral tolerancing is used, both the size and the plus-minus tolerance must have the same number of decimal places.

LIMIT DIMENSIONING When using limit dimensioning, both upper and lower limits must have the same number of decimal places.

$$\frac{.500}{.495} \quad \text{instead of} \quad \frac{.5}{.495}$$

Tolerance Expression: Metric

MILLIMETERS

$$\text{Unilateral tolerance:} \quad \frac{24\ 0}{-0.03} \quad \text{or} \quad \frac{24 + 0.03}{-0}$$

A single zero is shown without a plus or minus sign.

BILATERAL TOLERANCING

$$24 + 0.03 \qquad 23 - 0.03$$
$$\quad - 0.05 \qquad \quad + 0.05$$

Both the plus and minus tolerances must have the same number of decimal places.

LIMIT DIMENSIONING

$$\frac{24.00}{23.97} \quad \text{or} \quad \frac{24.03}{24.00}$$

Both the upper and lower limits must have the same number of decimal places.

How Tolerance Affects Mating Parts

When two parts mate or are interchanged in an assembly, tolerance becomes vitally important. Consider the following example (Figure B4.2): The shaft must fit the bearing and be able to turn freely. The diameter of the shaft is specified as 1.000 ± .001. This means that the maximum limit of the shaft is 1.001 and the minimum is .999. The tolerance is then .002 and bilateral.

The maximum limit of the bearing bore is also 1.001 and the minimum limit is .999. The tolerance is once again .002. Will the shaft made by one machine shop fit the bearing made by another machine shop using the tolerances specified? If the shaft is turned to the maximum limit of 1.001 and the bearing is bored to its minimum limit of .999, both parts would be within acceptable tolerance, but would not fit to each other since the shaft is .002 larger than the bearing. However, if the bearing bore was specified in limit form or unilateral tolerance of

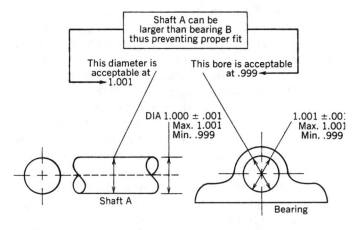

FIGURE B4.2 Tolerance overlap can prevent proper fit of mating parts.

$1.002 \begin{smallmatrix} +.002 \\ -.000 \end{smallmatrix}$ the parts would fit as intended. Even if the shaft was turned to the high limit of 1.001, it would still fit the bearing even though the bore was machined to the low limit of 1.002. Although a machinist is not usually concerned with establishing tolerance and limit specifications, you can easily see how fit problems can be created by overlapping tolerances discussed in this example.

Standard Tolerances

On many drawings you will use in the machine shop, tolerances will be specified at the dimensions. If no particular tolerances are specified at the dimension, accepted standard tolerances may be applied. These are often listed in part of the title block on the drawing and generally conform to the following:

Fractional dimensions $\pm\frac{1}{64}$
Two-place decimal fractions $\pm.010$
Three-place decimal fractions $\pm.005$
Four-place decimal fractions $\pm.0005$
Angles $\pm\frac{1}{2}$ degree

Always check any drawing carefully to determine if standard tolerances apply and what they might be for the particular job you are doing.

FITS

Fit refers to the amount or lack of clearance between two mating parts. Fits can range from free running or sliding, where a certain amount of clearance exists between mating parts, to **press** or **interference** fits where parts are forced together under pressure. Clearance fits can range from a few millionths of an inch, such as would be the case in the component parts of a ball or roller bearing, to a clearance of several thousandths of an inch, for a very low-speed drive or control lever application.

Many times a machinist is concerned with press or interference fits. In this case two parts are forced together usually by mechanical or hydraulic pressing. The frictional forces involved then hold the parts together without any additional hardware such as keys or set screws. Tolerances for press fits can become very critical because parts can be easily damaged by attempts to press fit them if there is an excessive difference in their mating dimensions. In addition, press fitting physically deforms the parts to some extent. This can result in damage, mechanical binding, or the need for a secondary resizing operation such as hand reaming or honing after the parts are pressed together.

A very typical example of press fit is when a ball bearing inner race is pressed onto a shaft or an outer race is pressed into a bore. Thus, the bearing is retained by friction and the free running feature is obtained within the bearing itself. Ball-to-race clearance is only a few millionths of an inch in precision bearings. If a bearing is pressed into a bore or onto a shaft with excessive force because pressing allowances are incorrect, the bearing may be physically deformed to the extent that mechanical binding is present. This will often cause excess friction and heat while in operation resulting in rapid failure of the part. On the other hand, insufficient frictional retention of the part resulting from a press fit that is not sufficient can result in the wrong part turning under load or some of the mechanism falling apart while operating.

Press Fit Allowances

Press fit allowances depend on a number of factors including length of engagements, diameter, material, particular components being pressed, and need for later disassembly of parts.

Soft materials such as aluminum can be pressed very successfully. However, soft materials may experience considerable **deformation** and these parts may not stand up to repeated pressings. Like metal parts pressed without the benefits of lubrication may **gall,** making them very difficult if not impossible to press apart. Very thin parts such as tubing may bend or deform to such a degree that the press retention is not sufficient to hold the parts together under design loads. The following general rule can be applied when determining the press allowance for cylindrical parts.

$$\text{Allowance} = .0015 \times \text{diameter of part in inches}$$

EXAMPLES

Determine the press allowance for a pin with a .250-in. diameter.

$$.0015 \times .250 = .000375$$
(slightly more than $\frac{3}{10,000}$ of an inch)

Determine the press allowance for a 4.250-in. diameter.

$$.0015 \times 4.250 = .00637$$
(slightly more than $\frac{6}{1000}$ of an inch)

Generally, pressing tolerances range from a few tenths to a few thousandths of an inch depending on the diameter of the parts and the other factors previously discussed. Proper measurement tools and techniques must be employed to make accurate determinations of the dimensions involved. For further specific dimensions on pressing allowances, consult a machinist's handbook.

Press Fits and Surface Finishes

The surface finish (texture) of parts being press fitted can also play an important part. Smooth-finished parts will press fit more readily than rough-finished parts. If the roughness height of the surface texture is large (64 μ inch (microinch) and higher), more frictional forces will be generated in the pressing operation and the chances for misalignment, galling, and seizing will be increased, especially if no lubricant is used. Lubrication will improve this situation to some extent. However, lubrication can be detrimental to press fit retention in some cases. A few molecules of lubricant between fit surfaces, especially if they are quite smooth, can result in the parts slipping apart when subjected to certain pull or push forces.

Shrink and Expansion Fits

Parts can be fitted by making use of the natural tendency of metals to expand or contract when heated and cooled. By heating a part, it will expand and can be then slipped on a mating part. Upon cooling, the heated part will contract and grip the mating part, often with tremendous force. Parts may also be mated by cooling one or the other so that it contracts, thus making it smaller. Upon warming to ambient temperature, it will expand to meet the mating part.

Shrink and **expansion** fits can have superior holding power over press fits, although special heating and cooling equipment may be necessary. Like press fits, however, allowances are extremely important. Consult a machinist's handbook for proper allowance specifications.

GEOMETRIC DIMENSIONING AND TOLERANCING

Equally important—and in many cases more important than controlling the size of a particular individual part—is controlling the form, orientation, location, profile, and runout of a part or assembly feature. This relates directly to the ability to interchange individualized parts and assemblies. For example, you have undoubtedly purchased standard replacement parts for your auto from many different sources. In many cases, these may be made by manufacturers other than the original maker of your auto. However, they fit and function exactly as the original equipment. To make this kind of interface possible, the manufacturing and engineering community has developed a system of geometric dimensioning and tolerancing that helps a manufacturer control form, orientation, location, profile, and runout of parts and assemblies. This system of geometric dimensioning and tolerancing—standardized in the American Society of Mechanical Engineers publication ASME Y14.5M-1994—is a complex subject and would require a great deal of time and space to cover completely. You will learn more about this as you go further into your training. For the present, the following discussion is intended to cover the basic concepts only.

Controlling Feature Location

You can see that to bolt the pump to the motor (Figure B4.3) it obviously is necessary to insure that the pattern of bolt holes in the pump matches the pattern of bolt holes in the motor. Also, the bore in the pump

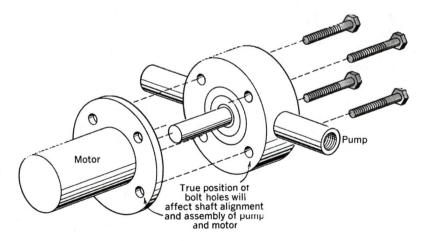

FIGURE B4.3 Pump and motor bolt hole patterns must match in position in order to accomplish assembly.

True position of bolt holes will affect shaft alignment and assembly of pump and motor

Motor

Pump

housing must match the boss on the motor so that the shaft will engage the motor with proper alignment. If the respective assemblies are made by different manufacturers, you can see that if either bolt pattern position deviates very far from the specified dimensions, the assemblies would be difficult or impossible to interface. As long as the two manufacturers work closely together, the assemblies will interface. However, if the pump manufacturer wants to start using a motor made by another manufacturer, the bolt hole pattern on the new pump motor will also have to interface with the pattern on the pump. This is an example where the *location* or the *true position* of the holes could be more critical than the size of the holes themselves. On drawings, the following symbols are used to indicate location control:

⊕ True position

◎ **Concentricity**

⹀ Symmetry

Controlling Orientation

Controlling orientation is equally important. Consider the pump and motor assembly in the previous example. The pump drive shaft must be perpendicular to the impeller case so that it can engage the motor without mechanical binding. Therefore, **perpendicularity** is one example of form that must be controlled during manufacturing. Table B4.1 shows symbols used on drawings to indicate geometric controls.

Datums and Basic Dimensions

Datums are reference points, lines, areas, and planes taken to be exact for the purpose of calculations and measurements. An initially machined surface on a casting, for example, may be selected as a datum surface and used as a reference from which to measure and locate other part features. Datums are usually not changed by subsequent machining operations and are identified by single or sometimes double letters (except *I*, *O*, and *Q*) inside a rectangular frame. For example,

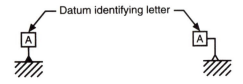

TABLE B4.1   Geometric characteristic symbols

Type of Tolerance		Characteristic	Symbol
For individual features	Form	Straightness Flatness Circularity (roundness) Cylindricity	— ▱ ○ ⌭
For individual or related features	Profile	Profile of a line Profile of a surface	⌒ ⌓
For related features	Orientation	Angularity Perpendicularity Parallelism	∠ ⊥ ∥
	Location	Position Concentricity Symmetry	⊕ ◎ ⹀
	Runout	Circular runout Total runout	↗* ↗↗*

*Arrowheads may or may not be filled.

SOURCE: Adapted from *ASME Y14.5M-1994*, Fig. 3–1, p. 42. New York: ASME International.

The term *basic dimension* represents a true, theoretically exact dimension describing location or shape of a part feature. Basic dimensions have no tolerance. The tolerance comes from the associated geometric tolerances used with the basic dimension. Basic dimensions are shown on drawings by enclosing a number in a rectangle:

$$\boxed{1.375}$$

MMC, LMC, and RFS

Three material condition modifier symbols that you will encounter on working drawings are:

MMC or Ⓜ for maximum material condition
LMC or Ⓛ for least material condition
RFS or Ⓢ for regardless of feature size

These identify at which size the geometric tolerances apply.

MMC refers to the maximum amount of material remaining on a feature or size. On an external cylindrical feature this would be the high limit of the feature tolerance. For example, a shaft with a diameter of .750 ±.010 would have an MMC diameter of .760 since this would leave maximum material remaining on the part. An internal cylindrical feature such as a hole with a diameter of .750 ±.010 would have an MMC diameter of .740 since this would leave maximum material remaining on the part.

LMC works in just the opposite way. The LMC of the shaft would be .740 diameter, and the LMC of the hole would be .760 diameter. These sizes would have the least amount of material remaining on the part.

RFS, or Regardless of Feature Size, means that the geometric tolerance applies no matter what the feature size is. The tolerance zone size remains the same, unlike MMC or LMC, which allow a tolerance zone size growth as the feature size changes.

RFS is the default condition for all geometric tolerances. Unless MMC or LMC is specified on the drawing by Ⓜ or Ⓛ, the tolerance zone size remains the same though the feature size changes. An example of RFS would be a hole located with a true position tolerance call out, where the size of the hole itself is not as important as the hole location.

Drawing Formats

The formats shown in Figure B4.4 are used to express some of the common geometric dimensions and tolerances on working drawings. Geometric dimensions and tolerance symbols appear on drawings in boxes such as those shown in the figure. The first box entry is the specific geometric call out. This will be either a form or position symbol. The next box entry will indicate any datum—a point, line, or plane—from which the geometric dimension is to be measured. The third box entry specifies the tolerance zone that applies to the form or position called out by the symbol.

SHOP TIP

For further information about geometric dimensions and tolerances, consult the current edition of publication ASME Y14.5M-1994.

STATISTICAL PROCESS CONTROL (SPC)

The major objective of any manufacturing is to insure that the end product meets design **specifications.** This is of the utmost importance when manufacturing the many discrete pieces that will ultimately become part of a subassembly or a complex finished product. In the modern world, many different manufacturers using a wide variety of processes are involved in producing a wide variety of parts for complex precision products such as automobile and aircraft engines, instruments of all types, and both the electronic and precision mechanical components of computers, to cite just a few examples.

The need to perfect manufacturing processes to a degree that will result in the maximization of good products is well known. Manufacturing at any level is expensive, so it is vital to maximize the number of product units that meet design specifications. Furthermore, there must be ways to quickly determine where there are problems within a process so that these may be solved quickly in order to minimize the production of scrap or otherwise unsuitable parts. The tools and techniques of statistical process control (SPC) are very effectively applied for this purpose.

Statistics may be used in at least two fundamental ways in manufacturing process control. First, this special branch of mathematics allows the manipulation of manufactured part inspection data in ways such that inferences can be made about large numbers of product units simply by selecting a representative sampling of the units, inspecting these for design specifications, and then manipulating the inspection data mathematically in order to make an inference about the suitability or lack thereof of the larger batch. This method saves time as it is not necessary or even possible to inspect each and every part from a given production run.

Second, data obtained from the inspection of product units can quickly reveal how consistent the manufacturing process is and can also reveal an exact point in the process where specifications are not being met. There are always imperfections in even the best of manufacturing equipment and processes. Machines and cutting tools wear out, operators and programmers may not be as attentive as they should, and many other factors may influence the outcome of a manufacturing process. A knowledge of information about specification discrepancies can greatly speed both the locating and correction of the problem. Having statistical data available allows the manufacturer to quickly determine where an adjustment or correction needs to be made in the process.

For parts made by the various machining processes, such as turning, milling, or grinding, dimensional specifications are the primary item of interest, since it is the size of the part features and often their geometry that control how well they will fit and perform with other parts. However, it should be noted that part dimensions and geometry are not the only specifications that need to be monitored during a manufacturing process. For example, the particular shade of paint, surface texture, or metallurgical

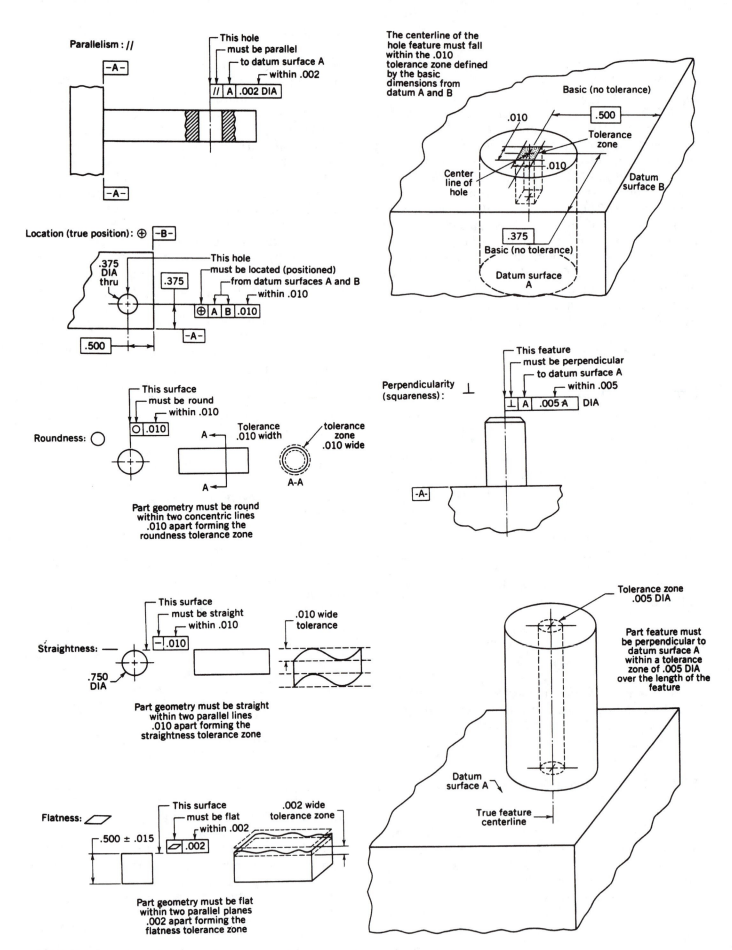

FIGURE B4.4 Drawing formats for geometric dimension and tolerances.

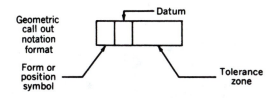

Geometric call out notation format

Datum

Form or position symbol

Tolerance zone

Examples:
Parallelism ‖

Top surface
— must be parallel
— to datum surface B
— within .002

Top surface

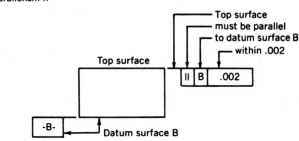

‖ | B | .002

-B-

Datum surface B

.002 DIA tolerance zone

Hole centerline must be parallel to datum A within .002 DIA tolerance zone

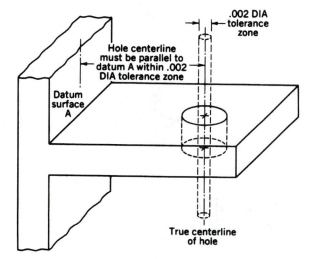

Datum surface A

True centerline of hole

FIGURE B4.4 *Continued*

specifications on a product may be equally as important as their dimensional specifications.

Various statistics are used in SPC to make determinations about product units. These statistics include arithmetic average, mean, mode, standard deviation, and correlations. Statistics may be used to measure central tendencies, average size distribution around a nominal dimension, and deviation of specifications from acceptable tolerance ranges.

A very basic example of this would be arithmetic average. For example, you just measured 10 parts produced on a CNC fuming center. Their nominal diameter is specified at .500 ±.010. You are interested in the average size of the 10 parts. Part dimensions are recorded as follows:

Part 1	.495	Part 6	.497
Part 2	.502	Part 7	.503
Part 3	.496	Part 8	.499
Part 4	.504	Part 9	.493
Part 5	.501	Part 10	.492

To calculate the average size of these parts and generate a statistic called *arithmetic average,* which is one measure of central tendency, apply the following formula:

$$\frac{\text{sum of part sizes}}{\text{number of parts}} = \text{arithmetic average}$$

$$.495 + .502 + .496 + .504 + .501 + .497 + .503$$
$$+ .499 + .493 + .492 = 4.982$$

$$\frac{4.982}{10 \text{ parts}} = \text{an average size of .4982}$$

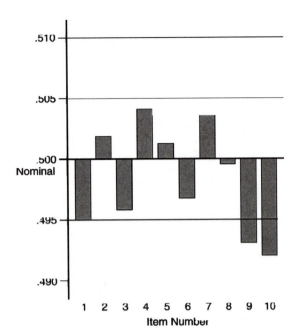

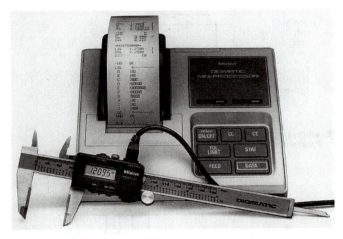

FIGURE B4.6 The dial caliper is an example of SPC equipment for dimensional measurement of machined parts. The attached microprocessor records, graphs, and prints part dimensions and generates appropriate statistical data. (MTI Corporation)

FIGURE B4.5 Measuring equipment for SPC can generate various types of bar, line, and scatter graphs showing part dimension distributions and other statistical data.

In another situation you might want to know how these parts are distributed around the theoretical exact size of .500. A type of graph called a *histogram* (Figure B4.5) may be developed for this purpose. The histogram is particularly useful because being able to see how the dimensions of the product units are distributed can quickly show those that are outside the specified tolerance. If many parts begin to fall outside the tolerance specifications, then immediate action must be taken to correct a deficiency in the process that is responsible for the discrepancy.

Tools for SPC

The advent of the microprocessor has permitted the development of many useful tools for SPC. For SPC functions in machining manufacturing, these measuring instruments look much the same as their conventional electronic counterparts and include calipers, micrometers, depths gages, and dial indicators (Figures B4.6 and B4.7). SPC equipment also includes microcomputers and suitable software able to perform a variety of SPC functions and to generate the appropriate statistical data.

By interfacing the measuring equipment with a microprocessor and suitable software, statistical data can be recorded, graphed, and then analyzed. The programmable capability of the SPC micro-

FIGURE B4.7 The outside micrometer is another example of SPC equipment for dimensional measurement of machined parts. The attached microprocessor records, graphs, and prints part dimensions and generates appropriate statistical data. (MTI Corporation)

processor allows various parameters to be preset for a given inspection task. For example, the computer may be programmed with the high and low tolerance limits of parts. As parts are produced and inspected by the attached electronic gaging device, statistics are generated immediately. These include maximum and minimum size, average size, and deviations of tolerance specifications from acceptable norms. SPC computer systems and software have the capability to record and compile statistical data for many different part specifications as well as from many different processes in the same or different manufacturing facilities. PC microcomputers, minis, or mainframes can accept data from different inspection stations or over local area computer networks (LANs). The computer system can then generate SPC reports for screen viewing, printouts, or permanent computer memory storage.

Role of the Machinist in SPC

Every machinist has a personal responsibility to produce machined parts that are within specified tolerances. In this capacity, some of your work may require that you inspect parts made by a production machine using SPC tools such as those discussed that both measure and generate statistical information about part dimensions. If the machining process that you are running is not producing parts to acceptable tolerances, you may also be responsible for determining the cause of the problem and to provide corrective actions by making appropriate adjustments to feeds, speeds, and cutting tools.

SELF-TEST

1. Why are tolerances important in manufacturing?
2. What are typical standard tolerances?
3. Name three geometric specifications called out on drawings.
4. What is the general rule for press fit allowances?
5. Describe shrink and expansion fits.
6. Describe SPC tools and activities.

Reading Shop Drawings

From earliest times, man has communicated his thoughts through drawings. The pictorial representation of an idea is a vital line of communication between the designer and the people who produce the final product. Technological design would be impossible were it not for the several different ways that an idea may be represented by a drawing. The drawing also provides an important testing phase for an idea. Many times an idea may be rejected at the drawing board stage before a large investment is made to equip a manufacturing facility and risk production of an item that does not meet all design requirements.

This does not mean that all design problems can be solved in the drafting room. Almost anything can be represented by a drawing, even to the extent that some designs can be quite impossible to manufacture. It is important that the designer be aware of the problems that confront the machinist. On the other hand, you must fully understand all of the symbols and terminology on the designer's drawing. You must then interpret these terms and symbols in order to transform the ideas of the designer into useful products.

OBJECTIVES

After completing this unit, you should be able to:

1. Identify symbols and terminology on working drawings.
2. Complete a simple orthographic drawing.
3. Read and interpret a typical working drawing for a machinist.

PICTORIAL REPRESENTATIONS

Perspective Drawings

The perspective drawing (Figure B5.1) is used when it is desired to show an object as it would appear to the eye. Perspective can be either one point, known as *parallel perspective,* or two point, known as *angular perspective.* In the perspective view, the lines of the object recede to a single point. The preception of depth is indicated.

Isometric Drawing

An isometric drawing (Figure B5.2) is also intended to represent an object in three dimensions. However, unlike the perspective, the object lines do not recede but remain parallel. Furthermore, isometric views are drawn about the three isometric axes that are 120 degrees apart.

Oblique Drawing

Object lines in the oblique drawing (Figure B5.3) also remain parallel. The oblique differs from the isometric in that one axis of the object is parallel to the plane of the drawing.

The perspective view is used mainly by artists and technical illustrators. You will seldom come in contact with a perspective drawing. However, you should be aware of its existence. Isometric and oblique are also not generally used as working drawings for the machinist. However, you may occasionally see them in the machine shop.

Exploded Drawings

The exploded drawing (Figure B5.4) is a type of pictorial drawing designed to show several parts in their proper location prior to assembly. Although the exploded view is not used as the working draw-

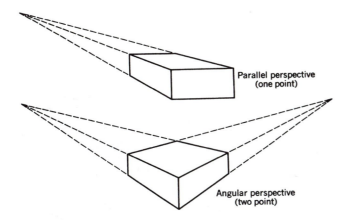

FIGURE B5.1 Perspective drawing.

Parallel perspective
(one point)

Angular perspective
(two point)

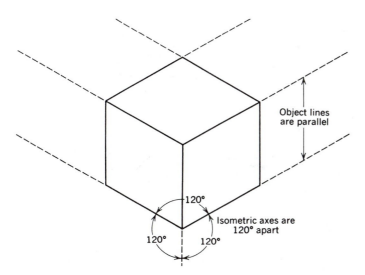

Object lines
are parallel

120°

Isometric axes are
120° apart

120°   120°

FIGURE B5.2 Isometric drawing.

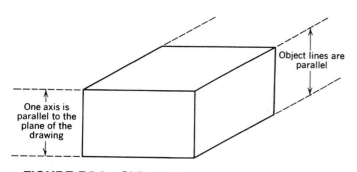

Object lines are
parallel

One axis is
parallel to the
plane of the
drawing

FIGURE B5.3 Oblique drawing.

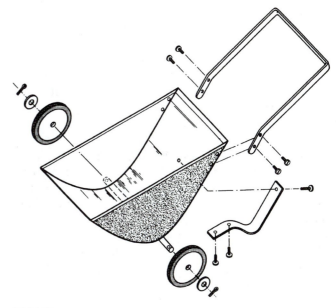

FIGURE B5.4 Exploded drawing.

ing for the machinist, it has an important place in mechanical technology. Exploded views appear extensively in manuals and handbooks that are used for repair and assembly of machines and other mechanisms.

Orthographic Drawings

THE ORTHOGRAPHIC PROJECTION DRAWING
In almost every case, the working drawing for the machinist will be in the form of the three-view or orthographic drawing. The typical orthographic format always shows an object in the three-view combination of side, end, and top (Figure B5.5). In some cases, an object can be completely shown by

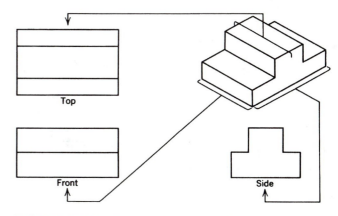

Top

Front

Side

FIGURE B5.5 Standard orthographic drawings.

a combination of only two orthographic views. However, any orthographic drawing must have a minimum of two views in order to show an object completely. The top view is referred to as the *plan* view. The front or side views are referred to as *elevation* views. The terms *plan* and *elevation* may appear on some drawings, especially those of large complex parts or assemblies.

HIDDEN LINES FOR PART FEATURES NOT VISIBLE Features that are not visible are indicated by dotted lines. These are called *hidden lines* as they indicate the locations of part features hidden from view. The plain bearing (Figure B5.6) is shown in a typical orthographic drawing. The front view is the only one in which the hole through the bearing, or bore, can be observed. In the side and top views, the bore is not visible. Therefore, it is indicated by dotted or hidden lines. The mounting holes through the base are visible only in the top view. They appear as hidden lines in the front and side views.

SECTIONED VIEWS When internal features are complex to the extent that indicating them as hidden lines would be confusing, a sectioned drawing may be employed. Two common styles of sections are used. In the full section (Figure B5.7*a*), the object has been cut completely through. In the half section (Figure B5.7*b*), one quarter of the object is removed. The section indicator line shows the plane at which the section is taken. For example (Figure B5.7*a*), the end view of the object shows the section line marked by the symbol "AA." The section line "BB" (Figure B5.7*b*) indicates the portion removed in the half section. An object may be sectioned at any plane as long as the section plane is indicated on the drawing.

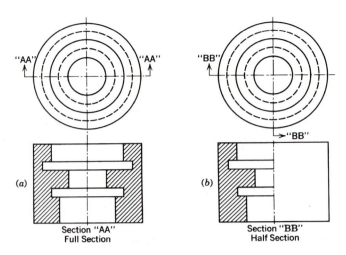

FIGURE B5.7 Sectioned drawings.

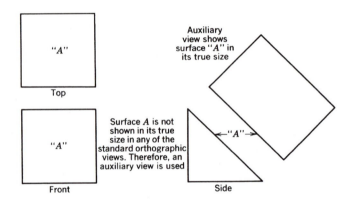

FIGURE B5.8 Auxiliary view.

AUXILIARY VIEWS One of the reasons for adopting the orthographic drawing is to represent an object in its true size and shape. This is not possible with the pictorial drawings discussed earlier. Generally, the orthographic drawing meets this requirement. However, the shape of certain objects is such that their actual size and shape are not truly represented. An auxiliary view may be required (Figure B5.8). On the object shown in the figure, surface "A" does not appear in its true size in any of the standard orthographic views. Therefore, surface "A" is projected to the auxiliary view, thus revealing its true size.

READING AND INTERPRETING DRAWINGS

Scale

In some cases, an object may be represented by a drawing that is the same size as the object. In other cases, an object may be too large to draw full

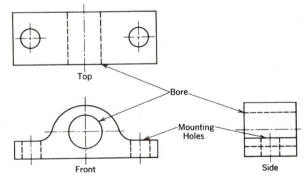

FIGURE B5.6 Hidden lines for part features not visible.

size, or a very small part may be better represented by a drawing that is larger. Therefore, all drawings are drawn to a specific scale. For example, when the drawing is the same size as the object, the scale is said to be full, or 1 = 1. If the drawing is one-half size, the scale is one-half, or $\frac{1}{2}$ = 1. A drawing twice actual size would be double scale, or 2 = 1. The scale used is generally indicated on the drawing.

Dimensioning of Detail and Assembly Drawings

You will primarily come in contact with the detail **drawing.** This is a drawing of an individual part and, in almost all cases, will appear in orthographic form. Depending on the type of work a machinist may be doing, he may also see an assembly drawing. The assembly drawing is a drawing of subassemblies or several individual parts assembled into a complete unit. For example, a drawing of a complete automobile engine would be an assembly drawing. In addition, a detail drawing of each engine component would also exist.

A detail drawing contains all of the essential information needed by you in order to make the part. Most important are the **dimensions.** Dimensioning refers generally to the sizes specified for the part and the locations of its features. Furthermore, dimensions reflect many design considerations, such as the fit of mating parts, that will affect the operating characteristics of all machines. Much of the effort that you expend performing the various machining operations will be directed toward controlling the dimensions specified on the drawing.

Several styles of dimensioning appear on drawings. The most common of these is the standard fractional inch notation (Figure B5.9). The outline of the part along with the several holes are dimensioned according to size and location. Generally, the units of the dimensions are not shown. Note that certain dimensions are specified to

come within certain ranges. This is known as **tolerance.**

Tolerance refers to an acceptable range of part size or feature location and is generally expressed in the form of a minimum and maximum limit. The bore (Figure B5.9) is shown to be 1.250 in. ±.005 in. This notation is called a *bilateral tolerance* because the acceptable size range is both above and below the nominal (normal) size of 1.250 in. The bored hole could be any size from 1.245 to 1.255 in diameter. The thickness of the part is specified as 1.000 + .000 and −.002. This tolerance is *unilateral* as all the range is on one side of nominal. Thus, the thickness could range from .998 in. to 1.000 in.

No tolerance is specified for the outside dimensions of the part or the locations of the various features. Since these dimensions are indicated in standard fractional form, the tolerance is taken to be plus or minus $\frac{1}{64}$ of an inch unless otherwise specified on the drawing. This range is known as *standard tolerance* and applies only to dimensions expressed in standard fractional form.

Another system of dimensioning used in certain industries is that of decimal fraction notation (Figure B5.10). In this case, tolerance is determined by the number of places indicated in the decimal notation.

2 places	.00 tolerance is ±.010 in.
3 places	.000 tolerance is ±.005 in.
4 places	.0000 tolerance is ±.0005 in.

Always remember that standard tolerances apply only when no other tolerance is specified on the drawing.

The coordinate or absolute system of dimensioning (Figure B5.11) may be found in special applications such as numerically controlled machining. In this system, all dimensions are specified from the same zero point. The figure shows the dimensions expressed in decimal form. Standard fraction notation may also be used. Standard tolerances apply unless otherwise specified.

With the increase in the use of the metric system in recent years, some industries have adopted a system of dual dimensioning of drawings with both metric and inch notation (Figure B5.12). Dual dimensioning has, in some cases, created a degree of confusion for the machinist. Hence, industry is constantly devising improved methods by which to differentiate metric and inch drawing dimensions. You must use caution when reading a dual-dimensioned drawing to insure that you are conforming your work to the proper system of measurement for your

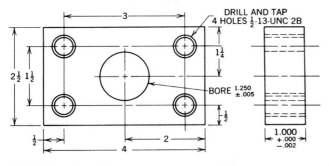

FIGURE B5.9 Fractional inch dimensioning.

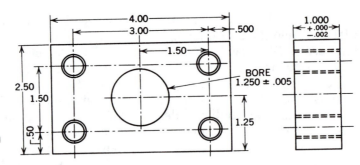

FIGURE B5.10 Decimal inch dimensioning.

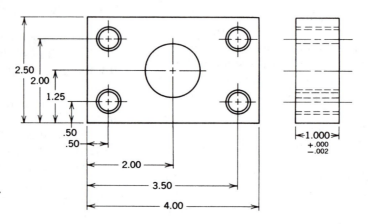

FIGURE B5.11 Absolute or co-ordinate dimensioning.

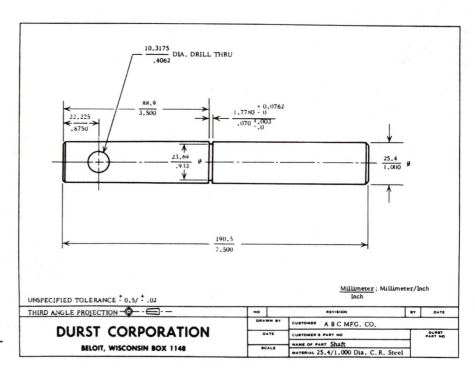

FIGURE B5.12 Dual dimensioning: metric and inch.

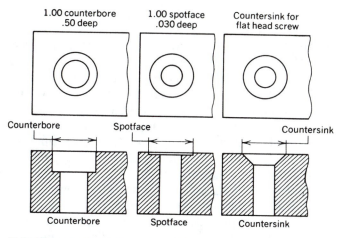

1.00 counterbore
.50 deep

1.00 spotface
.030 deep

Countersink for
flat head screw

Counterbore   Spotface   Countersink

Counterbore   Spotface   Countersink

FIGURE B5.13 Countersinking, counterboring, and spot-facing symbols.

TABLE B5.1   Abbreviations on drawings

Symbol	Definition
BHN	Brinell hardness number
B.C.	Bolt circle diameter
C'BORE	Counterbore
C'SINK	Countersink
C to C	Center to center
D, Dia., Diam.	Diameter
F.A.O.	Finish all over
Hardn & Gr	Harden and grind
I.D.	Inside diameter
O.D.	Outside diameter
R, Rad.	Radius
Rc	Rockwell, C scale
S'FACE	Spotface
S.H.C.S., Soc. Hd. Cp. Scr.	Socket head capscrew
Stl.	Steel
Scr.	Screw
T.I.R.	Total indicator reading
Typ.	Typical

tools. In the figure, metric dimensions appear above the line and inch dimensions appear below the line.

Abbreviations for Machine Operations

Working drawings contain several symbols and abbreviations that convey important information to the machinist (Figure B5.13). Certain machining operations may be abbreviated on a drawing. For example, countersinking is a machining operation in which the end of a hole is shaped to accept a flathead screw, and on a drawing countersinking may be abbreviated as C'SINK. The desired angle will also be specified. Table B5.1 gives some of the abbreviations used on mechanical drawings.

Finish Marks and Symbols

Very often you will perform work on a part that has already been partially shaped. An example of this might be a casting or forging. The finish mark, shaped like a letter f and sometimes used to indicate which surfaces are to be machined, is not used much today, being replaced by a v symbol. The point of the v touches the work where machining is needed. Surface finish symbols, as shown in Figure B5.14, tell the machinist what kind of finish is required in terms of microinches. For example, a finish mark notation of 4, 32, or 64 refers to a specific surface finish. A surface roughness comparator gage is sometimes used by the machinist to determine

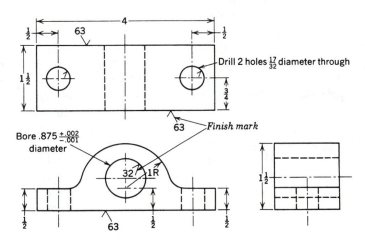

FIGURE B5.14 Finish marks.

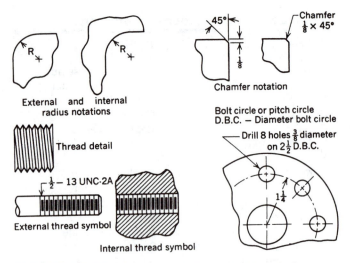

FIGURE B5.15 Other symbols and abbreviations.

relative smoothness. See Section B, Unit 1, Figure B1.7 for a surface roughness comparator gage.

Other Common Symbols and Abbreviations

External and internal radii are generally indicated by the abbreviation R and the specified size (Figure B5.15). **Chamfers** may be indicated by size and angle as shown in the figure. **Threads** are generally represented by symbols or they may be drawn in detail. Threads will also have a notation that indicates type, size, and fit. Consider the notation $\frac{1}{2}$—13 UNC 2A. This thread notation indicates the following:

$\frac{1}{2}$—Major thread diameter

13—Number of threads per inch

UNC—Shape and series of thread

2—Class of fit

A—External thread (internal is denoted B)

A specific bolt circle or pitch circle is often indicated by the abbreviation B.C. (or D.B.C.), meaning Bolt Circle diameter. The size of the diameter is indicated by normal dimensioning or with an abbreviation such as $2\frac{1}{2}$ B.C.

MECHANICAL DRAWING FORMATS

A designer's idea may at first appear as a freehand sketch perhaps in one of the pictorial forms discussed previously. After further discussion and examination, the decision may be made to have a part or an assembly manufactured. This necessitates suitable orthographic drawings that can be supplied to the machine shop. The original drawings produced by the drafting department are not used directly by the machine shop. These original drawings must be carefully preserved, as a great deal of time and money has been invested in them. Were they to be sent directly to the machine shop, they would soon be destroyed by constant handling. Therefore, a copy of the original drawing is made.

Several methods are employed to obtain copies of original drawings. An older method of copying is called **blueprinting**, where the drawing appears as white lines on a blue background. The process is no longer widely used, but the name has come to mean any of several processes of reproduction. The most commonly used method of reproducing drawings is sometimes called *blue line*, where blue lines are on a white background. Any number of prints may be made and distributed to the various departments of a manufacturing facility. For example, assembly prints are needed in the assembly area while prints of individual parts are required at the machine tool stations and in the inspection department.

The typical detail drawing format (Figure B5.16) contains a suitable title block. In many cases, the name of the firm appears in the title block, as shown in Figure B5.12. The block also contains the name of the part, specified tolerance, scale, and the initials of the draftsman. A finish mark notation may also appear. The print may also contain a change block. Often designs may be modified after an original drawing is made. Subsequent drawings will reflect any changes, and these will be noted in the change block. A drawing may also contain one or more general notes. The notes contain important information for the machinist. Therefore, you should always find and read any general notes appearing on a print.

One common note found on mechanical drawings is *break all corners*. This means the machinist must deburr sharp edges, usually with a file. Sharp edges can be hazardous to anyone handling the part, and a sharp edge is more easily damaged in handling than one that has been deburred.

A typical assembly drawing format contains essentially the same information as found on the detail drawing plan (Figure B5.17). However, assembly drawings generally show only those dimensions that pertain to the assembly. Dimensions of the individual parts are found on the detail plans. In addition to the normal information, a bill of materials appears on the assembly plan. This bill contains the part number, description, size, material, and required quantity of each piece in the assembly. Often, the source of a specific item that is not manufactured by

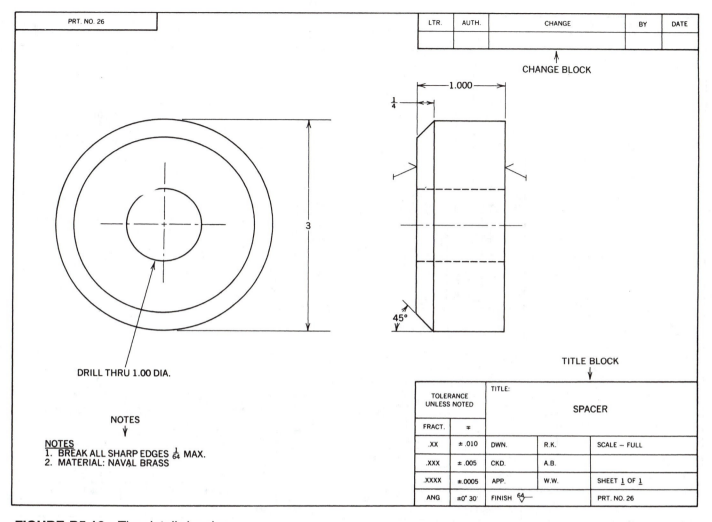

		LTR.	AUTH.	CHANGE	BY	DATE

CHANGE BLOCK

PRT. NO. 26

1.000

¼

3

45°

DRILL THRU 1.00 DIA.

NOTES
↓

NOTES
1. BREAK ALL SHARP EDGES 1/64 MAX.
2. MATERIAL: NAVAL BRASS

TITLE BLOCK
↓

TOLERANCE UNLESS NOTED		TITLE: SPACER		
FRACT.	≠			
.XX	± .010	DWN.	R.K.	SCALE – FULL
.XXX	± .005	CKD.	A.B.	
.XXXX	±.0005	APP.	W.W.	SHEET 1 OF 1
ANG	±0° 30'	FINISH 64 —		PRT. NO. 26

FIGURE B5.16   The detail drawing.

the assembler will be specified in the bill of material. An assembly drawing may also include a list of references of detail drawings of the parts in the assembly. Any general notes containing information regarding the assembly will also be included.

SHOP SKETCHING

It often becomes necessary to record pertinent information and dimensions when away from the shop or where drafting equipment is not available. Shop sketching is the method used to record this information for later use when assembling or machining the required parts.

The same standards used for drafting are used for shop drawings; that is, center, section, hidden, dimension, and break lines are all used. It is especially important to use **centerlines** since most dimensioning is from centerlines or from machined edges (datum or base line). Notes, where needed, should also be included. The most important thing to remember is to get *all* necessary dimensions for later use. In some cases, it may be impractical to return to the location to get an omitted dimension. Since mechanical drawings should not be scaled (measured on the drawing) and the part should be made only by given dimensions, a crude drawing will suffice to convey the necessary information if all of the needed dimensions are given.

All that is needed to make a shop drawing is a tablet and pencil and the measuring tools. A small rule will help to make straighter lines but it takes more time to use it. Circles are particularly hard for most people to make freehand. One method to sketch circles is shown in Figure B5.18. Centerlines are necessary for all circles as they serve as dimension lines. When one quadrant is drawn at a time, the job is easier.

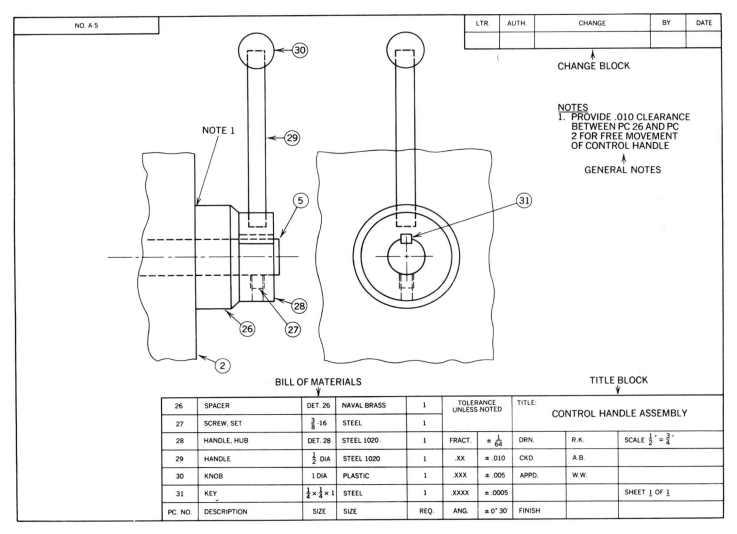

LTR.	AUTH.	CHANGE	BY	DATE

CHANGE BLOCK

NOTES
1. PROVIDE .010 CLEARANCE
   BETWEEN PC 26 AND PC
   2 FOR FREE MOVEMENT
   OF CONTROL HANDLE

GENERAL NOTES

NOTE 1

BILL OF MATERIALS

TITLE BLOCK

26	SPACER	DET. 26	NAVAL BRASS	1	TOLERANCE UNLESS NOTED		TITLE: CONTROL HANDLE ASSEMBLY		
27	SCREW, SET	$\frac{3}{8}$ -16	STEEL	1					
28	HANDLE, HUB	DET. 28	STEEL 1020 .	1	FRACT.	$\pm \frac{1}{64}$	DRN.	R.K.	SCALE $\frac{1}{2}'' = \frac{3}{4}''$
29	HANDLE	$\frac{1}{2}$ DIA	STEEL 1020	1	.XX	± .010	CKD.	A.B.	
30	KNOB	1 DIA	PLASTIC	1	.XXX	± .005	APPD.	W.W.	
31	KEY	$\frac{1}{4} \times \frac{1}{4} \times 1$	STEEL	1	.XXXX	± .0005			SHEET 1 OF 1
PC. NO.	DESCRIPTION	SIZE	SIZE	REQ.	ANG.	± 0° 30′	FINISH		

FIGURE B5.17 The assembly drawing.

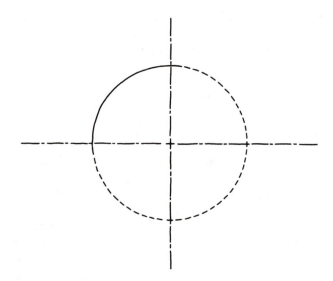

FIGURE B5.18 Freehand circles can be made one quadrant at a time.

An example of a large taper shaft to which a new gear must be bored and fitted in a machine shop is shown in Figure B5.19. The shaft cannot be removed for transport from its remote location. Armed with the correct measuring tools, the machinist goes to the location and makes a sketch. The taper per foot can later be calculated with the given dimensions, and the gear accurately bored to fit.

Shown in Figure B5.20 is a pipe flange for which a steel plate must be made in the shop for an end cap. The plate to be made is simply a steel circle with holes drilled in it for bolts, but the holes have to be in the right position, so someone must go to the pipe location and make a sketch.

Two- and three-view orthographic sketches are often made. Sometimes auxiliary views are necessary. If you are good at sketching, isometric-dimensioned drawings are very useful.

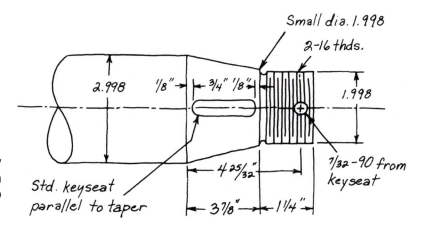

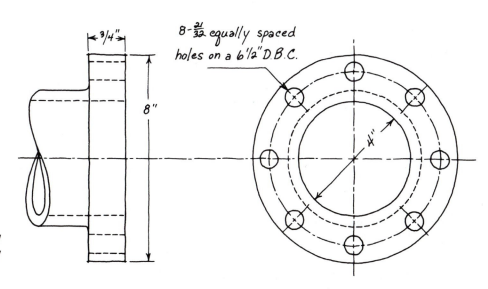

FIGURE B5.19 All necessary dimensions and information must be shown in shop sketches.

Small dia. 1.998

2-16 thds.

1.998

Std. keyseat parallel to taper

7/32-90 from keyseat

FIGURE B5.20 A two-view sketch is sufficient in many cases.

8-21/32 equally spaced holes on a 6½" D.B.C.

SELF-TEST

1. Sketch the object (Figure B5.21) as it would appear in correct orthographic form.

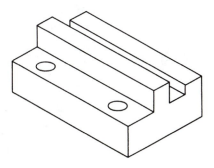

FIGURE B5.21

(For Problems 2 to 10 refer to Figure B5.22.)

2. What is the minimum size of the hole through the clevis head?
3. What length of thread is indicated on the drawing?
4. What is the tolerance of the slot in the clevis head?
5. What radius is specified where the shank and clevis head meet?
6. What is the total length of the part?
7. What is the width of the slot in the clevis head?
8. Name two machining operations specified on the drawing.
9. What is the size and angle of the chamfer on the thread end?
10. What does note 1 mean?

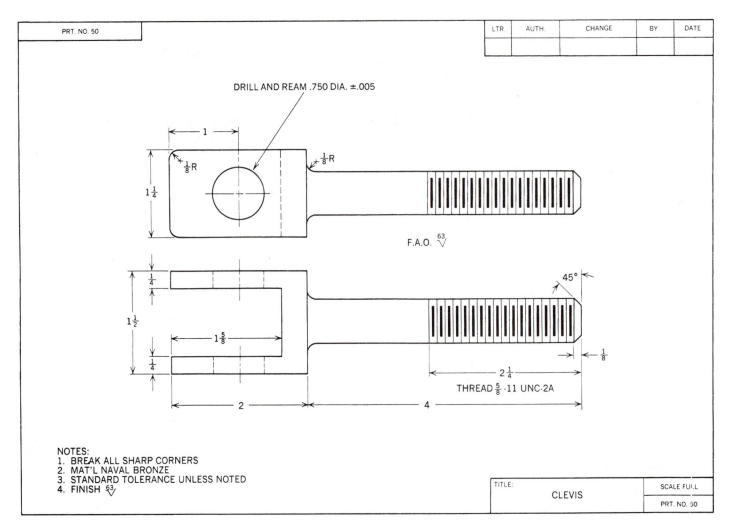

FIGURE B5.22

SECTION C
Preparations for Machining Operations

Before a manual machinist or a CNC operator turns on a machine, there are several principles that are absolutely essential that he or she should understand.

Probably the most important of these is cutting speeds, which, if a wrong setting is used, can quickly destroy the tooling and/or damage the workpiece. Feed rates that are too high or too low can also damage tooling and produce poor finishes.

The machine operator must also understand the use and purpose of cutting fluids. Higher production rates, longer tool life, and better finishes are a few of the advantages of using cutting fluids.

Choosing the right cutting tool material for the job is also very important. That information is provided in this text in the section that discusses the specific machine being used. For example, in the lathe section, cutting tools for the lathe are discussed, and in the milling machine section, milling cutters are illustrated.

Unit 1   Machinability and Chip Formation
Unit 2   Speeds and Feeds for Machine Tools
Unit 3   Cutting Fluids

UNIT 1   Machinability and Chip Formation

Cutting tool materials tend to get hot during the process of machining, that is, in cutting metals and producing chips. The heat rise on the cutting edge of a tool can be sufficient to burn the sharp edge and dull it (Figure C1.1). The tool must then be resharpened or a new sharp edge must be indexed in place. A certain amount of wear and tool breakdown is to be expected, but machining time lost while replacing or sharpening the tool must be considered. Generally, a trade-off between maximum production and tool breakdown or loss is accepted in machining operations.

There are several factors that cause or reduce tool heating. Some tool materials can withstand higher temperatures than others. The hardness and toughness of workpiece materials are also factors in tool breakdown caused by heating. Tool breakage can be caused by heavy cuts (excessive feed) and interrupted cuts or by accidentally reversing the machine movement or rotation while a cut is being made.

The speed of the cutting tool point or edge moving along the workpiece or of the workpiece moving past the tool is a major factor in controlling tool breakdown and must always be considered when cutting metal with a machine tool. This is true of all machining operations—sawing, turning, drilling, milling, and grinding. In all of these, cutting speed is of utmost importance.

The feed on a machine is the movement that forces the tool into or along the workpiece. Increments of feed are usually measured in thousandths of an inch. Excessive feed usually results in broken tools (Figure C1.2). Often this breakage ruins the tool, but in some cases the tool can be reground.

Cutting fluids are used in machining operations for two basic purposes: to cool the cutting tool and workpiece and to provide lubrication for easy chip flow across the toolface.

Some cutting fluids are basically **coolants.** These are usually water-based soluble oils or synthetics and their main function is to cool the work-tool interface. Cutting oils provide some cooling action but also act as lubricants and are mainly used where cutting forces are high as in die threading or tapping operations.

Tool materials range from hardened carbon steel to diamond, each type having its proper use whether in traditional metal cutting in machine tools or for grinding operations. This section will enable you to use correctly all of these cutting tools, to avoid undue damage to tools and workpieces, and to use the tools to the greatest advantage.

Machinability is the relative difficulty of a machining operation with regard to tool life, surface finish, and power consumption. In general, softer materials are easier to machine than harder ones. Chip removal and control are also major factors in the machining industry. These and other considerations concerning the use of machine tools are covered in this unit.

The previous discussion applies to all forms of metal cutting operations, whether they be manual machining or done by Computer Numerical Control (CNC). It is assumed that students using this book will go on to CNC training. Speeds, feeds, and coolants are especially important in automated machining.

OBJECTIVES

After completing this unit, you should be able to:

1. Determine how metal cutting affects the surface structures of metals.
2. Analyze chip formation, structures, and chip breakers.
3. Explain machinability ratings and machining behavior of metals.

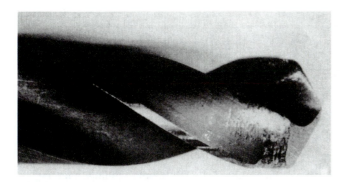

FIGURE C1.1 The end of this twist drill has been burned due to excessive speed.

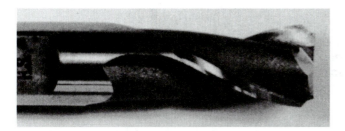

FIGURE C1.2 The sharp corners of this end mill have been broken off due to excessive feed.

SAFETY IN CHIP HANDLING

CAUTION

Certain metals, such as magnesium and zirconium, when divided finely as a powder or even when as coarse as machining chips, can ignite with a spark or just by the heat of machining. Such fires are difficult to extinguish, and if water or a water or water-based fire extinguisher is used, the fire will only increase in intensity. The greatest danger of fire occurs when a machine operator fails to clean up zirconium or magnesium chips on a machine when the job is finished. If the next operator cuts alloy steel, which can produce high temperatures in the chip or even sparks that can ignite the magnesium chips, a fire may destroy the entire machine if not the shop. Chloride-based powder fire extinguishers are commercially available. These are effective for such fires since they prevent water absorption and form an air-excluding crust over the burning metal. Sand is also used to smother fires in magnesium.

Metal chips from machining operations are very sharp and are a serious hazard. They should not be handled with bare hands. Gloves may be worn only when the machine is not running.

PRINCIPLES OF METAL CUTTING

In machining operations, either the tool material rotates or moves in a linear motion or the workpiece rotates or moves (Table C1.1). The moving or rotating tool must be made to move into the work material in order to cut a chip. This procedure is called **feed.** The amount of machine feed controls the thickness of the chip. The depth of cut is often called **infeed.** Besides the use of single-point tools with one cutting edge, as they are called in lathe, shaper, and planer operations, there are multiple-point tools such as milling machine cutters, drills, and reamers. In one sense, grinding wheels could be considered multiple-point cutters, since very small chips are removed by the many tiny cutting points on grinding wheels.

One of the most important aspects of cutting tool geometry is **rake** (Figure C1.3). Tool rake ranging from negative to positive has an effect on the formation of the chip and on the surface finish. **Zero-** and negative-**rake** tools are stronger and have a longer working life than positive-rake tools. Negative-rake tools produce poor finishes at low cutting speeds but give a good finish at high speeds. Positive-rake tools are freer cutting at low speeds and can produce good finishes when they are properly sharpened.

A common misconception is that the material splits ahead of the tool as wood does when it is being split with an axe. This is not true with metals; the metal is sheared off and does not split ahead of the chip (Figures C1.4 to C1.7). The metal is forced along in the direction of the cut and the grains are elongated and distorted ahead of the tool and forced along a shear plane, as can be seen in the micrographs. The surface is disrupted more with the negative-rake tool than with the tool with the positive rake in this case because it is moved at a slow speed. Negative-rake tools require more power than positive-rake tools.

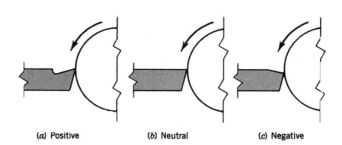

(a) Positive   (b) Neutral   (c) Negative

FIGURE C1.3 Side view of rake angles.

TABLE C1.1 Machining principles and operations

Operation	Diagram	Characteristics	Type of Machines
Turning		Work rotates, tool moves for feed	Lathe and vertical boring mill
Milling (horizontal)		Cutter rotates and cuts on periphery; work feeds into cutter and can be moved in three axes	Horizontal milling machine
Face milling		Cutter rotates to cut on its end and periphery of vertical workpiece	Horizontal mill, profile mill, and machining center
Vertical (end) milling		Cutter rotates to cut on its end and periphery, work moves on three axes for feed or position; spindle also moves up or down	Vertical milling machine, die sinker, machining center
Shaping		Work is held stationary and tool reciprocates; work can move in two axes; toolhead can be moved up or down	Horizontal and vertical shapers
Planing		Work reciprocates while tool is stationary; tool can be moved up, down, or crosswise; worktable cannot be moved up or down	Planer
Horizontal sawing (cutoff)		Work is held stationary while the saw cuts either in one direction, as in band sawing, or reciprocates while being fed downward into the work	Horizontal band saw, reciprocating cutoff saw

TABLE C1.1 Machining principles and operations (*continued*)

Operation	Diagram	Characteristics	Type of Machines
Vertical band sawing (contour sawing)		Endless band moves downward, cutting a **kerf** in the workpiece that can be fed into the saw on one plane at any direction	Vertical band saw
Broaching		Workpiece is held stationary while a multitooth cutter is moved across the surface; each tooth in the cutter cuts progressively deeper	Vertical broaching machine, horizontal broaching machine
Horizontal spindle surface grinding		The rotating grinding wheel can be moved up or down to feed into the workpiece; the table, which is made to reciprocate, holds the work and can also be moved crosswise	Surface grinders, specialized industrial grinding machines
Vertical spindle surface grinding		The rotating grinding wheel can be moved up or down to feed into the workpiece; the circular table rotates	Blanchard-type surface grinders
Cylindrical grinding		The rotating grinding wheel contacts a turning workpiece that can reciprocate from end to end; the wheelhead can be moved into the work or away from it	Cylindrical grinders, specialized industrial grinding machines

TABLE C1.1 Machining principles and operations (*continued*)

Operation	Diagram	Characteristics	Type of Machines
Centerless grinding		Work is supported by a workrest between a large grinding wheel and a smaller feed wheel	Centerless grinder
Drilling and reaming		Drill or reamer rotates while work is stationary	Drill presses, vertical milling machines
Drilling and reaming		Work turns while drill or reamer is stationary	Engine lathes, turret lathes, automatic screw machines
Boring		Work rotates, tool moves for feed on internal surfaces	Engine lathes, horizontal and vertical turret lathes, and vertical boring mills (on some horizontal and vertical boring machines the tool rotates and the work does not)

SOURCE: J. E. Neely and R. R. Kibbe, *Modern Materials and Manufacturing Processes*, John Wiley & Sons, Inc., New York, Copyright 1987.

Higher speeds give better surface finishes and produce less disturbance of the grain structure. This can be seen in Figures C1.8 and C1.9. At a lower speed of 100 surface feet per minute **(SFPM),** the metal is disturbed to a depth of .005 to .006 in., and the grain flow is moving in the direction of the cut. The grains are distorted and in some places the surface is torn. This condition can later produce **fatigue** failures and a shorter working life of the part than would a better surface finish. Tool marks, rough surfaces, and an insufficient internal radius at shaft shoulders can also cause early fatigue failure where there are high stress and frequent reversals of the stress. At 400 SFPM, the surface is less dis-

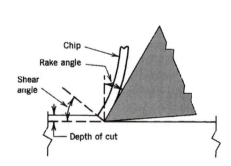

FIGURE C1.4 Metal cutting diagram.

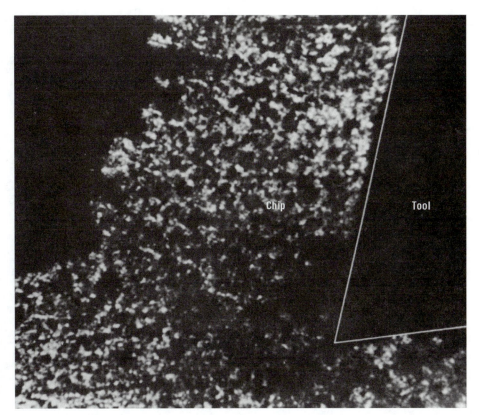

FIGURE C1.5 A chip from a positive-rake tool magnified 100 diameters at the point of the tool. The grain distortion is not as evident as in Figures 6 and 7. This is a continuous chip.

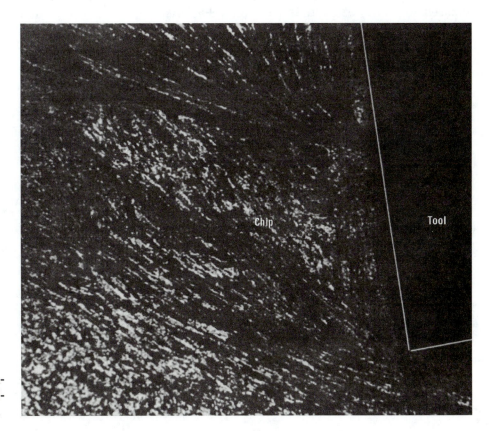

FIGURE C1.6 Point of a negative-rake tool magnified 100 diameters at the point of the tool.

FIGURE C1.7 A zero-rake tool at 100 diameters shows similar grain flow and distortion to that created by the negative-rake tool.

rupted and the grain structure is altered only to a depth of about .001 in. When the cutting speeds are increased to 600 SFPM and above, little additional improvement is noted.

There is a great difference between the surface finish of metals cut with coolant or lubricant and metals that are cut dry. This is because of the cooling effect of the cutting fluid and the lubricating action that reduces friction between tool and chip. The chip tends to curl away from the tool more quickly and the chip is more uniform when a cutting fluid is used. Also, the chip becomes thinner and pressure

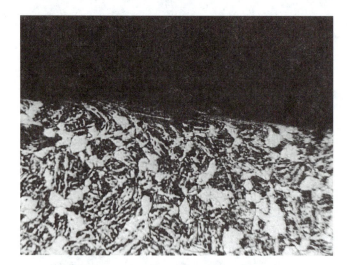

FIGURE C1.8 This micrograph shows the surface of the specimen that was turned at 100 SFPM. The surface is irregular and torn and the grains are distorted to a depth of approximately .005 to .006 in. (250×).

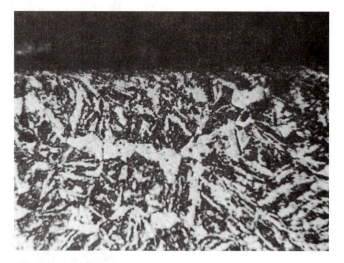

FIGURE C1.9 At 400 SFPM, this micrograph reveals that the surface is fairly smooth and the grains are only slightly distorted to a depth of approximately .001 in. (250×).

FIGURE C1.10 A continuous form chip is beginning to curl away from this positive-rake tool.

FIGURE C1.12 A discontinuous, thick chip is being formed with a zero-rake tool.

welding is reduced. When metals are cut dry, pressure welding is a definite problem, especially in the softer metals such as 1100 aluminum and low-carbon steels. Pressure welding produces a built-up edge (BUE) that causes a rough finish and a tearing of the surface of the workpiece. Built-up edge is also caused by speeds that are too slow; this often results in broken tools from excess pressure on the cutting edge. Figures C1.10 to C1.12 show chips formed with tools having positive, negative, and zero rakes. These chips were all formed at low surface speeds and consequently are thicker than they would have been at higher speeds. There is also more distortion at low speeds. The material was cut dry.

At high speeds, a crater begins to form on the top surface of the tool because of the wear of the chip against the tool; this causes the chip to begin to curl

(Figure C1.13). The crater makes an airspace between the chip and the tool, which is an ideal condition since it insulates the chip from the tool and allows the tool to remain cooler; the heat goes off with the chip. The crater also allows the cutting fluid to get under the chip.

Various metals are cut in different ways. Softer, more ductile metals produce a thicker chip and harder metals produce a thinner chip. A thin chip indicates a clean cutting action with a better finish.

A machine operator often notices that using certain tools having certain rake angles at greater speeds (several hundred surface feet per minute as compared with 50 to 100 SFPM) produces a better finish on the surface. The operator may also notice that when speeds are too low, the surface finish becomes rougher. Not only do speeds, feeds, tool shapes, and

FIGURE C1.11 A thick discontinuous chatter chip being formed at slow speed with a negative-rake tool.

FIGURE C1.13 The crater on the cutting edge of this tool was caused by chip wear at high speeds. The crater often helps to cool the chip.

depth of cut have an effect on finishes but the surface structure of the metal itself is disturbed and altered by these factors.

Tool materials are high-speed steel, carbide, ceramic, and diamond. Most manufacturing today is done with carbide tools. Greater amounts of materials may be removed and tool life extended considerably when carbides, rather than high-speed steels, are used. Much higher speeds can also be used with carbide tools than with high-speed tools.

ANALYSIS OF CHIP STRUCTURES

Machining operations performed on various machine tools produce chips of three basic types: the continuous chip, the continuous chip with built-up edge on the tool, and the discontinuous chip. The formation of the three basic types of chip can be seen in Figure C1.14. Various kinds of chip formations are shown in Figure C1.15. High cutting speeds produce thin chips, and tools with a large positive rake angle favor the formation of the continuous chip. Any circumstances that lead to a reduction of friction between the chip-tool interface, such as the use of cutting fluid, tend to produce a continuous chip. The continuous chip usually produces the best surface finish and has the greatest efficiency in terms of power consumption. Continuous chips create lower temperatures at the

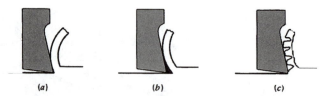

FIGURE C1.14 The three types of chip formation: *(a)* Continuous; *(b)* Continuous with built-up edge; *(c)* Discontinuous (segmented).

cutting edge, but at very high speeds there are higher cutting forces and very high tool pressures. Since there is less strength at the point of positive rake-angle tools than in the negative-rake tools, tool failure is more likely with large positive rake angles at high cutting speeds or with intermittent cuts.

Negative-rake tools are most likely to produce a built-up edge with a rough continuous chip and a rough finish on the work, especially at lower cutting speeds and with soft materials. Positive rake angles and the use of cutting fluid plus higher speeds decrease the tendency for a built-up edge on the tool. However, most carbide tools for turning machines have a negative rake. There are several reasons for this. Carbides are always used at high cutting speeds, lessening the tendency for a built-up edge. Better finishes are obtained at high speeds even with

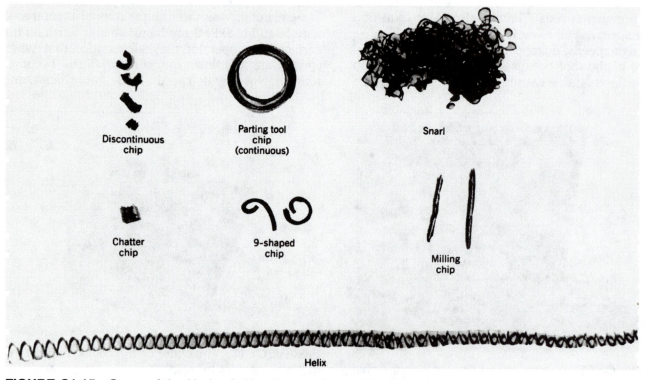

Discontinuous chip

Parting tool chip (continuous)

Snarl

Chatter chip

9-shaped chip

Milling chip

Helix

FIGURE C1.15 Some of the kinds of chips that are formed in machining operations.

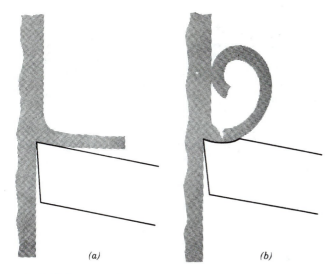

FIGURE C1.16 The action of a chip breaker to curl a chip and cause it to break off: *(a)* plain tool; *(b)* tool with chip breaker.

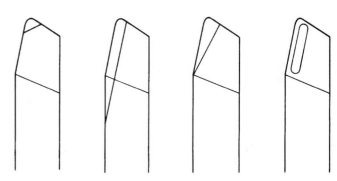

FIGURE C1.17 Types of chip breakers in high-speed tools.

negative rake and the tool can withstand greater shock loads than positive-rake tools. Another advantage of negative-rake tools is that an indexable insert can have 90-degree angles between its top surface and flank, making more cutting edges on both sides. With a negative rake the flank of the tool has relief even though the tool has square edges. More horsepower is needed for negative-rake tools than for positive-rake tools when all other factors are the same.

The discontinuous or segmented chip is produced when a brittle metal, such as cast iron or hard bronze, is cut. Some ductile metals can form a discontinuous chip when the machine tool is old or loose and a **chattering** condition is present, or when the tool form is not correct. The discontinuous chip is formed as the cutting tool contacts the metal and compresses it to some extent; the chip then begins to flow along the tool and, when more stress is applied to the brittle metal, it tears loose and a rupture occurs. This causes the chip to separate from the work material. Then a new cycle of compression, the tearing away of the chip, and its breaking off begins.

Low cutting speeds and a zero or negative rake angle can produce discontinuous chips. The discontinuous chip is more easily handled on the machine since it falls into the chip pan.

The continuous chip sometimes produces snarls or long strings that are not only inconvenient but dangerous to handle. The optimum kind of chip for operator safety and producing a good surface finish is the 9-shaped chip that is usually produced with a chip breaker.

Chip breakers take many forms and most carbide tool holders either have an inserted chip breaker or the chip breaker is formed in the insert tool itself (Figure C1.16). The chip breaker is designed to curl the chip against the work and then to break it off to produce the proper type of chip.

Machine operators usually must form the tool shape themselves on a grinder when using high-speed tools and may or may not grind a chip breaker on the tool. If they do not, a continuous chip is formed that usually makes a wiry tangle, but if they grind a chip breaker in the tool, depending on the feed and speed, they produce a more acceptable type of chip. A high rate of feed will produce a greater curl in the chip, and, often even without a chip breaker, a curl can be produced by adjusting the feed of the machine properly. The depth of cut also has an effect on chip curl and breaking. A larger chip is produced when the depth of cut is greater. This heavier chip is less springy than light chips and tends to break up into small chips more readily. Figure C1.17 shows how chip breakers may be formed in a high-speed tool.

MACHINABILITY OF METALS

Various ratings are given to metals of different types on a scale based on properly annealed carbon steel containing about .1 percent carbon (which is 100 on the scale). Much information has been published on the relative machinability of the various grades of carbon steel. They have been compared almost universally with B-1112, a free-machining steel, which was once the most popular screw machining stock. Machinability ratings by some manufacturers are now based on AISI 1212, which is rated at 100 percent and at 168 SFPM. Metals that are more easily machined have a number higher than 100 and materials that are more difficult to machine are numbered lower than 100. Table C1.2 gives the machinability of various alloys of steel based on AISI B-1112 as 100 percent. Of course, the machinability is greatly affected by the cutting tool material, tool geometry, and the use of cutting fluids.

In general, the machinist must select the type of tool, speeds, feeds, and kind of cutting fluid for the material being cut. The most important material property, however, is hardness. A machinist often shop tests the hardness of material with a file to determine relative machinability since hardness is related to machinability. Hardness is also a factor in producing good finishes. Soft metals such as copper, 1100 series aluminum, and low-carbon hot-rolled steel tend to have poor finishes when cut with ordinary tooling. Cutting tools tend to tear these soft metals away rather than cut them; however, very sharp tools and larger rake angles plus the use of cutting fluids can help to produce better finishes. Harder, tougher alloys almost always produce better finishes even with negative-rake carbide tools. For example, an AISI 1020 hot-rolled steel bar and an AISI 4140 steel bar of the same diameter cut at the same speed and feed with the same tooling would have markedly different finishes. The low-carbon HR steel would have a poor finish compared to the alloy steel.

The operator must also understand the effects of heat on normally machinable metals such as alloy steel, tool steel, or gray cast iron. Welding on any of these metals may harden the base metal near the weld and make it difficult to cut, even with carbide tools. To soften these hard areas, the entire workpiece must first be annealed.

Most alloy steel can be cut with high-speed tools but at relatively low cutting speeds, which produces a poor finish unless some back rake is used. These steels are probably best machined by using carbide tools at higher cutting speeds ranging from 200 to 350 SFPM, or even higher in some cases. Some nonferrous metals, such as aluminum and magnesium, can be machined with much higher speeds ranging from 500 to 8000 SFPM when the correct cutting fluid is used. Some alloy steels, however, are more difficult to machine because they tend to work harden. Examples of these are austenitic manganese steels (those used for wear resistance), Inconel, stainless steels (SAE 301 and others), and some tool steels.

TABLE C1.2 Machinability ratings for some commonly used steels

AISI Number	Machinability Index (% Relative Speed Based on AISI B1112 as 100%)
B1112	100
C1120	81
C1140	72
C1008	66
C1020	72
C1030	70
C1040	64
C1060 (annealed)	51
C1090 (annealed)	42
3140 (annealed)	66
4140 (annealed)	66
5140 (annealed)	70
6120	57
8620	66
301 (stainless)	36
302 (stainless)	36
304 (stainless)	36
420 (stainless)	36
440A (stainless)	24
440B (stainless)	24
440C (stainless)	24

SELF-TEST

1. In what way do tool rakes, positive and negative, affect surface finish?
2. Soft materials tend to pressure weld on the top of the cutting edge of the tool. What is this condition called and what is its result?
3. Which indicate a greater disruption of the surface material: thin, uniform chips or thick, segmented chips?
4. In metal cutting does the material split ahead of the tool? If not, what does it do?
5. Which tool form is stronger, negative or positive rake?
6. What effect does cutting speed have on surface finish? On surface disruption of grain structure?
7. How can surface irregularities caused by machining later affect the usefulness of the part?
8. Which property of metals is directly related to machinability? How do machinists usually determine this property?
9. Describe the type of chip produced on a machine tool that is the safest and easiest to handle.
10. Define *machinability*.

Speeds and Feeds for Machine Tools

Modern machine tools are very powerful and with modern tooling they are designed for a high rate of production. However, if the operator uses the incorrect feeds and speeds for the machine operation, a much lower production rate and an inferior product will result. Sometimes an operator will take a very light roughing cut when it is possible to take a greater depth of cut. Often the cutting speeds are too low to produce good surface finishes and for optimum metal removal. Because of the cost of labor, time is very important and removal of a larger amount of metal can shorten the time needed to produce a part and often improve its physical properties.

The importance of cutting speeds and machine feeds in machining operations cannot be overemphasized. The right speed can mean the difference between burning the end of a drill or other tool, causing lost time, or having many hours of cutting time between sharpenings or replacement. The correct feed can also lengthen tool life. Excessive feeds often result in tool breakage. This unit will prepare you to think in terms of correct feeds and speeds.

OBJECTIVES

After completing this unit, you should be able to:

1. Calculate correct cutting speeds for various machine tools and grinding machines.
2. Determine correct feeds for various machining operations.

CUTTING SPEEDS

Cutting speeds (CS) are normally given in tables for cutting tools and are based on surface feet per minute (FPM or SFPM). Surface feet per minute means that either the tool moves past the work or the work moves past the tool at a rate based on the number of feet that pass the tool in one minute. It can be determined from a flat surface or on the **periphery** of a cylindrical tool or workpiece. A small cylinder will have more revolutions per minute (RPM) than a large one with the same surface speed. For example, if a thin wire were pulled off a 1-in.-diameter spool at the rate of 20 FPM, the spool would rotate three times faster than one 3 in. in diameter with the wire being pulled off at the same rate of 20 FPM.

Since machine spindle speeds are given in revolutions per minute (RPM), they can be derived in the following manner:

$$\text{RPM} = \frac{\text{Cutting speed (CS: in feet per minute)} \times 12}{\text{Diameter of cutter (}D\text{; in inches)} \times \text{pi}}$$

If you use 3 to approximate pi (3.1416), the formula becomes

$$\frac{\text{CS} \times 12}{D \times 3} = \frac{\text{CS} \times 4}{D}$$

This simplified formula is certainly the most common one used in machine shop practice and it applies to the full range of machine tool operations, which include the lathe and the milling machine as well as the drill press. The simplified formula

$$\text{RPM} = \frac{\text{CS} \times 4}{D}$$

is used throughout this book. The formula is used, for example, as follows: where D = the diameter of the drill or other rotating tool in inches and CS = an assigned cutting speed for a particular material, for a $\frac{1}{2}$-in. drill in low-carbon steel the speed would be

$$\frac{90 \times 4}{\frac{1}{2}} = 720 \text{ RPM}$$

Table C2.1 gives cutting speeds and starting values for common materials. These may have to be varied up or down depending on the specific machining task. Always observe the cutting action carefully and make appropriate speed corrections as needed. Until you gain some experience in machining, use the lower values in the table when selecting cutting speeds. As you can see by the hardness values in the table, there is a general relationship between hardness and cutting speed. This is not always true, however. Some stainless steel is relatively soft, but it must be cut at a low speed.

Cutting speeds/RPM tables for various materials are available in handbooks and on wall charts. The cutting speed for carbide tools is normally three to four times that of high-speed cutting tools.

You will learn more about cutting tool materials in later sections in this book, but a good rule to remember concerning cutting speeds is: Generally, high speed cutting tools will wear out quickly at too-high cutting speeds. Carbide cutting tools will chip or break up quickly at too-low cutting speeds. When steel chips become blue colored, it is a sign of a higher temperature caused by high cutting speed and/or dull tools. Blue chips are acceptable and in fact desirable when using carbides but not when using high-speed steel tools. The chips should not be discolored at all with high-speed tools, especially when using cutting fluids.

Cutting speed constants are influenced by the cutting tool material, workpiece material, rigidity of the machine setup, and the use of cutting fluids. As a rule, lower cutting speeds are used to machine hard or tough materials or where heavy cuts are taken and it is desirable to minimize tool wear and thus maximize tool life. Higher cutting speeds are used in machining softer materials in order to achieve better surface finishes.

Cutting speeds for drills are the same whether the drill rotates or whether the work rotates and the drill remains stationary, as in lathe work. For a step drill where there are two or more diameters, the largest diameter is used in the formula.

Where the workpiece rotates and the tool is stationary, as in lathes, the outer diameter (D) of the workpiece is used in the speed formula. As the workpiece is reduced to a smaller diameter, the speed should be increased. However, there is some latitude in speed adjustments and 5 to 10 SFPM one way or another is usually not significant. Therefore, if a machine cannot be set at the calculated RPM, use the next-lower speed. If there is vibration or chatter, a lower speed will often eliminate it.

For multiple-point cutting, as in milling machines, D is the diameter of the milling cutter in inches. Cutting speed in milling is the rate at which a point on the cutter passes by a point on the workpiece in a given period of time.

If the cutting speed is sought and the RPM known, the following formula may be used:

$$\text{CS} = \frac{\text{RPM} \times D}{4}$$

FEEDS

The machine movement that causes a tool to cut into or along the surface of a workpiece is called *feed*. The amount of feed in metal cutting is usually measured in thousandths of an inch. Feeds are expressed in slightly different ways on various types of machine tools.

Drills on drill presses rotate, causing the cutting edges on the end of the drill to cut into the workpiece. When metal is cut, considerable force is required to feed the cutting edges of the drill into the workpiece. Drilling machines that have power feeds are designed to advance the drill a given amount for

TABLE C2.1 Cutting speeds and starting values for some commonly used materials

Work Material	Hardness (BHN)	Tool Material	
		High-Speed Steel	Cemented Carbide
Aluminum	60–100	300–800	1000–2000
Brass	120–220	200–400	500–800
Bronze (hard drawn)	220	65–130	200–400
Gray cast iron (ASTM 20)	110	50–80	250–350
Low-carbon steel	220	60–100	300–600
Medium-carbon alloy steel (AISI 4140)	229[a]	50–80	225–400
High-carbon steel	240[a]	40–70	150–250
Carburizing-grade alloy steel (AISI 8620)	200–250[a]	40–70	150–350
Stainless steel (type 410)	120–200[a]	30–80	100–300

[a] Normalized.

each revolution of the spindle. Therefore, a .006-in. feed means that the drill advances .006 in. for every revolution of the machine spindle. Thus feeds for drill presses are expressed in inches per revolution (IPR). The amount of feed varies with the drill size and the kind of work material (Table C2.2).

When drill presses have no power feed, the operator provides the feed with a handwheel or hand lever. These drilling machines are called *sensitive* drill presses because the operator can "sense" or feel the correct amount of feed. Also, the formation of the chip as a tightly rolled **helix** instead of a stringy chip is an indicator of the correct feed. Excessive feed on a drill can cause jamming or stopping of the machine, drill breakage, and a hazardous situation if the workpiece is not properly clamped.

Feeds on turning machines such as lathes are also expressed in inches per revolution (IPR) of the lathe spindle. The tool moves along the rotating workpiece to produce a chip. The depth of the cut is not related to feed. The **quick-change gearbox** on the lathe makes possible the selection of feeds in IPR in a range from approximately .001 in. to about .100 in., depending on size and machine manufacturer. Feed rates for small (10-in. swing) lathe operations can be as high as .015 IPR for roughing operations and usually .003 to .005 IPR for finishing. However, some massive turning machines may use .030 to .050 IPR feed with a .750-in. depth of cut. Coarser feeds, depending on the machine size, horsepower, and rigidity, necessary for rapid stock removal, are called *roughing.* As long as the roughing dimension is kept well over the finish size, feeds should be set to the maximum the machine will handle. In roughing operations, finish is not important and no attempt should be made to obtain a good finish; the only consideration is stock removal and, of course, safety. It is in the finishing cuts with fine feeds that dimensional accuracy and surface finish is of the greatest importance and stock removal is of little consideration.

Feeds on milling machines refer to the rate at which the workpiece material is advanced into the cutter by table movement or where the cutter is advanced into the workpiece in the manner of a drill cutting on its end. Since each tooth on a multitooth milling cutter makes a chip, the chip thickness depends on the amount of table feed. Feed rate in milling is measured in inches per minute (IPM) and is calculated by the formula

$$IPM = F \times N \times RPM, \text{ where}$$

$$
\begin{aligned}
IPM &= \text{feed rate, in inches per minute} \\
F &= \text{feed per tooth} \\
N &= \text{number of teeth in the cutter being used} \\
RPM &= \text{revolutions per minute of the cutter}
\end{aligned}
$$

Feeds for end mills used in vertical milling machines range from .001 to .002 in. feed per tooth for very small diameter cutters in steel work material to .010 in. feed per tooth for large cutters in aluminum workpieces. Since the cutting speed for mild steel is 90, the RPM for a $\frac{3}{8}$-in. high-speed, two-flute end mill is

$$RPM = \frac{CS \times 4}{D} = \frac{90 \times 4}{3/8} = \frac{360}{.375} = 960 \text{ RPM}$$

To calculate the feed rate, we will select .002 in. feed per tooth. (See Section F, Unit 2, for a table for end mill feed rates.)

$$IPM = F \times N \times RPM = .002 \times 2 \times 960 = 3.84$$

The cutter should rotate at 960 RPM and the table feed be adjusted to approximately 3.84 IPM.

Horizontal milling machines have a horizontal spindle instead of a vertical one, but the feed rate calculations are the same as for vertical mills. Horizontal milling cutters usually have many more teeth than end mills and are much larger. Cutting speeds and RPM are also calculated in the same way. (See Section F, Unit 4, Tables 1 and 2, for cutting speeds and feed rates for horizontal milling.)

Feeds for surface grinding are set for the crossfeed so that the workpiece moves at the end of each stroke of the table. The crossfeed dial is graduated in thousandths, but the feed is not usually calculated very precisely. As a rule, the crossfeed is set to move from one-quarter to one-third of the wheel width at each stroke. This coarse feed allows the wheel to break down (wear) more evenly than do fine feeds, which cause one edge of the wheel face to break down.

In plain cylindrical grinding where the rotating workpiece is traversed back and forth against a large grinding wheel, table feeds are used as in surface grinding. The feed or traverse setting must be adjusted for roughing or finishing to get optimum results.

TABLE C2.2   Drilling feed table

Drill Size Diameter (in.)	Feeds per Revolution (in.)
Under $\frac{1}{8}$	.001 to .002
$\frac{1}{8}$ to $\frac{1}{4}$	.002 to .004
$\frac{1}{4}$ to $\frac{1}{2}$	.004 to .007
$\frac{1}{2}$ to 1	.007 to .015
Over 1	.015 to .025

SELF-TEST

1. If the cutting speed (CS) for low-carbon steel is 90 and the formula for RPM is $(CS \times 4)/D$, what should the RPM of the spindle of a drill press be for a $\frac{1}{8}$-in.-diameter high-speed twist drill? For a $\frac{3}{4}$-in.-diameter drill?
2. If the two drills in Question 1 were used in a lathe instead of a drill press, what should the RPM of the lathe spindle be for both drills?
3. If, in Question 2, the small drill could not be set at the calculated RPM because of machine limitations, what could you do?
4. An alloy steel 2-in.-diameter cylindrical bar having a cutting speed of 50 is to be turned on a lathe using a high-speed tool. At what RPM should the lathe be set?
5. Are feeds on a drill press based on inches per minute or inches per revolution of the spindle?
6. Which roughing feed on a small lathe would work best, .100 or .010 in. per revolution?
7. What kind of machine tool bases feed on inches per minute instead of inches per revolution?
8. Approximately what should the feed rate be on a $\frac{3}{4}$-in.-wide grinding wheel on a surface grinder?

Cutting Fluids

Since the beginning of the Industrial Revolution, when metals first began to be cut on machines, cutting fluids have been an important aspect of machining operations. Lubricants in the form of animal fats were first used to reduce friction and cool the workpiece. These straight fatty oils tended to become rancid, had a disagreeable odor, and often caused skin rashes on machine operators. Although lard oil alone is no longer used to any great extent, it is still used as an additive in cutting oils. Plain water was sometimes used to cool workpieces, but water alone is a corrosive liquid and tends to rust machine parts and workpieces. Water is now combined with oil in an emulsion that cools but does not corrode. Many new chemical and petroleum-based cutting fluids are in use today, making possible the high rate of production in machining and manufacturing of metal products we presently enjoy. The reasons for this are that cutting fluids reduce machining time by allowing higher cutting speeds and they reduce tool breakdown and down time. This unit will prepare you to use cutting fluids properly when you begin to operate machine tools.

OBJECTIVES

After completing this unit, you should be able to:

1. List the various types of cutting fluids.
2. Explain the correct uses of and care in using several cutting fluids.
3. Describe several methods of cutting fluid application.

EFFECTS OF CUTTING FLUIDS

Cutting fluids is a generic term that covers a number of different products used in cutting operations. Basically, there are two major effects derived from cutting and grinding fluids: cooling and lubrication. Although water-based fluids are most effective for cooling the tool and workpiece, and oil-based fluids are the better lubricants, there is a considerable overlap between the two.

Cutting force and temperature rise are not generated so much by the friction of the chip sliding over the surface of the tool as by the shear flow of the workpiece metal just ahead of the tool (Figure C3.1). Some tool materials, such as carbide and ceramic, are able to withstand high temperatures. When these materials are used in high-speed cutting operations, the use of cutting fluids may actually be counterproductive. The increased temperatures tend to promote an easier shear flow and thus reduce cutting force. High-speed, high-temperature cutting also produces better finishes and disrupts the surface less than does lower-speed machining.

For this reason, machining with extremely high cutting speeds using carbide tools is often done dry. The use of coolant may cause a carbide tool to crack and break up due to thermal shock, since the flooding coolant can never reach the hottest point at the tip of the tool and so will not maintain an even temperature on the tool.

It is at the lower cutting speeds in which high-speed tools and even carbides are used that cutting fluids are used to the greatest advantage. Also, cutting fluids are a necessity in most precision grinding operations. Most milling machine, drilling, and many lathe operations require the use of cutting fluids.

TYPES OF CUTTING FLUIDS

Lubricants as well as coolants (which also provide some lubrication) are able to reduce cutting forces at lower cutting speeds (allowing somewhat higher speeds to be used). They also reduce or remove heat generation that tends to break down high-speed tools. Both lubricants and coolants extend tool life and improve workpiece finish. Another advantage in

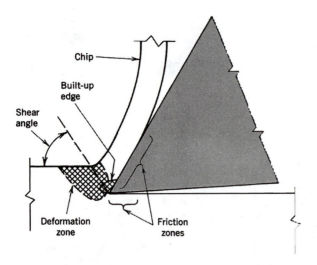

FIGURE C3.1 Metal-cutting diagram showing the shear flow ahead of the tool.

using cutting fluids is that the workpiece dimensions can be accurately measured on the cool workpiece. A hot workpiece must be cooled to room temperature to obtain a correct measurement.

Several types of cutting fluids are in use; their nomenclature is by no means universal since a single fluid is often called by many different names. Next we describe the names used in this book.

Synthetic Fluids

Sometimes called *chemical fluids,* these can be *true solutions,* consisting of inorganic substances dissolved in water, which include nitrites, nitrates, borates, or phosphates. As their name implies, these chemical fluids form true solutions in water unlike the oils in emulsions. These chemical substances are water **miscible** and vary in color from milky to transparent when mixed with water. This type is superior to others for machining and grinding titanium and for grinding operations in general. The *surface-active* type is a water solution that contains additives that lower the surface tension of the water. Some types have good lubricity and corrosion inhibitors.

Some of the advantages of these chemical fluids are their resistance to becoming rancid (especially those that contain no fats), good detergent (cleansing) properties, rapid heat dissipation, and the fact that they are easy to mix, requiring little agitation. There are also some disadvantages. A lack of lubrication (oiliness) in most types may cause sticking in some machine parts that depend on the cutting fluid for lubrication. Also, the high detergency has a tendency to remove the normal skin oils and irritate

workers' hands where there is a long continual exposure. Synthetic fluids generally provide less corrosion resistance than oilier types and some tend to foam to a certain extent. When improved lubricating qualities are needed for synthetic fluids, sulfur, chlorine, or phosphorus is added.

Semisynthetic Fluids

Sometimes called *semichemical fluids,* these cutting fluids are essentially a combination of synthetic fluids and emulsions. They have a lower oil content than the straight emulsion fluids, and the oil droplets are smaller because of a higher content of emulsifying molecules, thus combining the best qualities of both types. Since a small amount of mineral oil is added to these semisynthetic fluids, they possess enhanced lubrication qualities. In addition, they may contain other additives such as chlorine, sulfur, or phosphorus.

Emulsions

Also called *water-miscible fluids* or *water-soluble oils,* emulsions like ordinary soap cause oil to combine with water in a suspension of tiny droplets. Special soaps and other additives such as amine soaps, rosin soaps, petroleum sulfonates, and naphthenic acids are blended with a naphtha-based or paraffin mineral oil to emulsify it. Other additions, such as bactericides, help to reduce bacteria, fungi, and mold and to extend emulsion life. Without these additives, an emulsion tends to develop a strong, offensive odor because of bacterial action and must be replaced with a new mix.

These ordinary emulsified oils contain oil particles large enough to reflect light, and therefore they appear opaque or milky when combined with water. In contrast, many of the synthetic and semisynthetic emulsion particles are so small that the water remains clear or translucent when mixed with the chemical. The ratio of mixing oil and water varies with the job requirement and can range from 5 to 70 parts water to 1 part oil. However, for most machining and grinding operations, a mixture of 1 part oil to 20 parts water is generally used. A mixture that is too rich for the job can be needlessly expensive, and a mixture that is too lean may cause rust to form on the workpiece and machine parts. When mixing, oil should be poured into the water.

Some emulsions are designed for greater lubricating value, with animal or vegetable fats or oils providing a "superfatted" condition. Sulfur, chlorine, and phosphorus provide even greater lubricating value for metal cutting operations where extreme pressures are encountered in chip forming. These fluids are mixed in somewhat rich ratios: 1 part oil with 5 to 15 parts water. All of these water-soluble cutting fluids are considered to be in the category of coolants even though they do provide some lubrication.

Cutting Oils

Cutting oils are fluids that may be animal, vegetable, petroleum, or marine (fatty tissue of fish and other marine animals) oil, or a combination of two or more of these. The plain cutting oils (naphthenic and paraffinic) are considered to be lubricants and are useful for light-duty cutting on metals of high machinability, such as free-machining steels, brass, aluminum, and magnesium. Water-based cutting fluids should never be used for machining magnesium because of the fire hazard. The cutting fluid recommended for magnesium is an **anhydrous** (without water) oil that has a very low acid content. Fine magnesium chips can burn if ignited, and water tends to make it burn even more fiercely. A mineral-lard oil combination is often used in automatic screw machine practice.

Where high cutting forces are encountered and extreme pressure lubrication is needed, as in threading and tapping operations, certain oils, fats, waxes, and synthetic materials are added. The addition of animal, marine, or vegetable oil to petroleum oil improves the lubricating quality and the wetting action. Chlorine, sulfur, or phosphorus additives provide better lubrication at high pressures and high temperatures. These chlorinated or sulfurized oils are dark in color and are commonly used for thread-cutting operations.

Cutting oils also tend to become rancid and develop disagreeable odors unless germicides are added to them. Cutting oils tend to stain metals and, if they contain sulfur, may severely stain nonferrous metals such as brass and aluminum. In contrast, soluble oils generally do not stain workpieces or machines unless they are trapped for long periods of time between two surfaces, such as the base of a milling vise bolted to a machine table.

Tanks containing soluble oil on lathes, milling, and grinding machines tend to collect tramp oil and dirt or grinding particles. For this reason, the fluid should be removed periodically, the tank cleaned, and a clean solution put in the tank. Water-based cutting fluid can be contaminated quickly when a machinist uses a pump oilcan to apply cutting oil to the workpiece instead of using the coolant pump and nozzle on the machine. The tramp oil goes into the tank and settles on the surface of the coolant, creating an oil seal where bacteria can grow. This causes an odorous scum to form that quickly contaminates the entire tank.

Special vegetable-base lubricants have been developed that boast machining efficiency and tool life. Only very small amounts are used and messy coolant flooding is not necessary. They are being used in milling, sawing reaming, and grinding operations.

Gaseous Fluids

Fluids such as air are sometimes used to prevent contamination on some workpieces. For example, some reactive metals such as zirconium may be contaminated by water-based cutting fluids. The atmosphere is always, to some extent, a cutting fluid and is actually a coolant when cutting dry. Air can provide better cooling when a jet of compressed air is directed at the point of cut. Other gases such as argon, helium, and nitrogen are sometimes used to prevent oxidation of the chip and workpiece in special applications.

METHODS OF CUTTING FLUID APPLICATION

The simplest method of applying a cutting fluid is manually with a pump oilcan. This method is sometimes used on small drill presses that have no coolant pumping system. Another simple method is with a small brush that is dipped into an open pan of cutting oil and applied to the workpiece, as in lathe threading operations.

Most milling and grinding machines, large drill presses, and lathes have a built-in tank or reservoir containing a cutting fluid and a pumping system to deliver it to the cutting area. The cutting fluid used in these machines is typically a water-soluble synthetic or soluble oil mix, with some exceptions, as in machining magnesium or in certain grinding applications where special cutting oils or fluids are used. The cutting fluid in the tank is picked up by a low-pressure (5 to 20 PSI) pump and delivered through a tube or hose to a nozzle where the machining or grinding operation is taking place (Figure C3.2).

When this pumping system is used, the most common method of application is by flooding. This is done by simply aiming a single, round nozzle at the work-tool area and applying copious amounts of cutting fluid. This system usually works fairly well for most turning and milling operations. In lathe operations, the nozzle is usually above the workpiece and pointed downward at the tool, but it is rather inefficient. A better method of application would be above and below the tool, as shown in Figure C3.3. Internal lathe work, such as boring operations and

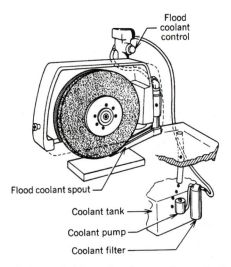

FIGURE C3.2 Fluid recirculates through the tank, piping, nozzle, and drains in a flood grinding system. (DoAll Company)

close chuck work, cause the coolant flow to be thrown outward by the spinning chuck. In this case, a chip guard is needed to contain the spray. Although a single nozzle is normally used to direct the coolant flow over the cutter and on the workpiece on milling machines, a better method is to use two nozzles (Figure C3.4) to make sure the cutting zone is completely flooded. Some lathe toolholders are now provided with through-the-tool pressurized coolant systems. Also, some end mills for vertical milling machines have coolant channels that direct the fluid at the cutting edge under pressure. Not only is this an extremely efficient method of coolant application, it also forces the chips away from the cutting area.

When a cutting fluid is properly applied, it can have a tremendous effect on finishes and surfaces. A

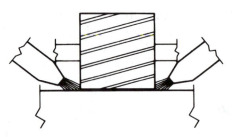

FIGURE C3.4 The use of two nozzles insures flooding of the cutting zone.

finish may be rough when machined dry but may be dramatically improved with lubrication. When carbide tools are used, whether on turning or milling work, and coolant is used, the work-tool area must be flooded to avoid intermittent cooling and heating cycles that create thermal shock and consequent cracking of the tool. The operator should not shut off the coolant while the cut is in progress to "see" the cutting operation.

There is no good way to cool and lubricate a standard twist drill, especially when drilling horizontally in a lathe. In deep holes, cutting forces and temperatures increase, often causing the drill to expand and bind. Even in vertical drilling on a drill press, the helical **flutes,** designed to lift out the chips, also pump out the cutting fluid. Some drills are made with oil holes running the length of the drill to the cutting lips that help to offset the problem. The fluid is pumped under relatively high pressure through a rotating gland. The cutting oil not only lubricates and cools the drill and workpiece, it also flushes out the chips. Gun drills designed to make very accurate and very deep holes have a similar arrangement, but the fluid pressures are as high as 1000 PSI. Shields must be provided for operator

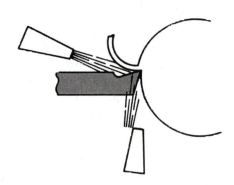

FIGURE C3.3 Although not always possible, this is the ideal way of applying cutting fluid to a cutting tool.

FIGURE C3.5 Mist grinding fluid application on a surface grinder. (DoAll Company)

safety when using these extremely high-pressure coolant systems.

Air-carried mist systems (Figure C3.5) are popular for some drilling and end milling operations. Usually, shop compressed air is used to make a spray of coolant drawn from a small tank on or near the machine. Since only a small amount of liquid ever reaches the cutting area, very little lubrication is provided. These systems are chosen for their ability to cool rather than to lubricate, which may reduce tool life. However, mist cooling has many advantages. No splash guards, chip pans, and return hoses are needed, and only small amounts of liquid are used. The high-velocity air stream also cools the spray further by evaporation. There are some safety hazards in using mist spray equipment from the standpoint of the operator's health. Conventional coolants are not highly toxic, but when sprayed in a fine mist they can be inhaled; this could affect certain people over a period of time. However, many operators find that breathing any of the mist is uncomfortable and offensive. Good ventilation systems or ordinary fans can remove the mist from the operator's area.

SELF-TEST

1. What are the two basic functions of cutting fluids?

2. Name the four types of cutting fluids (liquids).
3. When soluble oil coolants become odorous or cutting oil becomes rancid in the reservoir, what can be done to correct the problem?
4. When a machine such as a lathe has a coolant pump and tank that contains a soluble oil-water mix, why should a machine operator not use a pump can containing cutting oil on a workpiece?
5. Some of the synthetic cutting fluids tend to irritate some workers' hands. What causes this?
6. What method of applying cutting fluid is often used on a small drill press?
7. What kinds of cutting fluid accessories are provided on most machine tools, lathes, milling machines, drill presses, and grinding machines?
8. Describe the most common method of cutting fluid application from a nozzle.
9. Why is a single, round nozzle rather inefficient when used on a grinding wheel?
10. Spray-mist cooling systems work well for cooling purposes, but do not lubricate the work area and tool very well. Why is this?

SECTION D
Drilling Machines

In manual machining, holes are generally made in materials by means of a drill press. In CNC machining, computer-generated hole patterns are often made by a vertical milling machine or some similar device.

In both kinds of hole drilling, the tools are the same. However, for CNC the drills must be precision machine sharpened instead of hand ground on a pedestal grinder.

This section covers the principles of and tooling found in the drilling process. Clamps and workholding devices are essentially the same in conventional manual machining and in CNC.

UNIT 1   The Drill Press

Drilling holes is one of the most basic of machining operations and one that is very frequently done by machinists. Metal cutting requires considerable pressure of feed on the cutting edge. A drill press provides the necessary feed pressure either by hand or power drive. The primary use of the drill press is to drill holes, but it can be used for other operations such as countersinking, counterboring, spot facing, reaming, and tapping, which are processes that modify the drilled hole.

Before operating any machine, a machinist must know the names and the functions of all its parts. In this unit, therefore, you should familiarize yourself with the operating mechanism of several types of drilling machines.

There is a tendency among students to dismiss the dangers involved in drilling since most drilling done in a school shop is performed on small sensitive drill presses with small-diameter drills. This tendency, however, only increases the danger to the operator and has resulted in serious injuries from otherwise harmless situations. Clamps should be used to hold down workpieces, since hazards are always present even with small-diameter drills.

The safety habits you develop now will protect you from many injuries in the school shop and all through your career. This unit will alert you to the potential dangers that exist in drilling and help you avoid the hand injuries most likely to occur while working with the drill press.

OBJECTIVES

After completing this unit, you should be able to:

1. List 10 safety hazards involving drilling machines.
2. Describe and explain steps the operator can take to avoid safety hazards.
3. Identify three basic drill press types and explain their differences and primary uses.
4. Identify the major parts of the sensitive drill press.
5. Identify the major parts of the radial arm drill press.

DRILLING SAFETY

One example of a safety hazard on even a small-diameter drill is the "grabbing" of the drill when it breaks through the hole. If the operator is holding the workpiece by hand and the piece suddenly begins to spin, the hand will likely be injured, especially if the workpiece is thin.

Even if the operator were strong enough to hold the workpiece, the drill can break, continue turning, and with its jagged edge become an immediate hazard to the operator's hand. The sharp chips that turn with the drill can also cut the hand that holds the workpiece.

Poor work habits produce many injuries. Chips flying into unprotected eyes, heavy tooling, or parts dropping from the drill press onto toes, slipping on oily floors, and getting hair or clothing caught in a rotating drill are all hazards that can be avoided by safe work habits.

The following are safety rules to be observed around all types of drill presses:

1. Proper dress (Figure D1.1) does not include loose clothing, gloves, or neckties. Sleeves should be rolled above the elbows, and if hair is long, tie it back or keep it contained in an industrial hairnet. Wristwatches or rings should be taken off when operating a drilling machine. Painful injuries can

FIGURE D1.1 Drill press operator wearing correct attire. (Lane Community College)

FIGURE D1.2 Drilling operation on upright drill press with tools properly located on adjacent work table.

result if any part of you or your attire is caught in a rotating drill.

2. Tools to be used while drilling should never be left lying on the drill press table, but should be placed on an adjacent worktable (Figure D1.2).

3. Get help when lifting heavy vises or workpieces.

4. Workpieces should always be secured with bolts and strap clamps, C-clamps, or **fixtures**. A drill press **vise** should be used when drilling small parts (Figure D1.3). If a clamp should come loose and a "merry-go-round" result, don't try to stop it from turning with your hands. Turn off the machine quickly; if the drill breaks or comes out, the workpiece may fly off the table.

5. Safety glasses are a necessity and are required by federal safety regulations. The small chips that fly at high speed can penetrate the unprotected eye, though will not usually go through clothing or skin. Wearing safety glasses will minimize the

FIGURE D1.3 Properly clamped drill press vise holding work for hard-steel drilling.

danger to your eyes. **Wear safety glasses at all times in a machine shop.**

6. Never clean the **taper** in the spindle when the drill is running, since this practice could result in broken fingers or worse injuries.

7. Always remove the chuck key immediately after using it. A key left in the chuck will be thrown out at high velocity when the machine is turned on. It is a good practice to **never let the chuck key leave your hand when you are using it.** It should not be left in the chuck even for a moment. Some keys are spring loaded so they will automatically be ejected from the chuck when released. Unfortunately, very few of these keys are in use in the industry.

8. Never stop the drill press spindle with your hand after you have turned off the machine. Sharp chips often collect around the chuck or spindle. Do not reach around, near, or behind a revolving drill.

9. When removing taper shank drills with a drift, use a piece of wood under the drills so they will not drop on your toes. This will also protect the drill points.

10. Interrupt the feed occasionally when drilling to break up the chip so it will not be a hazard and will be easier to handle.

11. Use a brush instead of your hands to clean chips off the machine. **Never** use an air jet for removing chips as this will cause the chips to fly at a high velocity and cuts or eye injuries may result. Do not clean up chips or wipe up oil while the machine is running.

12. Keep the floor clean. Immediately wipe up any oil that spills or the floor will be slippery and unsafe.

13. Remove **burrs** from a drilled workpiece as soon as possible, since any sharp edges or burrs can cause severe cuts.

14. When you are finished with a drill or other cutting tool, wipe it clean with a shop towel and store it properly.

15. Oily shop towels should be placed in a closed metal container to prevent a cluttered work area and avoid a fire hazard.

16. When moving the head or table on sensitive drill presses, make sure a safety clamp is set just below the table or head on the column; this will prevent the table from suddenly dropping if the column clamp is prematurely released.

Proper drilling and clamping procedures are important to your safety. Reference to these and other safety procedures will be made in other units on drilling.

TYPES OF DRILL PRESSES AND THEIR USES

There are three basic types of drill presses used for general drilling operations: the sensitive drill press, the upright drilling machine, and the radial arm drill press. The sensitive drill press (Figure D1.4), as the name implies, allows the operator to "feel" the cutting action of the drill as he hand feeds it into the work. These machines are either bench or floor mounted. Since these drill presses are used for light-duty applications only, they usually have a maximum drill size of $\frac{1}{2}$-in. diameter. Machine capacity is measured by the diameter of work that can be drilled (Figure D1.5).

The sensitive drill press has four major parts, not including the motor: the head, column, table, and base. Figure D1.6 labels the parts of the drill press that you should remember. The spindle rotates within the quill, which does not rotate but carries the spindle up and down. The spindle shaft is driven by a stepped-vee pulley and belt (Figure D1.7) or by a variable-speed drive (Figure D1.8). *The motor must be running and the spindle turning when changing speeds with a variable-speed drive.*

FIGURE D1.4 A sensitive drill press. These machines are used for light-duty application. (Wilton Corporation)

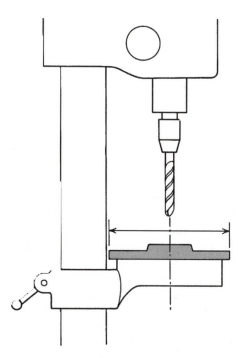

FIGURE D1.5 Drill presses are measured by the largest diameter of a circular piece that can be drilled in the center.

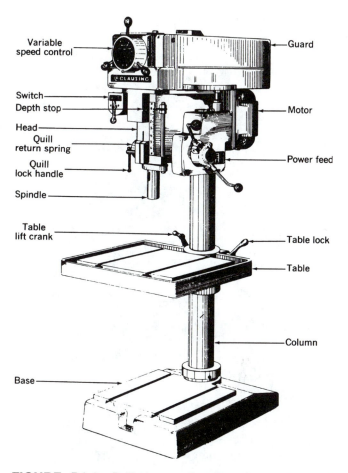

FIGURE D1.6 Drill press showing the names of major parts. (Clausing Machine Tools)

FIGURE D1.7 View of a vee-belt drive. Spindle speeds are highest when the belt is in the top steps and lowest at the bottom steps. (Clausing Machine Tools)

FIGURE D1.8 View of a variable-speed drive. The variable-speed selector should only be moved when the motor is running. With this drive it is possible to choose the exact speed for the drill size and material. (Clausing Machine Tools)

FIGURE D1.9 Upright drill press. (Wilton Corporation)

The upright drill press is very similar to the sensitive drill press, but it is made for much heavier work (Figure D1.9). The drive is more powerful and many types are *gear driven,* so they are capable of drilling holes to two inches or more in diameter. *The motor must be stopped when changing speeds on a gear drive drill press.* If it doesn't shift into the selected gear, turn the spindle by hand until it meshes. Since power feeds are needed to drill these large-size holes, these machines are equipped with power feed mechanisms that can be adjusted by the operator. The operator may either feed manually with a lever or handwheel or engage the power feed. A mechanism is provided to raise and lower the table.

As Figures D1.10, D1.11, and D1.12 show, the radial arm drill press is the most versatile drilling machine. Its size is determined by the diameter of the column and the length of the arm measured from the center of the spindle to the outer edge of the column. It is useful for operations on large castings that are too heavy to be repositioned by the operator for drilling each hole. The work is clamped to the table or base, and the drill can then be positioned where it is needed by swinging the arm and moving the head along the arm. The arm and head can be raised or lowered on the column and then locked in place. The radial arm drill press is used for drilling small to very large holes and for boring, reaming, counterboring, and countersinking. Like the upright machine, the radial arm press has a power feed mechanism and a hand feed lever.

The vertical milling machine (see Section F, Unit 1) evolved from the drill press. Both have a spindle in which the tool is inserted and a quill and hand lever for feeding the tool down into the work. In fact, the vertical mill can be used as a drill press as well as a milling machine. The drill press, however, cannot be used to make milling cuts since it has no moveable table and therefore the workpiece cannot be fed into the cutter.

On a drill press a hole circle or other pattern of holes can only be made by means of a fixture and drill **jig,** which consists of hardened **bushings** in a fixture that guide the drill into the workpiece. Otherwise, the hole pattern must be laboriously laid out and center punched. This is not a very accurate method and the tolerance is very broad. On the other hand, the CNC vertical mill can produce any pattern of holes accurately located and with close tolerance (Figure D1.13).

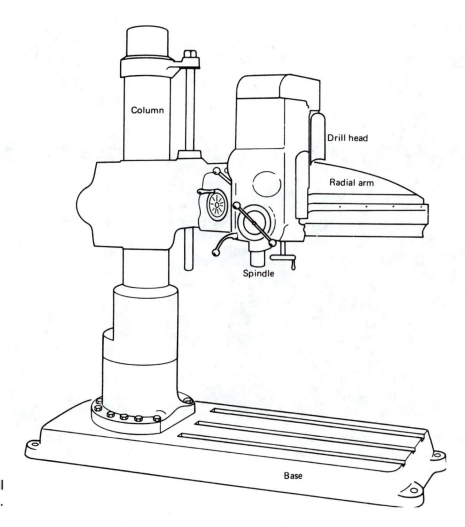

Column

Drill head

Radial arm

Spindle

Base

FIGURE D1.10 Radial arm drill press with names of major parts.

FIGURE D1.11 Small workpiece mounted on the side of the worktable for drilling.

FIGURE D1.12 Small holes are usually drilled by hand feeding on a radial arm drill. A workpiece is clamped on the tilting table so a hole may be drilled at an angle.

SELF-TEST

1. What clothing and personal styles could be considered a safety hazard? Explain in your answer.
2. Workpieces that are being drilled can be a danger to the operator. Why is this? How can this be corrected?
3. Why are your eyes in danger when using a drill press? How can you prevent this?
4. Morse tapers in drill press spindles must be clean and free from nicks or burrs. Machinists often use a finger to clean the taper and to feel for burrs. Should this be done with the machine running? Explain.
5. The chuck key should not be left in the drill press when you are finished with it. Is it acceptable to leave the chuck key in the chuck just for a moment while you are making your setup? Explain your answer.
6. Is slowing down a drill press spindle with your hand a good work habit to develop? Give the reason for your answer.

FIGURE D1.13 Machining an 11-hole pattern.

7. Large drills (up to $3\frac{1}{2}$-in. diameter) have a taper shank and are removed with a drift. This usually requires the use of both hands, leaving no way to hold the drill from falling. Often the drill is positioned over a hole in the table or, if it is a radial arm drill press, over the floor or workpiece. These drills are sometimes heavy and could cause an injury if dropped on a foot. How would you overcome this problem?

8. Describe a safe method of removing cutting chips from a drilling machine.

9. What are the dangers in having oil on the floor in the work area of a drilling machine?

10. Why should burrs be removed from holes and edges of workpieces be chamfered as soon as possible?

11. Identify these drill press types:
    a. List three basic types of drill presses and briefly explain their differences.
    b. Describe how the primary uses differ in each of these three drill press types.

12. Sensitive drill press. Fill in the correct letter in the space preceding the name of that part shown on Figure D1.14.

    _____ Spindle   _____ Base
    _____ Quill lock   _____ Power feed
           handle   _____ Motor
    _____ Column   _____ Variable speed
    _____ Switch   control
    _____ Depth stop   _____ Table lift crank
    _____ Head   _____ Quill return spring
    _____ Table   _____ Guard
    _____ Table lock

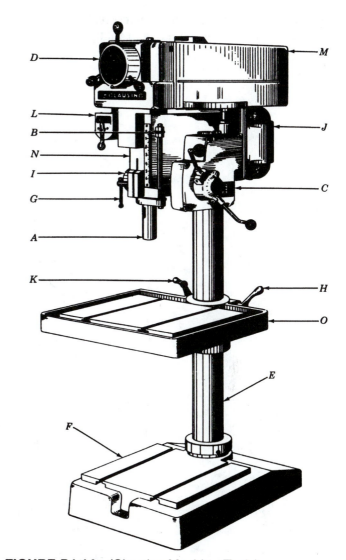

FIGURE D1.14 (Clausing Machine Tools)

13. Radial drill press. Fill in the correct letters in the space preceding the name of that part shown on Figure D1.15.

_____ Column

_____ Radial arm
_____ Spindle
_____ Base
_____ Drill head

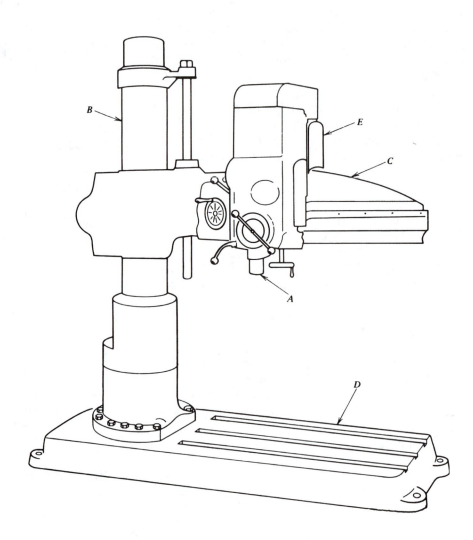

FIGURE D1.15

Drilling Tools

Before you learn to use drills and drilling machines, you will have to know of the great variety of drills and tooling available to the machinist. This unit will acquaint you with these interesting tools as well as show you how to select the one you should use for a given operation.

OBJECTIVES

After completing this unit, you should be able to:

1. Identify the various features of a twist drill.
2. Identify the series and size of 10 given decimal-equivalent drill sizes.

The drill is an end cutting rotary-type tool having one or more cutting lips and one or more flutes for the removal of chips and the passage of coolant. Drilling is the most efficient method of making a hole in metals softer than Rockwell 30. Harder metals can be successfully drilled, however, by using special drills and techniques.

In the past, all drills were made of carbon steel and would lose their hardness if they became too hot from drilling. Carbon steel drills are still made today for various purposes. Occasionally a machinist will pick one up thinking it is a high-speed drill and burn the cutting edge by turning it at an RPM suited to high-speed steel. As a rule of thumb, a carbon drill should be turned about half the RPM of a high-speed steel drill. Carbon steel drills can be identified by spark testing. (See Section A, Unit 2.) Today, however, many drills are made of high-speed steel. High-speed steel drills can operate at several hundred degrees Fahrenheit without breaking down and, when cooled, will be as hard as before. Carbide-tipped drills are used for special applications such as drilling abrasive materials and very hard steels. Other special drills are made from cast heat-resistant alloys.

TWIST DRILLS

The twist drill is by far the most common type of drill used today. These are made with two or more **flutes** and cutting lips and in many varieties of de-

sign. Figure D2.1 illustrates several of the most commonly used types of twist drills. The names of parts and features of a twist drill are shown in Figure D2.2.

The twist drill has either a straight or tapered **shank.** The taper shank has a Morse taper, a standard taper of about $\frac{5}{8}$ in. per foot, which has more driving power and greater rigidity than the straight shank types. Ordinary straight shank drills are typically held in drill chucks (Figure D2.3a). This is a friction drive, and slipping of the drill shank is a common problem. Straight shank drills with tang drives have a positive drive and are less expensive than tapered shank drills. These are held in special drill chucks with a Morse taper (Figure D2.3b).

Jobber drills have two flutes, a straight shank design, and a relatively short length-to-diameter ratio that helps to maintain rigidity. These drills are used for drilling in steel, cast iron, and nonferrous metals. Center drills and spotting drills are used for starting holes in workpieces. Oil hole drills are made so that coolant can be pumped through the drill to the cutting lips. This not only cools the cutting edges but also forces out the chips along the flutes. Core drills have from three to six flutes, making heavy stock removal possible. They are generally used for roughing holes to a larger diameter or for drilling out cores in castings. Left-handed drills are mostly used on multispindle drilling machines where some spindles are rotated in reverse of normal drill press rotation. The step drill generally has

FIGURE D2.3*a* Drill chucks such as this one are used to hold straight shank drills.

FIGURE D2.1 Various types of twist drills used in drilling machines. (The DoAll Company): *(a)* High helix drill; *(b)* Low helix drill; *(c)* Left-hand drill; *(d)* Three-flute drill; *(e)* Taper shank twist drill; *(f)* Standard helix jobber drill; *(g)* Center or spotting drill.

FIGURE D2.3*b* Tang drive drill chuck will fit into a Morse taper spindle. (Illinois/Eclipse, a Division of Illinois Tool Works, Inc.)

FIGURE D2.2 Features of a twist drill. (Bendix Industrial Tools Division)

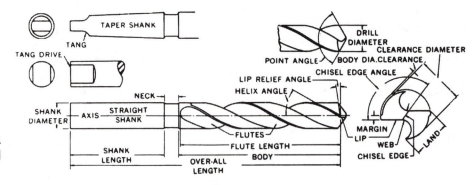

a flat or an angular cutting edge and can produce a hole with several diameters in one pass with either flat or countersunk shoulders.

Straight fluted drills are used for drilling brass and other soft materials because the zero rake angle eliminates the tendency for the drill to "grab" on breakthrough. For the same reason they are used on thin materials. Low helix drills, sometimes called *slow spiral* drills (Figure D2.4), are more rigid than standard-helix drills and can stand more **torque.** Like straight fluted drills, they are less likely to "grab" when emerging from a hole, because of the small rake angle. For this reason the low helix and straight flute drills are used primarily for drilling in brass, bronze, and some other nonferrous metals. Because of the low helix angle, the flutes do not remove chips very well from deep holes, but the large chip space allows maximum drilling efficiency in shallow holes. High helix drills, sometimes called *fast spiral* drills, are designed to remove chips from deep holes. The large rake angle makes these drills suitable for soft metals such as aluminum and mild steel. Spotting and centering drills (Figure D2.5) are used to accurately position holes for further drilling with regular drills. Centering drills are short and have little or no dead center. These characteristics

prevent the drills from wobbling. Lathe center drills are often used as spotting drills.

SPADE AND GUN DRILLS

Special drills such as spade and gun drills are used in many manufacturing processes. A spade drill is simply a flat blade with sharpened cutting lips. The spade bit, which is clamped in a holder (Figure D2.6), is replaceable and can be sharpened many times. Some types provide for coolant flow to the cutting edge through a hole in the holder or shank for the purpose of deep drilling. These drills are made with very large diameters of 12 in. or more (Figure D2.7) but can also be found as microdrills, smaller than a hair. Twist drills by comparison are rarely found with diameters over $3\frac{1}{2}$ in. Spade drills are usually ground with a flat top rake and with chipbreaker grooves on the end. A chisel edge and thinned web are ground in the dead center (Figure D2.8).

Some spade drills are made of solid tungsten carbide, usually only in a small diameter. Twist drills with carbide inserts (Figure D2.9) require a rigid drilling setup. Gun drills (Figure D2.10) are also carbide tipped and have a single vee-shaped flute in a steel tube through which coolant is pumped under

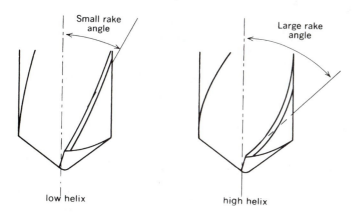

FIGURE D2.4   Low and high rake angles on drills.

FIGURE D2.5   Spotting drill. (The DoAll Company)

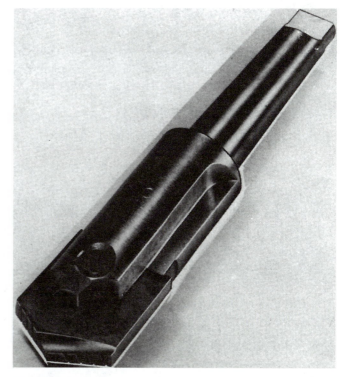

FIGURE D2.6   Spade drill clamped in holder. (The DoAll Company)

FIGURE D2.7 Large hole being drilled with spade drill. This 8-in. diameter hole, $18\frac{7}{8}$ in. deep, was spade drilled in solid SAE #4145 steel rolling mill drive coupling housing with a Brinell hardness of 200–240. The machine that did the job was a 6 ft 19 in. Chipmaster radial with a 25 HP motor. (Giddings & Lewis, Inc.)

FIGURE D2.9 Carbide-tipped twist drill. (The DoAll Company)

pressure. These drills are used in horizontal machines that feed the drill with a positive guide. Extremely deep precision holes are produced with gun drills.

DRILL SELECTION

The type of drill selected for a particular task depends upon several factors. The type of machine being used, rigidity of the workpiece, setup, and

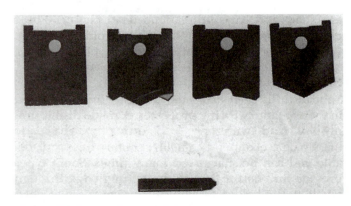

FIGURE D2.8 Spade drill blades showing various grinds. (The DoAll Company)

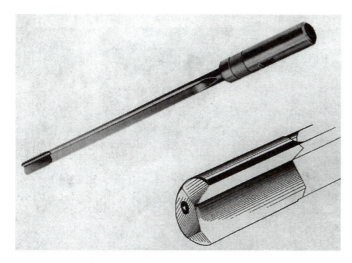

FIGURE D2.10 Single-flute gun drill with inset of carbide cutting tip. (The DoAll Company)

size of the hole to be drilled are all important. The composition and hardness of the workpiece are especially critical. The job may require a starting drill or one for secondary operations such as counterboring or spot facing, and it might need a drill for a deep hole or for a shallow one. If the drilling operation is too large for the size or rigidity of the machine, there will be chatter and the work surface will be rough or distorted.

A machinist also must make the selection of the size of the drill, the most important dimension of which is the diameter. Twist drills are measured

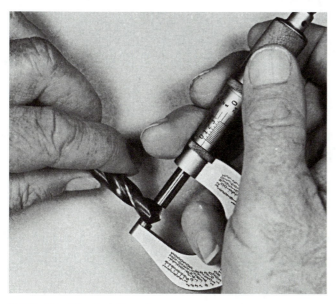

FIGURE D2.11 Drill being measured across the margins.

across the margins near the drill point (Figure D2.11). Worn drills measure slightly smaller here. Drills are normally tapered back along the margin, so they will measure a few thousandths of an inch smaller at the shank.

Drilling is basically a roughing operation. This provides a hole that is suitable for many uses such as for bolts or nuts, but holes must be finished by other operations, such as **boring** or reaming, for applications requiring more precision. Drills almost never make a hole smaller than their measured diameter, but often make a larger hole depending on how they have been sharpened. There are four drill size series: *fractional, number, letter,* and *metric* sizes. The fractional divisions are in $\frac{1}{64}$-in. increments, while the number, letter, and metric series have drill diameters that fall between the fractional inch measures. Together, the four series make up a long-running series in decimal equivalents, as shown in Table D2.1.

Identification of a small drill is simple enough as long as the number or letter remains on the shank. Most shops, however, have several series of drills, and individual drills often become hard to identify because the markings become worn off by the drill chuck. The machinist must then use a decimal equivalent table such as Table D2.1. The drill in question is first measured by a micrometer, the decimal reading is located in the table, and the equivalent fraction, number, letter, or metric size is found and noted.

Morse taper shanks on drills and Morse tapers in drill press spindles vary in size and are numbered from 1 to 6; for example, the smaller, light-duty drill press has a number 2 taper. Steel sleeves (Figure D2.12) have a Morse taper inside and outside with a slot provided at the end of the inside taper to facilitate removal of the drill shank. A **sleeve** is used for enlarging the taper end on a drill to fit a larger spindle taper. Steel **sockets** (Figure D2.13) function in the reverse manner of sleeves, as they adapt a smaller spindle taper to a larger drill. The tool used to remove a taper shank drill is called a **drift.** The drift (Figure D2.14) is made in several sizes and is used to remove drills or sleeves. The drift is placed round side up, flat side against the drill (Figure D2.15), and is struck a light blow with a hammer. A block of wood should be placed under the drill to keep it from being damaged and from being a safety hazard.

Machinists often hand grind drills on a pedestal grinder. Such sharpening procedures can never produce a perfectly symmetrical drill point. The result of offhand grinding often produces an oversize hole. The causes of this are shown in Figure D2.16.

TABLE D2.1 Decimal equivalents for drills

Decimals of an Inch	Inch	Wire Gage	Milli-meter	Decimals of an Inch	Inch	Wire Gage	Milli-meter
.0135		80		.0630			1.6
.0145		79		.0635		52	
.0156	$\frac{1}{64}$			.0650			1.65
.0157			.4	.0669			1.7
.0160		78		.0670		51	
.0180		77		.0689			1.75
.0197			.5	.0700		50	
.0200		76		.0709			1.8
.0210		75		.0728			1.85
.0217			.55	.0730		49	
.0225		74		.0748			1.9
.0236			.6	.0760		48	
.0240		73		.0768			1.95
.0250		72		.0781	$\frac{5}{64}$		
.0256			.65	.0785		47	
.0260		71		.0787			2
.0276			.7	.0807			2.05
.0280		70		.0810		46	
.0293		69		.0820		45	
.0295			.75	.0827			2.1
.0310		68		.0846			2.15
.0313	$\frac{1}{32}$			.0860		45	
.0315			.8	.0866			2.2
.0320		67		.0886			2.25
.0330		66		.0890		43	
.0335			.85	.0906			2.3
.0350		65		.0925			2.35
.0354			.9	.0935		42	
.0360		64		.0938	$\frac{3}{32}$		
.0370		63		.0945			2.4
.0374			.95	.0960		41	
.0380		62		.0966			2.45
.0390		61		.0980		40	
.0394			1	.0984			2.5
.0400		60		.0995		39	
.0410		59		.1015		38	
.0413			1.05	.1024			2.6
.0420		58		.1040		37	
.0430		57		.1063			2.7
.0433			1.1	.1065		36	
.0453			1.15	.1083			2.75
.0465		56		.1094	$\frac{7}{64}$		
.0469	$\frac{3}{64}$			.1100		35	
.0472			1.2	.1102			2.8
.0492			1.25	.1110		34	
.0512			1.3	.1130		33	
.0520		55		.1142			2.9
.0531			1.35	.1160		32	
.0550		54		.1181			3
.0551			1.4	.1200		31	
.0571			1.45	.1220			3.1
.0591			1.5	.1250	$\frac{1}{8}$		
.0595		53		.1260			3.2
.0610			1.55	.1280			3.25
.0625	$\frac{1}{16}$			.1285		30	

TABLE D2.1 *(continued)*

Decimals of an Inch	Inch	Wire Gage	Millimeter
.1299			3.3
.1339			3.4
.1360		29	
.1378			3.5
.1405		28	
.1406	$\frac{9}{64}$		
.1417			3.6
.1440		27	
.1457			3.7
.1470		26	
.1476			3.75
.1495		25	
.1496			3.8
.1520		24	
.1535			3.9
.1540		23	
.1563	$\frac{5}{32}$		
.1570		22	
.1575			4
.1590		21	
.1610		20	
.1614			4.1
.1654			4.2
.1660		19	
.1673			4.25
.1693			4.3
.1695		18	
.1719	$\frac{11}{64}$		
.1730		17	
.1732			4.4
.1750		16	
.1772			4.5
.1780		15	
.1811			4.6
.1800		14	
.1820		13	
.1850			4.7
.1870			4.75
.1875	$\frac{3}{16}$		
.1890			4.8
.1850		12	
.1910		11	
.1929			4.9
.1935		10	
.1960		9	
.1969			5
.1990		8	
.2008			5.1
.2010		7	
.2031	$\frac{13}{64}$		
.2040		6	
.2047			5.2
.2055		5	
.2067			5.25
.2087			5.3

Decimals of an Inch	Inch	Wire Gage	Millimeter
.2090		4	
.2126			5.4
.2130		3	
.2165	$\frac{7}{32}$		5.5
.2188			
.2205			5.6
.2210		2	
.2244			5.7
.2264			5.75
.2280		1	
.2283			5.8

Decimals of an Inch	Inch	Letter Sizes	Millimeter
.2323			5.9
.2340		A	
.2344	$\frac{15}{64}$		
.2362			6
.2380		B	
.2402			6.1
.2420		C	
.2441			6.2
.2460		D	
.2461			6.25
.2480			6.3
.2500	$\frac{1}{4}$	E	
.2520			6.4
.2559			6.5
.2570		F	
.2598			6.6
.2610		G	
.2638			6.7
.2656	$\frac{17}{64}$		
.2657			6.75
.2660		H	
.2677			6.8
.2717			6.9
.2720		I	
.2756			7
.2770		J	
.2795			7.1
.2810		K	
.2812	$\frac{9}{32}$		
.2835			7.2
.2854			7.25
.2874			7.3
.2900		L	
.2913			7.4
.2950		M	
.2953			7.5
.2969	$\frac{19}{64}$		
.2992			7.6
.3020		N	
.3031			7.7

Decimals of an Inch	Inch	Letter Sizes	Milli-meter	Decimals of an Inch	Inch	Milli-meter
.3051			7.75	.4528		11.5
.3071			7.8	.4531	$\frac{29}{64}$	
.3110			7.9	.4688	$\frac{15}{32}$	
.3125	$\frac{5}{16}$			.4724		12
.3150			8	.4844	$\frac{31}{64}$	
.3160		O		.4921		12.5
.3189			8.1	.5000	$\frac{1}{2}$	
.3228			8.2	.5118		13
.3230		P		.5156	$\frac{33}{64}$	
.3248			8.25	.5313	$\frac{17}{32}$	
.3268			8.3	.5315		13.5
.3281	$\frac{21}{64}$			.5469	$\frac{35}{64}$	
.3307			8.4	.5512		14
.3320		Q		.5625	$\frac{9}{16}$	
.3346			8.5	.5709		14.5
.3386			8.6	.5781	$\frac{37}{64}$	
.3390		R		.5906		15
.3425			8.7	.5938	$\frac{19}{32}$	
.3438	$\frac{11}{32}$			.6094	$\frac{39}{64}$	
.3345			8.75	.6102		15.5
.3465			8.8	.6250	$\frac{5}{8}$	
.3480		S		.6299		16
.3504			8.9	.6406	$\frac{41}{64}$	
.3543			9	.6496		16.5
.3580		T		.6563	$\frac{21}{32}$	
.3583			9.1	.6693		17
.3594	$\frac{23}{64}$			.6719	$\frac{43}{64}$	
.3622			9.2	.6875	$\frac{11}{16}$	
.3642			9.25	.6890		17.5
.3661			9.3	.7031	$\frac{45}{64}$	
.3680		U		.7087		18
.3701			9.4	.7188	$\frac{23}{32}$	
.3740			9.5	.7283		18.5
.3750	$\frac{3}{8}$			.7344	$\frac{47}{64}$	
.3770		V		.7480		19
.3780			9.6	.7500	$\frac{3}{4}$	
.3819			9.7	.7656	$\frac{49}{64}$	
.3839			9.75	.7677		19.5
.3858			9.8	.7812	$\frac{25}{32}$	
.3860		W		.7874		20
.3898			9.9	.7969	$\frac{51}{64}$	
.3906	$\frac{25}{64}$			.8071		20.5
.3937			10	.8125	$\frac{13}{16}$	
.3970		X		.8268		21
.4040		Y		.8281	$\frac{53}{64}$	
.4063	$\frac{13}{32}$			.8438	$\frac{27}{32}$	
.4130		Z		.8465		21.5
.4134			10.5	.8594	$\frac{55}{64}$	
.4219	$\frac{27}{64}$			.8661		22
.4331			11	.8750	$\frac{7}{8}$	
.4375	$\frac{7}{16}$			.8858		22.5

Decimals of an Inch	Inch	Milli-meter
.8906	$\frac{57}{64}$	
.9055		23
.9063	$\frac{29}{32}$	
.9219	$\frac{59}{64}$	
.9252		23.5
.9375	$\frac{15}{16}$	
.9449		24
.9531	$\frac{61}{64}$	
.9646		24.5
.9688	$\frac{31}{32}$	
.9843		25
.9844	$\frac{63}{64}$	25.003

SOURCE: *Bendix Cutting Tool Handbook,* "Decimal Equivalents—Twist Drill Sizes," The Bendix Corporation, Industrial Tools Division, 1972.

FIGURE D2.12 Morse taper drill sleeve. (The DoAll Company)

FIGURE D2.13 Morse taper drill socket. (The DoAll Company)

FIGURE D2.14 The drift in use. This tool is one of several sizes used to remove taper shanks, drills, and sleeves.

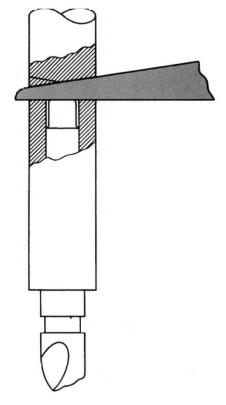

FIGURE D2.15 Cutaway of a drill and sleeve showing a drift in place.

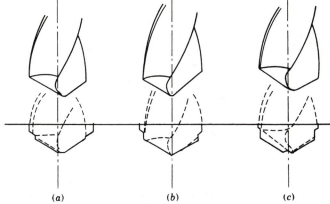

(a)   *(b)*   *(c)*

FIGURE D2.16 Causes of oversize drilling: *(a)* Drill lips ground to unequal lengths; *(b)* Drill lips ground to unequal angles; *(c)* Unequal angles and lengths.

Offhand drill sharpening is not practical for CNC machines. The reason is that ordinary drills have a wide dead center. They tend to wobble and run off center unless a small pilot hole has been previously drilled (Figure D2.17). A CNC machine usually feeds a drill into the workpiece without a pilot

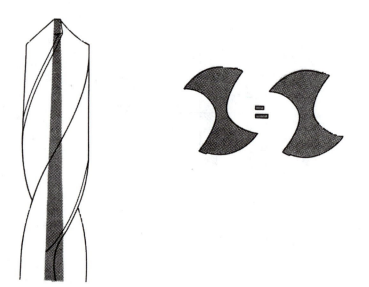

FIGURE D2.17 The tapered web of twist drills. (Bendix Industrial Tools Division)

hole or center punch mark. The drill must not wobble; it should begin drilling precisely where it contacts the workpiece. This can be done with split-point drills because they have no wide dead center; in fact, they have a zero-width dead center (Figure D2.18).

Drill Point Precision Sharpening

Drill-point grinding machines are available in several varieties and levels of complexity, ranging from those used for microdrilling to the automatic machines used by drill manufacturers. Some machines have a pivot so that the operator can shape the drill point to the proper geometry, and others use cams to produce a generated drill-point geometry (Figure D2.19). One kind uses the side of a grinding wheel, while another uses the circumference.

All of these point grinding machines have one thing in common: precision. Producing an accurate drill point is the major advantage of the drill sharpening machine over hand grinding. Other advan-

FIGURE D2.19 Dupoint drill pointer. (Courtesy of Mohawk Tools, Inc., Machine Tool Division)

tages are: Drill failures are reduced, drills stay sharp longer because both flutes always cut evenly, and the hole size is controllable within close limits.

Where drill jigs (drill guides) have been replaced by CNC machines, the drills used must be very precisely sharpened so that they will not wander or make an irregular hole. The way this is accomplished may be seen in Figure D2.20. However, the more complex and more expensive grinding machines are usually used to make split-point grind.

Less specialized and less expensive drill grinding machines are most likely to be found in a small ma-

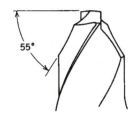

55°

FIGURE D2.18 Split-point design of a drill point. (Bendix Industrial Tools Division)

FIGURE D2.20 Cutting and chip-forming characteristics of split-point drills. (Courtesy of Mohawk Tools, Inc., Machine Tool Division)

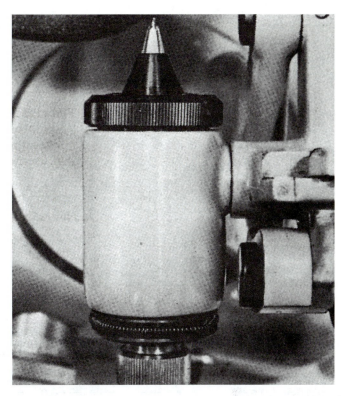

FIGURE D2.22 Rotating sleeve and drill holding unit. (Courtesy of Mohawk Tools, Inc., Machine Tool Division)

FIGURE D2.21 Precision drill grinder. (Courtesy of Mohawk Tools, Inc., Machine Tool Division)

chine shop. A machine of this type will produce accurate standard drill points on two-, three-, or four-flute drills from $\frac{1}{8}$ to $1\frac{1}{4}$ in. in diameter.

Microdrills, smaller than .008 in. or about .2 mm, are often considered expendable when dull. The most common failure is drill breakage. However, some precision drill grinders are capable of pointing a drill as small as .008 in. (Figure D2.21). This model has a magnifying eyepiece, a light, and a tiny four-jaw universal chuck to hold the drill (Figure D2.22). Split-point or four-facet drill grinds are used on these tiny drills to make them freer cutting.

SELF-TEST

1. Write the correct letters from the drawing in the space provided.

____ Web		____ Chisel edge angle	
____ Margin		____ Body clearance	
____ Drill point angle		____ Helix angle	
____ Cutting lip		____ Axis of drill	
____ Flute		____ Shank length	
____ Body		____ Tang	
____ Lip relief angle		____ Taper Shank	
____ Land		____ Straight shank	

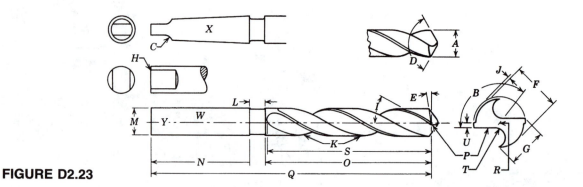

FIGURE D2.23

2. Determine the letter, number, fractional, or metric equivalents of the 10 following decimal measurements of drills:

	Decimal Diameter	Fractional Size	Number Size	Letter Size	Metric Size
a.	.0781				
b.	.1495				
c.	.272				
d.	.159				
e.	.1969				
f.	.323				
g.	.3125				
h.	.4375				
i.	.201				
j.	.1875				

3. State the basic advantage of drill-point sharpening machines over hand sharpening drills.

4. An acceptable tolerance for ordinary drilled holes in machine shop work is given as .010 in. × diameter of the drill. Is this acceptable for CNC applications and production drilling?

5. The chisel end on the dead center of a drill will wander and make an irregularly shaped hole if a center punch mark and spotting drill have not been used. How is this problem overcome on CNC machines that use no pilot or spotting drills?

6. Can drills as small as .008 in. be sharpened with a split-point grind?

Work Locating and Holding Devices on Drilling Machines

Workpieces of a great many sizes and shapes are drilled by machinists. In order to hold these parts safely and securely while they are drilled, several types of workholding devices are used. In this unit you will learn how to properly set up these devices for a drilling operation.

OBJECTIVES

After completing this unit, you should be able to:

1. Identify and explain the correct uses for several workholding and locating devices.
2. Set up for drilling holes; align and start a tap using the drill press.

WORK HOLDING

Because of the great forces applied by the machines in drilling, some means must be provided to keep the workpiece from turning with the drill or from climbing up the flutes after the drill breaks through. This is not only necessary for safety's sake but also for workpiece rigidity and good workmanship.

One method of workholding is to use strap clamps (Figure D3.1) and T-bolts (Figure D3.2). The clamp must be kept parallel to the table by the use of step blocks (Figure D3.3) and the T-bolt should be kept as close to the workpiece as possible (Figure D3.4). Parallels (Figure D3.5) are placed under the work at the point where the clamp is holding. This provides a space for the drill to break through without making a hole in the table. A thin or narrow workpiece should not be supported too far from the drill, however, since it will spring down under the pressure of the drilling. This can cause the drill on breakthrough to suddenly "grab" more material than it can handle. The result is often a broken drill (Figure D3.6). Thin workpieces or sheet metal should be clamped over a wooden block to avoid this problem. C-clamps of various sizes are used to hold workpieces on drill press tables and on angle plates (Figure D3.7).

Angle plates facilitate the holding of odd-shaped parts for drilling. The angle plate is either bolted or clamped to the table and the work is fastened to the angle plate. For example, a gear or wheel that requires a hole to be drilled into a projecting hub could be clamped to an angle plate.

Drill press vises (Figure D3.8) are very frequently used for holding small workpieces of regular shape and size with parallel sides. Vises provide the quickest and most efficient setup method for parallel work but should not be used if the work does not have parallel sides. The workpiece must be supported so the drill will not go into the vise. If precision parallels are used for support, they and the drill can be easily damaged since they are both hardened (Figure D3.9). For rough drilling, however, cold finished (CF) keystock would be sufficient for supporting the workpiece. Angular vises can pivot a workpiece to a given angle so that angular holes can be drilled (Figure D3.10). Another method of drilling angular holes is by tilting the drill press table. If there is no angular scale on the vise or table, a protractor head with a level may be used to set up the correct angle for drilling. Angle plates are also sometimes used for drilling angular holes (Figure D3.11a). The drill press table must be level (Figure D3.11b).

Work Locating

Vee-blocks come in sets of two, often with clamps for holding small size rounds (Figure D3.12). Larger-size round stock is set up with a strap clamp over the vee-blocks (Figure D3.13). The hole to be

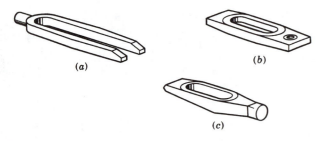

FIGURE D3.1 Strap clamps: *(a)* U-clamp; *(b)* Straight clamp; and *(c)* Finger clamp.

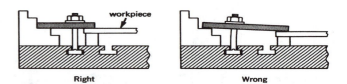

FIGURE D3.4 Right and wrong setup for strap clamps. The clamp bolt should be as close to the workpiece as possible.

FIGURE D3.2 T-bolts.

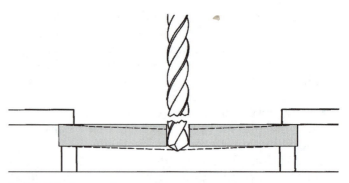

FIGURE D3.5 Parallels of various sizes.

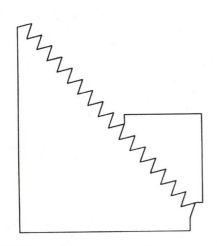

FIGURE D3.3 Adjustable step blocks.

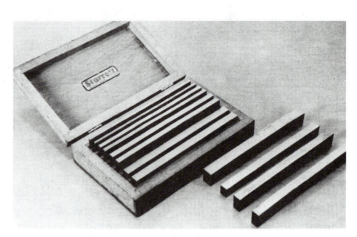

FIGURE D3.6 Thin, springy material is supported too far from the drill. Drilling pressure forces the workpiece downward until the drill breaks through, relieving the pressure. The work then springs back and the remaining "fin" of material is more than the drill can cut in one revolution. The result is drill breakage.

FIGURE D3.7 C-clamp being used on an angle plate to hold work that would be difficult to safely support in other ways.

FIGURE D3.9 Part setup in vise with parallels under it.

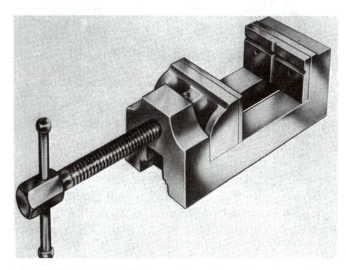

FIGURE D3.8 Drill press vise. Small parts are held for drilling and other operations with the drill press vise. (Wilton Corporation)

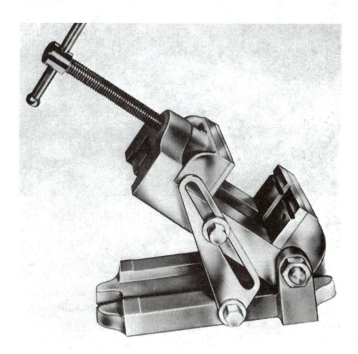

FIGURE D3.10 Angle vise. Parts that must be held at an angle to the drill press table while being drilled are held with this vise. (Wilton Corporation)

FIGURE D3.11a View of an adjustable angle plate on a drill press table using a protractor to set up.

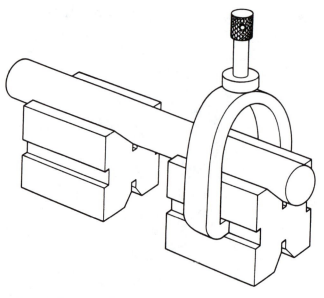

FIGURE D3.12 Set of vee-blocks with a vee-block clamp.

FIGURE D3.11b Checking the level of the table.

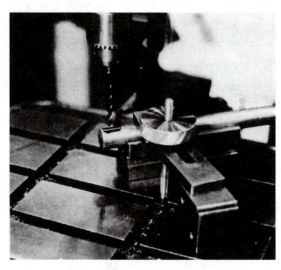

FIGURE D3.13 Setup of two vee-blocks and round stock with strap clamp between the vee-blocks.

cross drilled is first laid out and center punched. The workpiece is lightly clamped in the vee-blocks and the punch mark is centered, as shown in Figure D3.14. The clamps are tightened, and the drill chuck is located precisely over the punch mark by means of a **wiggler**. A wiggler is a tool that can be put into a drill chuck to locate a punch mark to the exact center of the spindle.

The wiggler is clamped into a drill chuck and the machine is turned on (Figure D3.15a). Push on the knob near the end of the pointer with a 6-in. rule or other piece of metal until it runs with no wobble (Figure D3.15b). With the machine still running, bring the pointer down into the punch mark. If the pointer begins to wobble again, the mark is not centered under the spindle and the workpiece will have

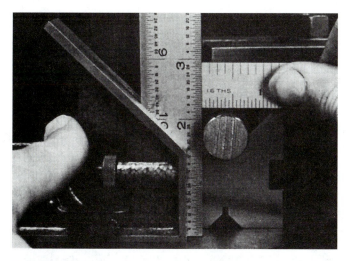

FIGURE D3.14 Round stock in vee-blocks. One method of centering the layout line or punch mark using a combination square and rule.

to be shifted. When the wiggler enters the punch mark without wobbling, the workpiece is centered.

After the work is centered, use a spotting or center drill to start the hole. Then, for larger holes, use a pilot drill, which is always a little larger than the web of the next drill size used. Pilot drills are not commonly used in industrial applications; only the spotting drill (if used) and the full-size drill are used. Use the correct cutting speed and coolant. Chamfer both sides of the finished hole with a countersink or chamfering tool. Round stock can also be cross drilled when held in a vise, using the same technique as with vee-blocks.

A **tap** may be started straight in the drill press by hand. After tap drilling the workpiece and without removing any clamps, remove the tap drill from the chuck and replace it with a straight shank center. (An alternative method is to clamp the shank of the tap directly in the drill chuck and turn the spindle by hand two or three turns.) Insert a tap in the work and attach a tap handle. Then put the center

FIGURE D3.15a Wiggler set in offset position.

FIGURE D3.15b Wiggler centered.

into the tap but *do not turn on the machine.* Apply sulfurized cutting oil and start the tap by turning the tap handle a few turns with one hand while feeding down with the other hand. Release the chuck while the tap is still in the work and finish the job of tapping the hole.

Jigs and fixtures are specially made tools for production work. In general, a **fixture** references a part to the cutting tool. **Jigs** guide the cutting tool such as a drill. They both hold and support the part (Figure D3.16). The use of a jig assures exact positioning of the hole pattern in duplicate and eliminates layout work on every part.

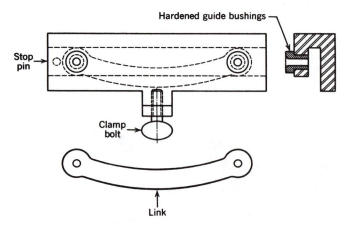

FIGURE D3.16 Simple box jig for drilling link. The link is shown below. Hardened guide bushings in the jig are used to limit wear.

Workholding devices are similar on most machines. Strap clamps, vises, and fixtures found on drill presses are also used on horizontal and vertical milling machines and on CNC machines. The most important thing about hold-down devices is just how much you need to tighten them. Too much torque on a hold-down bolt or nut can strip threads or break the bolt, whereas too little torque will cause the vise or work material to slip or move. That can ruin it and break the tooling.

SELF-TEST

1. What is the main purpose for using workholding devices on drilling machines?
2. List the names of all the workholding devices that you can remember.
3. Explain the uses of parallels for drilling setups.
4. Why should the support on a narrow or thin workpiece be as close to the drill as possible?
5. Angle drilling can be accomplished in several ways. Describe two methods. How would this be done if no angular measuring devices were mounted on the equipment?
6. What shape of material is the vee-block best suited to hold for drilling operations? What do you think its most frequent use would be?
7. What is the purpose of using a wiggler?
8. Why would you ever need an angle plate?
9. What is the purpose of starting a tap in the drill press?
10. Do you think jigs and fixtures are used to any great extent in small machine shops? Why?

Operating Drilling Machines

You have already learned many things about drilling machines and tooling. You should now be ready to learn some very important facts about the use of these machines. How fast should the drill run? How much feed should be applied? Which kind of coolant should be used? These and other questions are answered in this unit.

OBJECTIVES

After completing this unit, you should be able to:

1. Determine the correct drilling speeds for five given drill diameters.
2. Determine the correct feed in steel by chip observation.
3. Set up the correct feed on a machine by using a feed table.

After the workpiece is properly clamped and operator safety is assured, the most important considerations for drilling are speeds, feeds, and coolants. Of these, the control and setting of speeds will have the greatest effect on the tool and the work.

CUTTING SPEEDS

Cutting speeds (CS) are normally given for high-speed steel cutting tools and are based on surface feet per minute (FPM or SFM). Surface feet per minute means that either the tool moves past the work or the work moves past the tool at a rate based on the number of feet that passes a tool in one minute, whether it be on a flat surface or on the periphery of a cylindrical tool or workpiece. Since machine spindle speeds are given in revolutions per minute (RPM), cutting speed can be derived in the following manner:

$$\text{RPM} = \frac{\text{Cutting speed (in feet per minute)} \times 12}{\text{Diameter of cutter (in inches)} \times \text{pi}}$$

If you use 3 to approximate π (3.1416), then the formula becomes

$$\frac{\text{CS} \times 12}{D \times 3} = \frac{\text{CS} \times 4}{D}$$

This simplified formula is certainly the most common one used in machine shop practice, and it applies to the full range of machine tool operations, which include the lathe and the milling machine as well as the drill press. Throughout this text the simplified formula RPM = (CS $\times$ 4)/D will be used, where D = the diameter of the drill and CS = an assigned cutting speed for a particular material. For example, for a $\frac{1}{2}$-in. drill in low-carbon steel the speed would be

$$\frac{90 \times 4}{\frac{1}{2}} = 720 \text{ RPM}$$

See Table D4.1 for cutting speeds for some metals.

Cutting speeds/RPM tables for various materials are available in handbooks and as wall charts. Excessive speeds can cause the outer corners and margins of the drill to break down. This will in turn cause the drill to bind in the hole, even if the speed

TABLE D4.1 Drilling speed table

Material	Cutting Speed (CS)
Low-carbon steel	90
Aluminum	300
Cast iron	70
Alloy steel	50
Brass and bronze	120

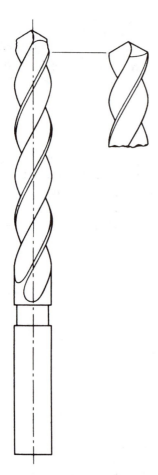

FIGURE D4.1 Broken-down drill corrected by grinding back to full diameter margins and regrinding cutting lips.

is corrected and more cutting oil is applied. The only cure is to grind the drill back to its full diameter (Figure D4.1) using methods discussed in Unit 3 of this section.

A blue chip from steel indicates the speed is too high. The tendency with very small drills, however, is to set the RPM of the spindle too slow. This gives the drill a very low cutting speed and very little chip is formed unless the operator forces it with an excessive feed. The result is often a broken drill.

CONTROLLING FEEDS

The feed may be controlled by the "feel" of the cutting action and by observing the chip. A long, stringy chip indicates too much feed. The proper chip in soft steel should be a tightly rolled helix in both flutes (Figure D4.2). Some materials such as cast iron will produce a granular chip. Drilling machines that have power feeds are arranged to advance the drill a given amount for each revolution

FIGURE D4.2 Properly formed chip.

of the spindle. Therefore, .006-in. feed means that the drill advances .006 in. every time the drill makes one full turn. The amount of feed varies according to the drill size and the work material. See Table D4.2.

It is a better practice to start with smaller feeds than those given in tables. Materials and setups vary, so it is safer to start low and work up to an optimum feed. You should stop the feed occasionally to break the chip and allow coolant to flow to the cutting edge of the drill.

There is generally no breakthrough problem when using power feed, but when hand feeding, the drill may catch and "grab" while coming through the last $\frac{1}{8}$ in. or so of the hole. Therefore, the operator should let up on the feed handle near this point and ease the drill through the hole. This "grabbing" tendency is especially true of brass and some plastics, but it is also a problem in steels and other materials. Large upright drill presses and radial arm

TABLE D4.2 Drilling feed table

Drill Size Diameter (in.)	Feeds per Revolution (in.)
Under $\frac{1}{8}$	.001 to .002
$\frac{1}{8}$ to $\frac{1}{4}$	.002 to .004
$\frac{1}{4}$ to $\frac{1}{2}$	.004 to .007
$\frac{1}{2}$ to 1	.007 to .015
over 1	.015 to .025

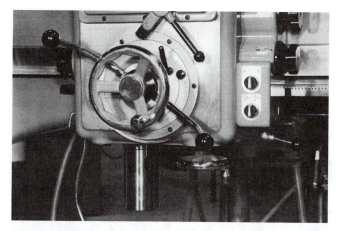

FIGURE D4.3 Feed clutch handle. The power feed is engaged by pulling the handles outward. When the power feed is disengaged, the handles may be used to hand feed the drill.

FIGURE D4.4a Speed and feed control dials.

FIGURE D4.4b Large speed and feed plates on the front of the head of the upright press can be read at a glance. (Giddings & Lewis, Inc.)

drills have power feed mechanisms with feed clutch handles (Figure D4.3) that also can be used for hand feeding when the power feed is disengaged. Both feed and speed controls are set by levers or dials (Figure D4.4a). Speed and feed tables are often found on plates on large drilling machines (Figure D4.4b).

Tapping with small taps is often done on a sensitive drill press with a tapping attachment (Figure D4.5) that has an adjustable friction clutch and reverse mechanism that screws the tap out when you raise the spindle. Large-size taps are power driven on upright or radial drill presses. These machines provide for spindle reversal (sometimes automatic) to screw the tap back out.

COOLANTS AND CUTTING OILS

A large variety of coolants and cutting oils are used for drilling operations on the drill press. **Emulsifying or soluble oils** (either mineral or synthetic) mixed in water are used for drilling holes where the main requirement is an inexpensive cooling medium. Operations that tend to create more friction and, hence, need more lubrication to prevent galling (abrasion due to friction) require a cutting oil. Animal or mineral oils with sulfur or chlorine added are often used. Reaming, counterboring, countersinking, and tapping all create friction and require the use of cutting oils, of which the sulfurized type is most used. Cast iron and brass are usually drilled dry, but water-soluble oil can be

FIGURE D4.5 Tapping attachment.

used for both. Aluminum can be drilled with water-soluble oil or kerosene for a better finish. Both soluble and cutting oils are used for steel.

DRILLING PROCEDURES

Deep hole drilling requires sufficient drill length and quill stroke to complete the needed depth. A high helix drill helps to remove the chips, but sometimes the chips bind in the flutes of the drill and, if drilling continues, will cause the drill to jam in the hole (Figure D4.6). A method of avoiding this problem is called *pecking:* the hole is drilled a short distance and then the drill is taken out of the hole, allowing the accumulated chips to fly off. The drill is again inserted into the hole, a similar amount is drilled, and the drill is again removed. This pecking is repeated until the required depth is reached.

A depth stop is provided on drilling machines to limit the travel of the quill so that the drill can be made to stop at a predetermined depth (Figure D4.7). The use of a depth stop makes drilling several holes to the same depth quite easy. **Spotfacing** and counterboring should also be set up with the depth stop. **Blind holes** (holes that do not go through the piece) are measured from the edge of the drill margin to the required depth (Figure D4.8). Once measured, the depth can be set with the stop and drilling can proceed. One of the most important uses of a depth stop, from a maintenance standpoint, is that of setting the depth so that the machine table or drill press vise will not be drilled full of holes.

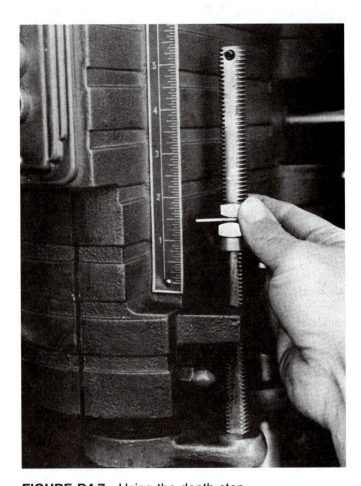

FIGURE D4.7   Using the depth stop.

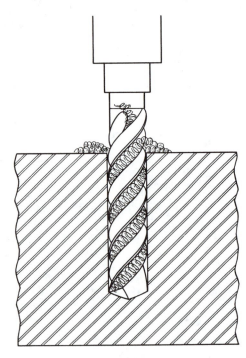

FIGURE D4.6   Drill jammed in hole because of packed chips.

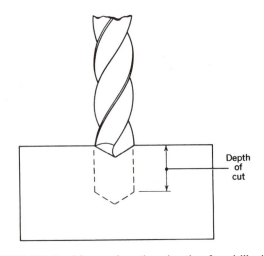

FIGURE D4.8   Measuring the depth of a drilled hole.

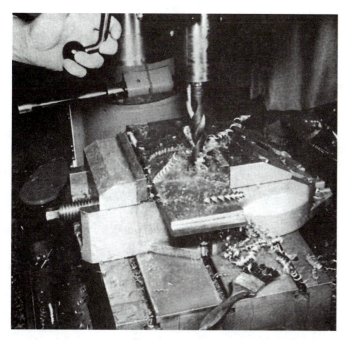

FIGURE D4.9 Heavy-duty drilling on a radial drill press.

Heavy-duty drilling should be done on an upright or radial drill press (Figure D4.9). The workpiece should be made very secure since high drilling forces are used with the larger drill sizes. The work should be well clamped or bolted to the work table. The head and column clamp should always be locked when drilling is being done on a radial drill press. When several operations are performed in a continuing sequence, quick-change drill and tool-holders are often used. Coolant is necessary for all heavy-duty drilling.

SELF-TEST

1. Name three important things to keep in mind when using a drill press (not including operator safety and clamping work).
2. If RPM = (CS × 4)/D and the cutting speed for low-carbon steel is 90, what would the RPMs be for the following drills:
   a. $\frac{1}{4}$-in. diameter _____
   b. 2-in. diameter _____
   c. $\frac{3}{4}$-in. diameter _____
   d. $\frac{3}{8}$-in. diameter _____
   e. $1\frac{1}{2}$-in. diameter _____
3. What are some of the results of excessive drilling speed? What corrective measures can be taken?
4. Explain what can happen to small-diameter drills when the cutting speed is too slow.
5. How can an operator tell by observing the chip if the feed is about right?
6. In what way are power feeds designated?
7. Name two differing cutting fluid types.
8. Such operations as counterboring, reaming, and tapping create friction that can cause heat. This can ruin a cutting edge. How can this situation be helped?
9. How can "jamming" of a drill be avoided when drilling deep holes?
10. Name three uses for the depth stop on a drill press.
11. Obtain three twist drills that have been previously sharpened from your instructor or the toolroom in $\frac{1}{4}$ in., $\frac{1}{2}$ in., and $\frac{3}{4}$ in. sizes. Check each of them for correct size using a micrometer.

    Drill these three hole sizes in a piece of mild steel scrap, using the appropriate drill press and the correct speeds, feeds, and coolant. Wear eye protection and make sure the hold-down clamps are sufficiently secure.

    Evaluation:
    a. Measure the drilled hole sizes. How close to the nominal drill diameter are the holes?
    b. Did the drilling operation go smoothly? If not, what was the problem: speeds, feeds, coolant, or dull drill?
    c. Write an analysis of the operation and submit it to your instructor. Do you think you would find the same circumstances while drilling with a CNC machine with the same drills and on mild steel?

WORKSHEET: DRILL AND HOLE GAGE

Objectives

1. Learn how to safely use power cutoff tools and hand cutting tools in bench work.
2. Learn to accurately measure and lay out workpieces.
3. Be able to correctly select and measure drills from any of several series.
4. Be able to determine if twist drills are accurately sharpened.
5. Be able to accurately drill holes with a drill press.

Materials

A $\frac{1}{8} \times 2 \times 6\frac{1}{4}$ in. cold finished flat bar, hacksaw, files, layout tools, drills, and countersink or chamfering tool.

Procedure for Safety Instruction

It is very necessary for you to study the units on safety *before* starting on the project. Also, Section D, units 1, 2, 3, and 4 must be studied and related tests completed before doing this project.

Procedure for Using the Steel Rule

You must be able to measure accurately before you start layout.

Procedure for Cutoff

1. Cut off the $\frac{1}{8} \times 2$ in. C.F. flat stock $\frac{1}{16}$ in. longer than the finish size. This would make the length $6\frac{1}{16}$ in.
2. Deburr the sharp edges caused by sawing.
3. Clean the metal piece in solvent and dry off.

Procedure for Layout

Using a steel rule and a scriber, locate the hole positions and center punch them in preparation for drilling (Figure D4.10). Do this after the sawing and filing operations are finished. Plan the spacing between the holes in a logical manner.

Procedure for Hacksawing

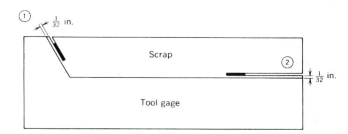

FIGURE D4.11

1. Make the angular saw cut 1, leaving $\frac{1}{32}$ inch of stock for filing. Hold the work so that the saw cut is in a vertical position. Do not saw to the end of the line. Leave $\frac{1}{16}$ inch (Figure D4.11).
2. Clamp the material so that the saw cut 2 is vertical. Keep the cut as close to the vise jaws as possible to prevent vibration. Saw as far as possible with the saw blade in the normal position. When the limit of the frame has been reached, turn the blade 90 degrees and continue sawing.
3. Leave $\frac{1}{32}$ in. outside the layout line for filing. If the kerf begins to wander in or out and you cannot correct it by angling the saw, cut off the scrap piece from the outside edge to the end of the saw cut and start a new cut.
4. Continue the cut until it joins the angular cut. Be careful and do not undercut at the corner where it meets the other cut.

Procedure for Filing

1. Now the drill gage must be filed to the layout line to produce flat, smooth surfaces. Produce a

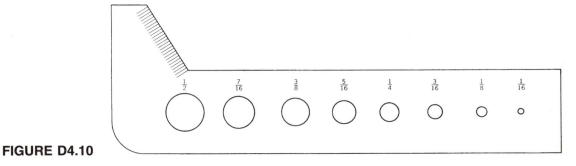

FIGURE D4.10

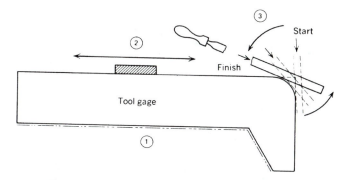

FIGURE D4.12

true radius by filing on the outside corner as shown in Figure D4.12.

2. Rough filing on the sawed surface, side 1, may be done by cross filing. If a large amount of stock needs to be removed, use a double cut file. Leave $\frac{1}{64}$ inch for finish filing.

3. Using a single cut file, all the straight surfaces may be finished by draw filing, side 2. Check frequently with a rule or straight edge to make sure that the surface is flat. The high spots are easily seen and should be filed until almost no light is seen between the rule and the part.

4. Rough file the radius, side 3, with a double cut file leaving $\frac{1}{32}$ inch for finishing.

5. With a single cut file, begin as shown at side 3. As you file with a forward stroke, move the handle downward.

6. Check frequently with a $\frac{1}{2}$ inch radius gage. Continue to file on high spots until almost no light is seen between the gage and the part.

7. The reference marks used for checking lip lengths may now be put on with a scriber and rule, a specially prepared stamp, or a scriber and division setup on a milling machine. Check with your instructor for the method you should use.

8. Your drill gage may be used at this point to check twist drill lip lengths and angles.

Procedure for Drilling Holes

1. From your tool room or instructor, select a No. 2 center drill plus the following drills: $\frac{1}{2}$, $\frac{1}{16}$, $\frac{3}{8}$, $\frac{5}{16}$, $\frac{1}{4}$, $\frac{3}{16}$, $\frac{1}{8}$, $\frac{1}{16}$ inch. Also obtain a countersink, drill press vise, parallels, and cutting oil.

2. Test drill a hole for each of the twist drills you will use in a piece of scrap metal before using them to drill holes in your drill and hole gage. Have them sharpened if necessary.

3. After laying out and center punching for the hole location, set up the drill gage in the vise with parallels arranged under each edge so that the largest drill can clear when you drill through. Drill the hole sizes as shown in Figure D4.13.

4. Place the center drill in the chuck and tighten. With the machine off, spot it in the punch mark for the $\frac{1}{2}$ in. hole. Clamp the vise without moving it.

5. Set the correct speed on the drill press and drill into the workpiece so there is a slight chamfer. Do not spot drill for the $\frac{1}{16}$ in. hole because you might enlarge it oversize. Spot drill all the other holes.

6. Change to the $\frac{1}{16}$ in. drill and drill that hole.

7. Change to the $\frac{1}{8}$ in. drill and drill the $\frac{1}{8}$ in. hole.

8. Continue drilling all the remaining holes, taking care to change to the correct speeds for each drill size.

9. Using a countersink, or chamfering tool, lightly chamfer both sides of each hole.

Evaluation

Turn in the finished project for your instructor's evaluation before going on to a new project or exercise.

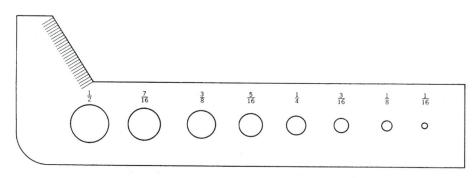

FIGURE D4.13

Countersinking and Counterboring

In drill press work it is often necessary to make a recess that will leave a bolt head below the surface of the workpiece. These recesses are made with countersinks or counterbores. When holes are drilled into rough castings or angular surfaces, a flat surface square to these holes is needed, and spot facing is the operation used. This unit will familiarize you with these drill press operations.

OBJECTIVES

After completing this unit, you should be able to:

1. Identify tools for countersinking and counterboring.
2. Select speeds and feeds for countersinking and counterboring.
3. Countersink and counterbore holes.

COUNTERSINKS

A countersink is a tool used to make a conical enlargement of the end of a hole. Figures D5.1 and D5.2 illustrate two types of single-flute countersinks, both of which are designed to produce smooth surfaces, free from chatter marks. A countersink is used as a chamfering or deburring tool to prepare a hole for reaming or tapping. Unless a hole needs to have a sharp edge, it should be chamfered to protect the end of the hole from nicks and burrs. A chamfer from $\frac{1}{32}$- to $\frac{1}{16}$-in. wide is sufficient for most holes.

A hole made to receive a flathead screw or rivet should be countersunk deep enough for the head to be flush with the surface or up to .015 in. below the surface. A flathead fastener should never project above the surface. The included angles on commonly available countersinks are 60, 82, 90, and 100 degrees. Most flathead fasteners used in metalworking have an 82-degree head angle, except for the aircraft industry where the 100-degree angle is prevalent. The cutting speed used when countersinking should always be slow enough to avoid chattering.

A combination drill and countersink with a 60-degree angle (Figure D5.3) is used to make center holes in workpieces for machining on lathes and grinders. The illustration shows a bell-type center drill that provides an additional angle for a chamfer of the center, protecting it from damage. The combination drill and countersink, known as a *center drill,* is used for spotting holes when using a drill press or milling machine since it is extremely rigid and will not bend under pressure.

COUNTERBORES

Counterbores are tools designed to enlarge previously drilled holes, much like countersinks, and are guided into the hole by a pilot to assure the concentricity of the two holes. A multiflute counterbore is shown in Figure D5.4, and a two-flute counterbore is shown in Figure D5.5. The two-flute counterbore has more chip clearance and a larger rake angle than the counterbore in Figure D5.4 so it is freer cutting and better suited for soft and ductile materials. Counterbored holes have flat bottoms, unlike the angled edges of countersunk holes, and are often used to recess a bolt head below the surface of a workpiece. Solid counterbores, such as shown in Figures D5.4 and D5.5, are used to cut recesses for socket head capscrews or filister head screws (Figure D5.6). The diameter of the counterbore is usually $\frac{1}{32}$ in. larger than the head of the bolt.

When a variety of counterbore and pilot sizes is necessary, a set of interchangeable pilot counter-

FIGURE D5.1 Single-flute countersink. (The DoAll Company)

FIGURE D5.2 (right) Chatterfree countersink. (The DoAll Company)

FIGURE D5.3 Center drill or combination drill and countersink. (The DoAll Company)

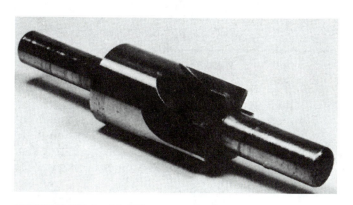

FIGURE D5.4 Multiflute counterbore.

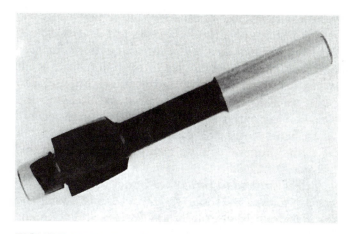

FIGURE D5.5 Two-flute counterbore.

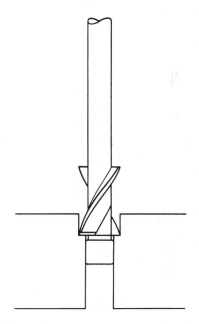

FIGURE D5.6 The counterbore is an enlargement of a hole already drilled.

bores is available. Figure D5.7 shows a counterbore in which a number of standard or specially made pilots can be used. A pilot is illustrated in Figure D5.8.

Counterbores are made with straight or tapered shanks to be used in drill presses, milling machines, or even lathes. When counterboring a recess for a hex head bolt, remember to measure the diameter of the socket wrench so the hole will be large enough to accommodate it. For most counterboring operations the pilot should have from .002- to .005-in. clearance in the hole. If the pilot is too tight in the hole, it may seize and break. If there is too much clearance between the pilot and the hole, the

FIGURE D5.7 Interchangeable pilot counterbore.

FIGURE D5.8 Pilot for interchangeable pilot counterbore.

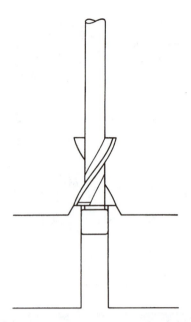

FIGURE D5.9 Spot facing on a raised boss.

counterbore will be out-of-round and will have an unsatisfactory surface finish.

It is very important that the pilot be lubricated while counterboring. Usually this lubrication is provided if a cutting oil or soluble oil is used. When cutting dry, which is often the case with brass and cast iron, the hole and pilot should be lubricated with a few drops of lubricating oil.

Counterbores or spot facers are often used to provide a flat bearing surface for nut or bolt heads on rough castings or a raised **boss** (Figure D5.9). This operation is called *spot facing*. Because these rough surfaces may not be at right angles to the pilot hole, great strain is put on the pilot and counterbore and can cause breakage of either one. To avoid breaking the tool, be very careful when starting the cut, especially when hand feeding. Prevent **hogging** into the work by tightening the spindle clamps slightly to remove possible **backlash.**

Recommended power feed rates for counterboring are shown in Table D5.1. The feed rate should be great enough to get under any surface scale quickly, thus preventing rapid dulling of the counterbore. The speeds used for counterboring are one-third less than the speeds used for twist drills of corresponding diameters. The choice of speeds and feeds is very much affected by the condition of the equipment, the power available, and the material being counterbored.

Before counterboring a workpiece, it should be securely fastened to the machine table or tightly held in a vise because of the great cutting pressures encountered. Workpieces also should be supported on parallels to allow for the protrusion of the pilot. To obtain several equally deep countersunk or coun-

TABLE D5.1 Feeds for counterboring

$\frac{3}{8}$-in. diameter up to .004 in. per revolution
$\frac{5}{8}$-in. diameter up to .005 in. per revolution
$\frac{7}{8}$-in. diameter up to .006 in. per revolution
$1\frac{1}{4}$-in. diameter up to .007 in. per revolution
$1\frac{1}{2}$-in. diameter up to .008 in. per revolution

terbored holes on the drill press or milling machine, the spindle depth stop can be set.

Counterbores can be used on a lathe to rough out a hole before it is finished-bored. This is often more efficient than using a single cutting edge boring bar. It permits the use of a larger-diameter boring bar for a more rigid setup.

SELF-TEST

1. When is a countersink used?
2. Why are countersinks made with varying angles?
3. What is a center drill?
4. When is a counterbore used?
5. What relationship exists between pilot size and hole size?
6. Why is lubrication of the pilot important?
7. As a rule, how does cutting speed compare between a counterbore and twist drill of equal size?
8. What affects the selection of feed and speed when counterboring?
9. What is spot facing?
10. What important points should be considered when a counterboring setup is made?

Reaming in the Drill Press

Many engineering requirements involve the production of holes having smooth surfaces, accurate location, and uniform size. In many cases, holes produced by drilling alone do not entirely satisfy these requirements. For this reason, the reamer was developed for enlarging or finishing previously formed holes. This unit will help you properly identify, select, and use machine reamers.

OBJECTIVES

After completing this unit, you should be able to:

1. Identify commonly used machine reamers.
2. Select the correct feeds and speeds for commonly used materials.
3. Determine appropriate amounts of stock allowance.
4. Identify probable solutions to reaming problems.

Reamers are tools used mostly to precision finish holes, but they are also used in the heavy construction industry to enlarge or align existing holes.

COMMON MACHINE REAMERS

Machine reamers have straight or taper shanks; the taper usually is a standard Morse taper. The parts of a machine reamer are shown in Figure D6.1 and the cutting end of a machine reamer is shown in Figure D6.2.

Chucking reamers (Figures D6.3, D6.4, and D6.5) are efficient in machine reaming a wide range of materials and are commonly used in drill presses, turret lathes, and screw machines. Helical flute reamers have an extremely smooth cutting action that finishes holes accurately and precisely. Chucking reamers cut on the chamfer at the end of the flutes. This chamfer is usually at a 45-degree angle.

Jobbers reamers (Figure D6.6) are used where a longer flute length than those of chucking reamers is needed. The additional flute length gives added guide to the reamer, especially when reaming deep holes.

The rose reamer (Figure D6.7) is primarily a roughing reamer used to enlarge holes to within .003 to .005 in. of finish size. The teeth are slightly backed off, which means that the reamer diameter is smaller toward the shank end by approximately .001 in./in. of flute length. The lands (outside surfaces of flutes) on these reamers are ground cylindrically without *radial relief* (backing off the land to provide a cutting edge), and all cutting is done on the end of the reamer. This reamer will remove a considerable amount of material in one cut.

Shell reamers (Figure D6.8) are finishing reamers. They are more economically produced, especially in larger sizes, than solid reamers because a much smaller amount of tool material is used in making them. Two slots in the shank end of the reamer fit over matching driving lugs on the shell or box reamer (Figure D6.9). The hole in the shell reamer has a slight taper ($\frac{1}{8}$ in./ft) in it to assure exact alignment with the shell reamer **arbor.** Shell reamers are made with straight or helical flutes and are commonly produced in sizes from $\frac{3}{4}$ to $2\frac{1}{2}$ in. in diameter. Shell reamer arbors come with matching straight or tapered shanks and are made in designated sizes from numbers 4 to 9.

Morse taper reamers (Figure D6.10), with straight or helical flutes, are used to finish ream-tapered holes in drill sockets, sleeves, and machine tool spindles. Helical taper pin reamers (Figure D6.11) are especially suitable for machine reaming of taper

FIGURE D6.1 The parts of a machine reamer. (Bendix Industrial Tools Division)

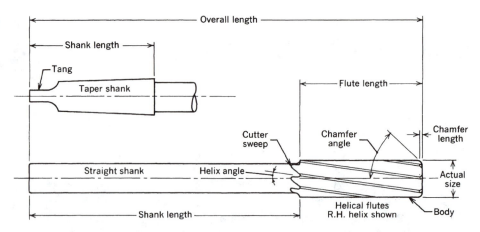

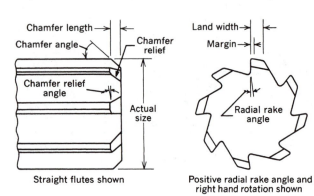

Straight flutes shown

Positive radial rake angle and right hand rotation shown

FIGURE D6.2 The cutting end of a machine reamer. (Bendix Industrial Tools Division)

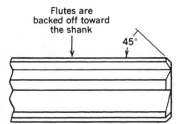

FIGURE D6.7 Rose reamer.

FIGURE D6.3 Straight shank straight flute chucking reamer. (TRW Inc.)

FIGURE D6.4 Straight shank helical flute chucking reamer. (TRW Inc.)

FIGURE D6.5 Taper shank helical flute chucking reamer. (TRW Inc.)

FIGURE D6.6 Taper shank straight flute jobbers reamer. (TRW Inc.)

FIGURE D6.8 Shell reamer helical flute. (TRW Inc.)

FIGURE D6.9 Taper shank shell reamer arbor. (TRW Inc.)

FIGURE D6.10 Morse taper reamer. (TRW Inc.)

FIGURE D6.11 Helical taper pin reamer. (TRW Inc.)

pin holes. There is no packing of chips in the flutes, which reduces the possibility of breakage. These reamers have a free-cutting action that produces a good finish at high cutting speeds. Taper pin reamers have a taper of $\frac{1}{4}$ in. per foot of length and are manufactured in 18 different sizes, ranging from the smallest at number 8/0 (eight naught) to the largest at number 10.

Taper bridge reamers (Figure D6.12) are used in structural iron or steel work, bridge work, and ship construction where extreme accuracy is not required. They have long tapered pilot points for easy entry into the out-of-line holes often encountered in structural work. Taper bridge reamers are made with straight and helical flutes to ream holes with diameters from $\frac{1}{4}$ to $1\frac{5}{16}$ in.

Carbide-tipped chucking reamers (Figure D6.13) are often used in production setups, particularly where abrasive materials or sand and scale as in castings are encountered. The right-hand helix chucking reamer (Figure D6.14) is recommended for ductile materials or highly abrasive materials or when machining blind holes. The carbide-tipped left-hand helix chucking reamer (Figure D6.15) will produce good finishes on heat-treated steels and other hard materials, but should be used on through holes only. All expansion reamers (Figure D6.16) after becoming worn can be expanded and resized by grinding. This feature offsets normal wear from abrasive materials and provides for a long tool life. These tools should not be adjusted for reaming size by loosening or tightening the expansion plug, but only by grinding.

Reaming is intended to produce accurate and straight holes of uniform diameter. The required ac-

FIGURE D6.12 Helical flute taper bridge reamer. (TRW Inc.)

FIGURE D6.13 Carbide-tipped straight flute chucking reamer. (TRW Inc.)

FIGURE D6.14 Carbide-tipped, helical flute chucking reamer, right-hand helix. (TRW Inc.)

FIGURE D6.15 Carbide-tipped, helical flute chucking reamer, left-hand helix. (TRW Inc.)

FIGURE D6.16 Carbide-tipped expansion reamer. (TRW Inc.)

curacy depends on a high degree of surface finish, tolerance on diameter, roundness, straightness, and absence of **bell mouth** at the ends of holes. To make an accurate hole it is necessary to use reamers with adequate support for the cutting edges; an adjustable reamer may not be adequate. Machine reamers are often made of either high-speed steel or cemented carbide. Reamer cutting action is controlled to a large extent by the cutting speed and feed used.

SPEED

The most efficient cutting speed for machine reaming depends on the type of material being reamed, the amount of stock to be removed, the tool material being used, the finish required, and the rigidity of the setup. A good starting point, when machine reaming, is to use $\frac{1}{2}$ to $\frac{1}{3}$ of the cutting speed used for drilling the same materials. Table D6.1 may be used as a guide.

Where conditions permit the use of carbide reamers, the speeds may often be increased over those recommended for HSS (high-speed steel) reamers. The limiting factor is usually an absence of rigidity in the setup. Any chatter, which is often caused by too high a speed, is likely to chip the cutting edges of a carbide reamer. Always select a speed that is slow enough to eliminate chatter. Close tolerances and fine finishes often require the use of considerably lower speeds than those recommended in Table D6.1.

FEEDS

Feeds in reaming are usually two to three times greater than those used for drilling. The amount of feed may vary with different materials, but a good starting point would be twice the feed rates given in Table D4.2 in Unit 4, "Operating Drilling Machines."

TABLE D6.1 Reaming speeds

Aluminum and its alloys	130–200[a]
Brass	130–200
Bronze, high tensile	50– 70
Cast iron	
Soft	70–100
Hard	50– 70
Steels	
Low-carbon	50– 70
Medium-carbon	40– 50
High-carbon	35– 40
Alloy	35– 40
Stainless steel	
AISI 302	15– 30
AISI 403	20– 50
AISI 416	30– 60
AISI 430	30– 50
AISI 443	15– 30

[a]Cutting speeds in surface feet per minute (FPM or SFM) for reaming with an HSS reamer.

Too low a feed may **glaze** the hole, which has the result of work hardening the material, causing occasional chatter and excessive wear on the reamer. Too high a feed tends to reduce the accuracy of the hole and the quality of the surface finish. Generally, it is best to use as high a feed as possible to produce the required finish and accuracy.

When a drill press that has only a hand feed is used to ream a hole, the feed rate should be estimated just as it would be for drilling. About twice the feed rate should be used for reaming as would be used for drilling in the same setup when hand feeding.

STOCK ALLOWANCE

The stock removal allowance should be sufficient to assure a good cutting action of the reamer. Too small a stock allowance results in **burnishing** (a slipping or polishing action), or it wedges the reamer in the hole and causes excessive wear or

TABLE D6.2 Stock allowance for reaming

Reamer Size (in.)	Allowance (in.)
$\frac{1}{32}$ to $\frac{1}{8}$	.003 to .006
$\frac{1}{8}$ to $\frac{1}{4}$	.005 to .009
$\frac{1}{4}$ to $\frac{3}{8}$	.007 to .012
$\frac{3}{8}$ to $\frac{1}{2}$	.010 to .015
$\frac{1}{2}$ to $\frac{3}{4}$	$\frac{1}{64}$ or $\frac{1}{32}$
$\frac{3}{4}$ to 1	$\frac{1}{32}$

TABLE D6.3 Coolants used for reaming

Material	Dry	Soluble Oil	Kerosene	Sulfurized Oil	Mineral Oil
Aluminum		x	x		
Brass	x	x			
Bronze	x	x			x
Cast iron	x				
Steels					
Low-carbon		x		x	
Alloy		x		x	
Stainless		x		x	

breakage of the reamer. The condition of the hole before reaming also has an influence on the reaming allowance since a rough hole will need a greater amount of stock removed than an equal-size hole with a fairly smooth finish. See Table D6.2 for commonly used stock allowance for reaming. When materials that work harden readily are reamed, it is especially important to have adequate material for reaming.

CUTTING FLUIDS

To ream a hole to a high degree of surface finish, a cutting fluid is needed. A good cutting fluid will cool the workpiece and tool and will also act as a lubricant between the chip and the tool to reduce friction and heat buildup. Cutting fluids should be applied in sufficient volume to flush the chips away. Table D6.3 lists some coolants used for reaming different materials.

REAMING PROBLEMS

Chatter is often caused by the lack of rigidity in the machine, workpiece, or the reamer itself. Corrections may be made by reducing the speed, increasing the feed, putting a chamfer on the hole before reaming, using a reamer with a pilot (Figure D6.17), or reducing the clearance angle on the cutting edge of the reamer. Carbide-tipped reamers especially cannot tolerate even a momentary chatter at the start of a hole, as such a vibration is likely to chip the cutting edges.

Oversize holes can be caused by inadequate workpiece support, worn guide bushings, worn or loose spindle bearings, or a bent reamer shank. When reamers gradually start cutting larger holes, it may be because of the work material galling or forming a built-up edge on reamer cutting surfaces (Figure D6.18). Mild steel and some aluminum alloys are particularly troublesome in this area.

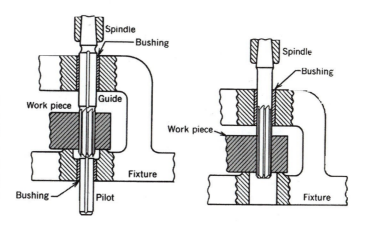

FIGURE D6.17 Use of pilots and guided bushings on reamers. Pilots are provided so that the reamer can be held in alignment and can be supported as close as possible while allowing for chip clearance. (Bendix Industrial Tools Division)

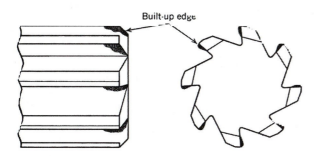

FIGURE D6.18 Reamer teeth having built-up edges.

Changing to a different coolant may help. Reamers with highly polished flutes, margins, and relief angles or reamers that have special surface treatment may also improve the cutting action.

Bell-mouthed holes are caused by misalignment of the reamer with the hole. The use of accurate bushings or pilots may correct bell mouth, but in many cases the only solution is the use of floating holders. A floating holder will allow movement in some directions while restricting it in others. A poor finish can be improved by decreasing the feed, but this will also increase the wear and shorten the life of the reamer. A worn reamer will never leave a good surface finish as it will score or groove the finish and often produce a tapered hole.

Too fast a feed will cause a reamer to break. Too large a stock allowance for finish reaming will produce a large volume of chips with heat buildup and will result in a poor hole finish. Too small a stock allowance will cause the reamer teeth to rub as they cut, not cut freely, which will produce a poor finish and cause rapid reamer wear. Coolant applied in insufficient quantity may also cause rough surface finishes when reaming.

SELF-TEST

1. How is a machine reamer identified?
2. What is the difference between a chucking and a rose reamer?
3. What is a jobbers reamer?
4. Why are shell reamers used?
5. How does the surface finish of a hole affect its accuracy?
6. How does the cutting speed compare between drilling and reaming for the same material?
7. How does the feed rate compare between drilling and reaming?
8. How much reaming allowance will you leave on a $\frac{1}{2}$-in. hole?
9. What is the purpose of using a coolant while reaming?
10. What can be done to overcome chatter?
11. What will cause oversize holes?
12. What causes a bell-mouthed hole?
13. How can poor surface finish be overcome?
14. When are carbide-tipped reamers used?
15. Why is vibration harmful to carbide-tipped reamers?

SECTION E
Turning Machines

The discussion of turning machines in this section is confined to horizontal spindle lathes. Many processes such as taper turning and threading are more easily and accurately produced on CNC lathes. However, a person must understand the principles of manual lathe machining in order to comprehend the operation of CNC machines.

In general, turning machines are those that rotate work material from which a chip is cut by a single-point tool. The spindle may be horizontal or vertical, but the process is the same.

The difference between the engine lathe and numerically controlled lathes and chuckers is simply a means of moving the slides and controlling the rotational speeds by a remote operating system. This was first accomplished with punched tape but now the programming is done by computer.

In this section, you will learn how to manually control the engine lathe. Later, as a CNC machinist you will still control the machine, but only by entering coordinates and data into a computer.

The Engine Lathe

The power-drive lathe or engine lathe is truly the father of all machine tools. With suitable attachments, the engine lathe may be used for turning, threading, boring, drilling, reaming, facing, spinning, and grinding, although many of these operations are preferably done on specialized machinery. Sizes range from the smallest jeweler's or precision lathes (Figure E1.1) to the massive lathes used for machining huge forgings (Figure E1.2).

Engine lathes (Figure E1.3) are used by machinists to produce one-of-a-kind parts or a few pieces for a short-run production. They are also used for toolmaking, machine repair, and maintenance.

Some lathes have a vertical spindle instead of a horizontal one with a large rotating table on which the work is clamped. These huge vertical boring mills (Figure E1.4) are the largest of our machine tools. A 25-foot diameter table is not unusual. Huge turbines, weighing many tons, can be placed on the table and clamped in position to be machined. The machining of such castings would be impractical on a horizontal spindle lathe.

Production lathes that are fully automatic or semiautomatic are generally specialized machines. These include turret lathes (Figure E1.5), automatic screw machines (Figure E1.6), tracer lathes (Figure E1.7).

All of these production machines are rapidly being replaced by CNC machinery. For example, the chucking lathe (Figure E1.8) can replace turret lathes and automatic screw machines; the tracer lathe has been replaced by numerically controlled lathes (Figure E1.9).

In the following units you will be learning how to use the engine lathe, that is, the power-driven nonspecialized machine. Modern engine lathes are highly accurate and complex machines capable of performing a great variety of operations. Before attempting to operate a lathe, you should familiarize yourself with its principal parts and their operation.

OBJECTIVE

After completing this unit, you should be able to:

Identify the most important parts of a lathe and their functions.

A lathe is a device in which the work is rotated against a cutting tool. As the cutting tool is moved lengthwise and crosswise to the axis of the workpiece, the shape of the workpiece is generated.

Figure E1.10 shows a lathe and its most important parts. A lathe consists of the following major component groups: headstock, bed, carriage, tailstock, quick-change gearbox, and a base or pedestal. The headstock is fastened on the left side of the bed. It contains the spindle that drives the various work-holding devices. The spindle is supported by spindle bearings on each end. If they are sleeve-type bearings, a thrust bearing is also used to take up end play. Tapered roller spindle bearings are often used on modern lathes. Spindle speed changes are also made in the headstock, either with belts or with gears (Figures E1.11a and E1.11b). The threading and feeding mechanisms of the lathe are also powered through the headstock.

Most belt-driven lathes use backgears for a slow speed range. Slow RPMs are obtained by engaging the backgears with the backgear lever. When higher RPMs are needed, the procedure is reversed by disengaging the backgears.

The spindle is hollow, which allows long, slender workpieces to pass through. The spindle end facing the tailstock is called the *spindle nose*. Spindle noses usually are one of three designs: a long-taper key

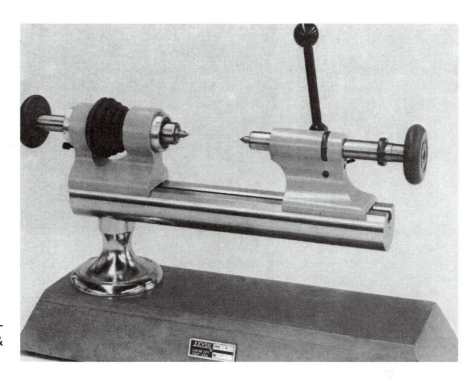

FIGURE E1.1 Jeweler's or instrument lathe. (Louis Levin & Son, Inc.)

drive (Figure E1.12), a camlock type (Figure E1.13), or a threaded spindle nose (Figure E1.14). Lathe chucks and other workholding devices are fastened to and driven by the spindle nose. The hole in the spindle nose typically has a standard Morse taper. The size of this taper varies with the size of the lathe.

The bed (Figure E1.15) is the foundation and backbone of a lathe. Its rigidity and alignment affect the accuracy of the parts machined on it. Therefore, lathe beds are constructed to withstand the stresses created by heavy machining cuts. The ways are on top of the bed, which usually consist of two inverted vees and two flat bearing surfaces. The ways of the

FIGURE E1.2 A massive forging being machined to exact specifications on a lathe. (Bethlehem Steel Corporation)

FIGURE E1.3 The engine lathe. (Clausing Machine Tools)

FIGURE E1.4 Vertical boring mill. (El-Jay Inc., Eugene, Oregon)

FIGURE E1.5 Saddle-type turret lathe. (The Warner & Swasey Company)

FIGURE E1.6 Small precision automatic screw machine in operation. (Sweetland Archery Products)

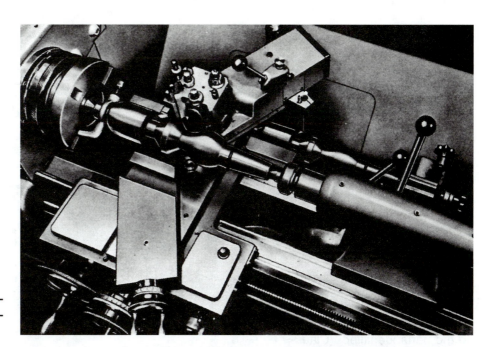

FIGURE E1.7 Tracer attachments on an engine lathe. (Clausing Machine Tools)

FIGURE E1.8 Numerically controlled chucking lathe with turret. (The Monarch Machine Tool Company, Sidney, Ohio)

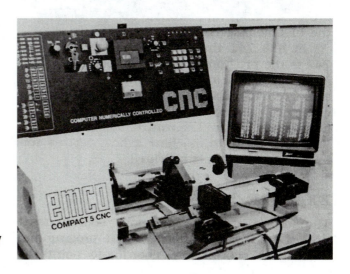

FIGURE E1.9 EMCO CNC Lathe. (Lane Community College)

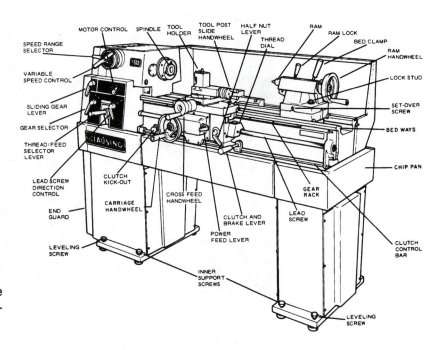

FIGURE E1.10 Engine lathe with the parts identified. (Clausing Machine Tools)

lathes are very accurately machined by grinding or by milling and hand scraping. Many modern lathes have hardened and ground ways. Wear or damage to the ways will affect the accuracy of workpieces machined on them. A gear rack is fastened below the front way of the lathe. Gears that link the carriage handwheel to this rack make the lengthwise movement of the carriage possible by hand-turning the carriage handwheel.

The carriage (Figure E1.16) is made up of the saddle and the apron. The saddle rides on top of the ways and carries the cross slide and the compound rest. The cross slide is moved perpendicular to the axis of the lathe by manually turning the cross-feed screw handle or by engaging the cross-feed lever (or clutch knob) for automatic power feed. On some lathes a feed change lever (or plunger) on the apron is used to direct power from the feed mechanism to either the longitudinal (lengthwise) travel of the carriage or to the cross slide. In some other lathes, two separate levers or knobs are used to transmit motion to the carriage or cross slide. The direction of move-

FIGURE E1.11a Geared headstock for heavy-duty lathe. (Lodge & Shipley Company)

FIGURE E1.11b Spindle drive showing gears and shifting mechanism located in the headstock. (Lodge & Shipley Company)

FIGURE E1.12 Long-taper key drive spindle nose. (Lane Community College)

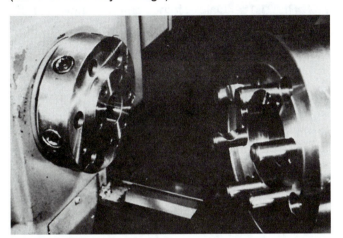

FIGURE E1.13 Camlock spindle nose.

FIGURE E1.14 Threaded spindle nose. (Lane Community College)

FIGURE E1.15 Lathe bed. (Clausing Machine Tools)

FIGURE E1.16 Lathe carriage. The arrow is pointing to the apron. (The Monarch Machine Tool Company, Sidney, Ohio)

ment is controlled by the feed reverse lever that is usually mounted on the headstock. It has the function of changing the direction of the lead screw rotation and consequently the feed direction. The compound rest is mounted on the cross slide and can be swiveled to any angle horizontal with the lathe axis in order to produce bevels and tapers. The compound rest can only be moved manually by turning the compound rest feed screw handle. Cutting tools

FIGURE E1.17 Tailstock. (The Monarch Machine Tool Company, Sidney, Ohio)

FIGURE E1.18 Quick-change gearbox showing index

are fastened on a tool post that is fastened to the compound rest.

The apron is the part of the carriage facing the operator. It contains the gears and feed clutches that transmit motion from the feed rod or lead screw to the carriage and cross slide. The thread dial is fastened to the apron, which indicates the exact moment to engage the half-nuts while thread cutting. The half-nut lever is used *only* while cutting threads. The entire carriage can be moved along the lathe bed manually by turning the carriage handwheel or under power by engaging the power feed controls on the apron. Once in position, the carriage can be clamped to the bed by tightening the carriage lock screw.

The tailstock (Figure E1.17) is used to support one end of a workpiece for machining or to hold various cutting tools such as drills, reamers, and taps. The tailstock has a sliding spindle that is operated by a handwheel and locked in position with a spindle clamp lever. The spindle is bored to receive a standard Morse taper shank. The tailstock consists of an upper and lower unit and can be adjusted to make tapered workpieces by turning the adjusting screws in the base unit, which offset the upper unit.

The **quick-change gearbox** (Figure E1.18) is the link that transmits power between the spindle and the carriage. By using the gear shift levers on the quick-change gearbox, you can select different feeds. Power is transmitted to the carriage through a feed rod or, on smaller lathes, through the lead screw with a keyseat in it. The index plate on the quick-change gearbox indicates the feed in thousandths of an inch or as threads per inch for the lever positions.

The base of the machine is used to support the lathe and to secure it to the floor. The lathe motor is usually mounted in the base. Figure E1.19 shows how the lathe is measured.

SELF-TEST

At a lathe in the shop, identify the following parts and describe their functions. *Do not* turn on the lathe until you get permission from your instructor.

THE HEADSTOCK

1. Spindle
2. Spindle speed-changing mechanism
3. Spindle nose
4. What kind of spindle nose is on your lathe?
5. Feed reverse lever

THE BED

1. The ways
2. The gear rack

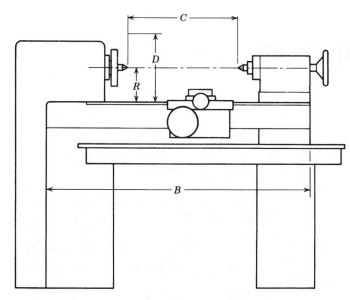

FIGURE E1.19 Measuring a lathe for size. *C* = maximum distance between centers; *D* = maximum diameter of workpiece over ways—swing of lathe; *R* = radius, one-half swing; *B* = length of bed. *(Machine Tools and Machining Practices)*

THE CARRIAGE

1. The cross slide
2. The compound rest
3. Saddle
4. Apron
5. Power feed lever
6. Feed change lever
7. Half-nut lever
8. Thread dial
9. Carriage handwheel
10. Carriage lock

THE TAILSTOCK

1. Spindle and spindle clamping lever
2. Tapered spindle hole and the size of its taper
3. The tailstock adjusting screws

THE QUICK-CHANGE GEARBOX

1. The lead screw
2. Shift the levers to obtain feeds of .005 and .013 in. per revolution. Rotating the chuck aids in shifting these levers.
3. Set the levers to obtain 4 threads/in. and then 12 threads/in.
4. Measure the lathe and record its size.

UNIT 2   Turning Machine Safety

Before 1910, safety programs in industry were virtually unknown and machines were not safeguarded. Lathes had exposed gears and pulleys. Accidents were very frequent and always considered the fault of careless workers. Since that time, many laws have been passed for accident prevention and guards and other safety devices are required on machines for worker safety. But this is not enough! Without your cooperation in wearing required protective clothing and devices and in following safe procedure, accidents will still happen.

This unit will make you aware of the many hazards that exist in lathe operations and help you to avoid them. Your responsibility as a student and as a worker is clear: Protect yourself and the people around you. Observe the safety rules.

OBJECTIVE

After completing this unit, you should be able to:

Describe and use 10 safety rules or procedures for lathe operations.

The lathe can be a safe machine only if the machinist is aware of the hazards involved in its operation. In the machine shop as anywhere, you must always keep your mind on your work in order to avoid accidents. Develop safe work habits in the use of setups, chip breakers, guards, and other protective devices. Standards for safety have been established as guidelines to help you eliminate unsafe practices and procedures on lathes. Some of the hazards are as follows:

1. **Pinch points due to lathe movement.** Fingers can be caught, for example, in gears or between the compound rest and a chuck jaw. The rule is to keep your hands away from such dangerous positions when the lathe is operating.
2. **Hazards associated with broken or falling components.** Heavy chucks or workpieces can be dangerous when accidentally dropped. Care must be used when handling them. If a threaded spindle is suddenly reversed, the chuck can come off and fly out of the lathe. A chuck wrench left in the chuck can become a missile when the machine is turned on. Always remove the chuck wrench immediately after using it (Figures E2.1a and E2.1b).

3. **Hazards resulting from contact with high-temperature components.** Burns usually result from handling hot chips (which can be up to 800°F or even more) or a hot workpiece. Gloves may be worn when handling hot chips or workpieces, but never when the machine is running.
4. **Hazards resulting from contact with sharp edges, corners, and projections.** These are perhaps the most common cause of hand injuries in lathe work. Dangerous sharp edges may be found many places: on a long stringy chip, on a tool bit, or on a burred edge of a turned or threaded part. Shields should be used for protection from flying chips and coolant. These shields are usually made of clear plastic and are hinged over the chuck or clamped to the carriage of engine lathes. Stringy chips must not be removed with bare hands; wear heavy gloves and use hook tools or pliers. Always turn off the machine before attempting to remove chips. Chips should be broken and should form a figure 9 rather than be a stringy mass or a long wire (Figure E2.2). Chip breakers on tools and correct feeds will help to produce safe, easily handled chips. Burred edges must be removed before the workpiece is removed from the lathe. Always

FIGURE E2.1a A chuck wrench left in the chuck is a danger to everyone in the shop.

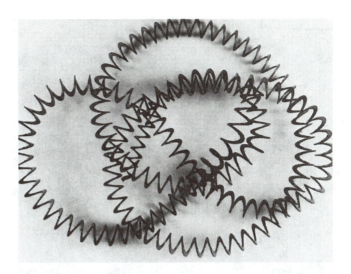

FIGURE E2.2 Unbroken lathe chips are sharp and hazardous to the operator. (Lane Community College)

FIGURE E2.1b A safety-conscious lathe operator will remove the chuck wrench when he finishes using it.

remove the tool bit when setting up or removing workpieces from the lathe.

5. **Hazards of workholding devices or driving devices.** When workpieces are clamped, their components often extend beyond the outside diameter of the holding device. Guards, barriers, and warnings such as signs or verbal instructions are all used to make you aware of the hazards. On power chucking devices you should be aware of potential pinch points between workpiece and workholding devices. Make certain sufficient gripping force is exerted by the jaws to safely hold the work. Never run a geared scroll chuck without having something gripped in the jaws. Centrifugal force on the jaws can cause the scroll to unwind and the jaws to come out of the chuck. Keep tools, files, and micrometers off the machine. They may vibrate off into the revolving chuck or workpiece.

6. **Electrical hazards.** Only qualified electricians should install any electric wiring. Enclosures for electrical and power transmission equipment are restricted to authorized personnel only.

7. **Employee's responsibility.** Setup checks should be made during each shift. Safety glasses and industrial hairnets or hair ties should be worn where required (Figure E2.3). Jewelry and clothing that will present a hazard must not be worn. An orderly work area must be maintained by the operator. The spindle or workpiece should never be slowed or stopped by hand gripping or by using a pry bar. Always use machine controls to stop or slow it.

8. **Extended workpieces.** Workpieces extending out of the lathe should be supported by a stock tube (Figure E2.4). If a slender workpiece is allowed to extend beyond the headstock spindle a foot or so without support, centrifugal force can cause it to fly outward. The piece will not only be bent but it will present a very great danger to anyone standing near.

These and other safety considerations are the responsibility of both employer and employee. Also the manufacturer, reconstructor, and modifier share this responsibility by providing proper safety devices for machines. Additional safety precautions you should take are as follows:

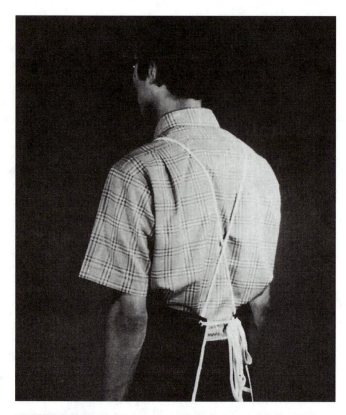

FIGURE E2.3 Correct attire for machining on a lathe. The apron is tied in back with a light cord so that it will break easily in case the apron catches in the machine.

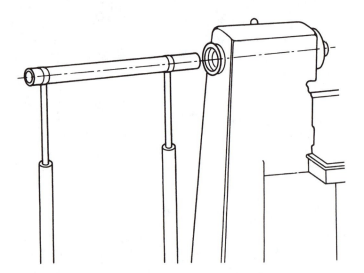

FIGURE E2.4 Stock tube is used to support long workpieces that extend out of the headstock of a lathe.

- Never overtighten locking bolts or set screws on toolholders. An overstressed part or bolt may fail suddenly and endanger the operator.
- Always keep the compound rest over its slide unless it is being used for angular cutting.
- Never operate a machine with a guard open or off.
- Keep your hands and fingers away from revolving work or machine parts.
- When cutting to a shoulder, finish the cut with hand feed.
- When setting up work, rotate the spindle by hand to see if the lathe dog or workpiece clears the carriage or compound rest.
- When setting up work between centers, make sure the dead center (tailstock center) is properly adjusted and the binding lever is tightened.
- Work that extends from the chuck more than five times its diameter should be supported by a steady rest or tailstock.
- Hold one end of abrasive cloth strips in each hand when polishing rotating work. Don't let either hand get closer than a few inches from the work (Figure E2.5).
- Keep rags, brushes, and fingers away from rotating work, especially when knurling. Roughing cuts tend to quickly drag in and wrap up rags, clothing, neckties, abrasive cloth, and hair (Figure E2.6).
- Move the carriage back out of the way and cover the tool with a cloth when checking boring work.
- When removing or installing chucks or heavy workpieces, use a board on the ways (a part of the lathe bed) so the piece can be slid into place. To lift a heavy chuck or workpiece (larger than an 8-in. diameter chuck) get help or use a crane (Figures E2.7a and E2.7b). Remove the tool or turn it out of the way during this operation.
- Always clean and oil a machine before you use it, and clean up the machine and work area when you are finished.
- Do not shift gears or try to take measurements while the machine is running and the workpiece is in motion.
- When grinding tool bits, eye protection must be worn, grinder tables or rests must be adjusted within safe limits (about $\frac{1}{16}$ in. from the wheel), and guards must be in place.
- Never use a file without a handle, as the file **tang** can quickly cut your hand or wrist if the file has been struck by a spinning chuck jaw or lathe dog. Left-hand filing is considered safest in the lathe; that is, the left hand grips the handle while the right hand holds the tip end of the file (Figure E2.8).
- Never hurry or become distracted from your work. If someone wishes to talk to you, stop your machine until you are finished talking.

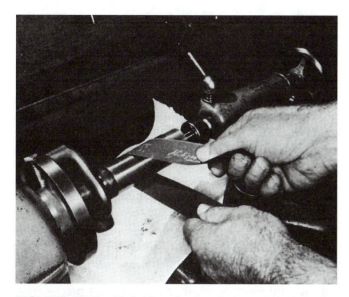

FIGURE E2.5 Polishing in the lathe with abrasive cloth.

FIGURE E2.6 Loose hair that is not kept confined or tied back may get caught in machinery with results such as this. (Photo courtesy of John Allan, Jr.)

FIGURE E2.7a One way of hoisting materials and equipment into the lathe is a lathe-mounted crane for handling chucks and workpieces. This one is adapted for mounting on quick-change toolholders on the compound. (Syclone Products Inc.)

FIGURE E2.7b The skyhook in use bringing a large chuck into place for mounting. (Syclone Products Inc.)

- Do not use an air hose to blow away chips. The flying chips are hazardous to your eyes as well as the eyes of others. Air pressure will drive small chips and dirt into bearing surfaces and ruin machinery.

- Don't try to fix a machine unless you are authorized and trained to do so.
- Avoid at all times carelessness, horseplay, and acting without thinking, especially when operating a machine. Remember, it is your own attitude

FIGURE E2.8 Left-hand filing in the lathe. (Lane Community College)

that will ultimately either prevent accidents or cause you to suffer pain and loss of working time or worse.

SELF-TEST

1. On whom does the responsibility for worker safety rest?

2. What is a pinch point?

3. How would it be possible for a chuck to come out of a lathe? A chuck wrench?

4. Is it possible for a lathe operator to get burns on his body? How?

5. How can a lathe operator receive cuts on the hands and arms?

6. Because of the nature and use of chucking devices, it is impossible to adequately guard them. How then can the operator avoid danger in this area?

7. The headstock gears are being changed by a mechanic in the lathe next to yours. He accidentally pulls the switch off for your lathe as well as the one he is working on when he went into the electrical enclosure. What should you do?

8. A student needs to machine several pieces from a $\frac{3}{8}$-in. rod. Since the rod is about 4 ft long and he is reluctant to cut it, he allows it to extend through the spindle and out about 30 in. The machining speed is about 1000 RPM. What do you think will result?

9. Overtightening bolts and set screws can sometimes be more hazardous than undertightening them. Why is this so?

10. When a person knows all the safety rules and procedures and still has accidents, what do you think is wrong?

Toolholders and Toolholding for the Lathe

For lathe work, cutting tools must be supported and fastened securely in the proper position to machine the workpiece. There are many different types of toolholders available to satisfy this need. Anyone working with a lathe should be able to select the best toolholding device for the operation performed.

OBJECTIVES

After completing this unit, you should be able to:

1. Identify quick-change and turret-type toolholders mounted on a lathe carriage.
2. Identify toolholding for the lathe tailstock.

QUICK-CHANGE TOOLING

A cutting tool is supported and held in a lathe by a toolholder that is secured in the tool post of the lathe with a clamp screw.

A **quick-change tool post** (Figure E3.1), so called because of the speed with which tools can be interchanged, is more versatile than the standard post. The toolholders used on it are accurately held because of the dovetail construction of the post. This accuracy makes for more exact repetition of setups. Tool height adjustments are made with a micrometer adjustment collar, and the height alignment will remain constant through repeated tool changes.

A three-sided quick-change tool post (Figure E3.2) has the added ability to mount a tool on the tailstock side of the tool post. These tool posts are securely clamped to the compound rest. The tool post in Figure E3.2 uses double vees to locate the toolholders, which are clamped and released from the post by turning the top lever.

Toolholders for the quick-change tool posts include those for turning (Figure E3.3), threading (Figures E3.4a and E3.4b), and holding drills (Figure E3.5). The drill holder makes it possible to use the carriage power feed when drilling holes instead of the tailstock hand feed. Figure E3.6 shows a boring bar toolholder in use; the boring bar is very rigidly supported.

An advantage of the quick-change tool post toolholders is that cutting tools of various shank thicknesses can be mounted in the toolholders (Figure E3.7). **Shims** are sometimes used when the shank is too small for the set screws to reach. Another example of quick-change tool post versatility is shown in Figure E3.8, where a tailstock turret is in use. Figure E3.9 shows a cutoff tool mounted in a toolholder. Figure E3.10 is a combination knurling tool and facing toolholder. A four-tool turret toolholder (Figure E3.11) can be set up with several different tools such as turning tools, facing tools, and threading or boring tools. Often one tool can perform two or more operations, especially if the turret can be indexed in 30-degree intervals. A facing operation (Figure E3.11), a turning operation (Figure E3.12), and chamfering of a bored hole (Figure E3.13) are all performed from this turret. Tool height adjustments are made by placing shims under the tool.

The toolholders studied so far are all intended for use on the carriage of a lathe. Toolholding is also done on the tailstock. Figure E3.14 shows how the tailstock spindle is used to hold Morse taper shank tools. One of the most common toolholding devices used on a tailstock is the drill chuck (Figure E3.15). A drill chuck is used for holding straight-shank drilling tools. When a series of operations must be performed and repeated on several workpieces, a tailstock turret (Figure E3.16) can be used. The tail-

FIGURE E3.1 Quick-change tool post, dovetail type. (Aloris Tool Company, Inc.)

FIGURE E3.2 Three-sided quick-change tool post. (Lane Community College)

FIGURE E3.3 Turning toolholder in use. (Aloris Tool Company, Inc.)

FIGURE E3.4a Threading toolholder, using the top of the blade. (Aloris Tool Company, Inc.)

FIGURE E3.4b Threading is accomplished with the bottom edge of the blade with the lathe spindle in reverse. This assures cutting right-hand threads without hitting the shoulders. (Aloris Tool Company, Inc.)

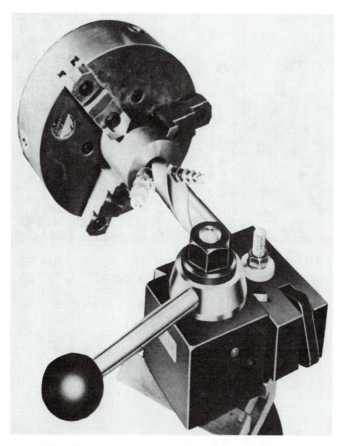

FIGURE E3.5 Drill toolholder in the tool post. Mounting the drill in the tool post makes drilling with power feed possible. (Aloris Tool Company, Inc.)

FIGURE E3.7 Toolholders are made with wide or narrow slots to fit tools with various shank thicknesses.

FIGURE E3.8 Tailstock turret used in quick-change toolholder. (Enco Manufacturing Company)

FIGURE E3.6 Boring toolholder. This setup provides good boring bar rigidity. (Aloris Tool Company, Inc.)

FIGURE E3.9 Quick-change cutoff toolholder. (Enco Manufacturing Company)

FIGURE E3.10 Quick-change knurling and facing toolholder. (Enco Manufacturing Company)

FIGURE E3.12 Turning cut with a turret-type toolholder. (Enco Manufacturing Company)

FIGURE E3.11 Facing cut with a turret-type toolholder. (Enco Manufacturing Company)

FIGURE E3.13 Chamfering cut with a turret-type toolholder. (Enco Manufacturing Company)

stock turret illustrated has six tool positions, one of which is used as a workstop. The other positions are for center drilling, drilling, reaming, counterboring, and tapping. Tailstock tools are normally fed by turning the tailstock handwheel.

In all cases, whether the tool is fastened in a toolholder as a single-point cutting tool, as is a boring bar, the amount of overhang is extremely important. Any tool or bar should be placed as close to the holder as possible. The further the tool is extended,

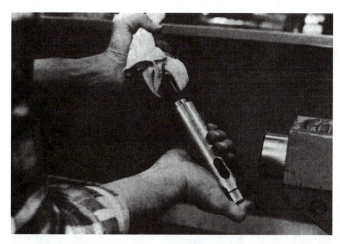

FIGURE E3.14 Taper shank just in front of tailstock spindle hole. (Lane Community College)

FIGURE E3.16 Tailstock turret. (Enco Manufacturing Company)

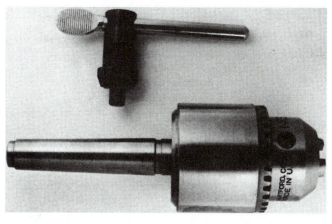

FIGURE E3.15 Drill chuck with Morse taper shank.

the more likely it is to produce chatter (vibration) or to dig in and damage the work.

SELF-TEST

At a lathe in the shop, identify various toolholders and their functions.

1. What is the purpose of a toolholder?
2. How are tool height adjustments made on a quick-change toolholder?
3. Explain how tool height adjustments are made on a turret-type toolholder.
4. How does tool or toolholder overhang affect the turning operations?
5. What kind of tools can be used in the lathe tailstock?
6. By what means are tools fastened in the tailstock?
7. When is a tailstock turret used?

Cutting Tools for the Lathe

A machinist must fully understand the purpose of cutting tool geometry, since it is the lathe tool that removes the metal from the workpiece. Whether this is done safely, economically, and with quality finishes depends to a large extent upon the shape of the point, the rake and relief angles, and the nose radius of the tool. In this unit, you will learn this tool geometry and also how to shape a lathe tool on a pedestal grinder.

Almost all cutting tools used in manufacturing and in machine shops are made of very hard materials that can resist breakdown even at high temperatures. These tool materials range from tungsten carbide to diamond. High-speed steel was once the only available high-temperature cutting tool, but it has been largely replaced except in the making of some special tool shapes. High-speed steel cutting tools can be easily shaped on a pedestal grinder. They are still used, though not often, by both manual and CNC machinists.

OBJECTIVES

After completing this unit, you should be able to:

1. Explain the purpose of rake and relief angles, chip breakers, and form tools.
2. Grind an acceptable right-hand roughing tool.

On a lathe, metal is removed from a workpiece by turning it against a single-point cutting tool. This tool must be very hard and it should not lose its hardness from the heat generated by machining. High-speed steel is used for many tools because it fulfills these requirements and is easily shaped by grinding. It should be noted, however, that the use of high-speed steel tools is limited since most production machining today is done with carbide tools. High-speed steel tools are often used for making specially shaped cutting tools. They are also useful for finishing operations, especially on soft metals.

The most important aspect of a lathe tool is its geometric form: the side and back rake, front and side clearance or **relief angles,** and chip breakers.

Figure E4.1 shows the parts and angles of the tool according to a commonly used industrial tool signature. The terms and definitions follow (the angles given are only examples and they could vary according to the application):

Back rake	BR	12°
Side rake	SR	12°
End relief	ER	10°
Side relief	SRF	10°
End cutting-edge angle	ECEA	30°
Side cutting-edge angle	SCEA	15°
Nose radius	NR	$\frac{1}{32}$ in.

TOOL SIGNATURE

1. The tool shank is that part held by the tool-holder.
2. Back rake is very important to smooth chip flow, which is necessary for a uniform chip and a good finish.
3. The side rake directs the chip flow away from the point of cut and provides for a keen cutting edge.
4. The end relief angle prevents the front edge of the tool from rubbing on the work.

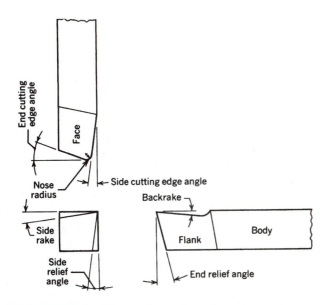

FIGURE E4.1 The parts and angles of a tool.

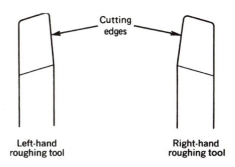

FIGURE E4.3 Left-hand and right-hand roughing tools.

5. The side relief angle provides for cutting action by allowing the tool to feed into the work material.
6. The cutting edge angle may vary considerably (from 5 to 32 degrees). For roughing, it should almost be square (5 degrees off 90 degrees), but tools used for squaring shoulders or for other light machining could have angles from 15 to 32 degrees.
7. The side cutting-edge angle, which is usually 10 to 20 degrees, directs the cutting forces back into a stronger section of the tool point. It helps to direct the chip flow away from the workpiece. It also affects the thickness of the cut (Figure E4.2).
8. The nose radius will vary according to the finish required.

Grinding a tool provides both a sharp cutting edge and the shape needed for the cutting operation. When the purpose for the rake and relief angles on a tool is clearly understood, then a tool suitable to the job may be ground. Left-hand tools are shaped just the opposite to right-hand tools (Figure E4.3). The right-hand tool has the cutting edge on the left side and cuts to the left or toward the headstock. The hand of the lathe tool can be easily determined by looking at the cutting end of the tool from the opposite side of the lathe; the cutting edge is to the right on a right-hand tool.

Tools are given a slight **nose radius** to strengthen the tip. A larger nose radius will give a better finish (Figure E4.4) but will also promote chattering (vibration) in a nonrigid setup. All lathe tools require some nose radius, however small. A sharp pointed tool is very weak at the point and will usually break off in use, causing a rough finish on the work. A facing tool (Figure E4.5) for shaft ends and **mandrel** work has very little nose radius and an included angle of 58 degrees. This facing tool is not used for chucking work, however, as it is a relatively weak tool. A right-hand

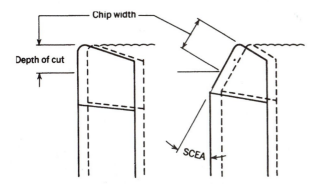

FIGURE E4.2 The change in chip width with an increase of the side cutting edge angle. A large SCEA can sometimes cause chatter (vibration of work or tool).

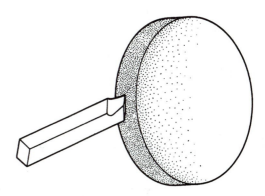

FIGURE E4.4 One method of grinding the nose radius on the point of the tool.

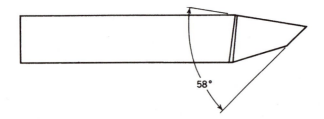

FIGURE E4.5 A right-hand facing tool showing point angles. This tool is not suitable for roughing operations because of its acute point angle.

(RH) or left-hand (LH) roughing or finishing tool is often used for facing in chuck-mounted workpieces. Some useful tool shapes are shown in Figure E4.6. These are used for general lathe work.

Tools that have specially shaped cutting edges are called *form* tools (Figure E4.7). These tools are plunged directly into the work, making the full cut in one operation.

Parting or cutoff tools are often used for necking or undercutting, but their main function is cutting off material to the correct length. The correct and incorrect ways to grind a cutoff tool are shown in Figure E4.8. Note that the width of the cutting edge becomes narrower than the blade as it is ground lower, which causes the blade to bind in a groove that is deeper than the sharpened end.

However, tools are sometimes specially ground for parting very soft metals or specially shaped grooves (Figure E4.9). The end is sometimes ground on a slight angle when a series of small hollow pieces is being cut off (Figure E4.10). This helps to eliminate the burr on small parts. This procedure is not recommended for deep parting.

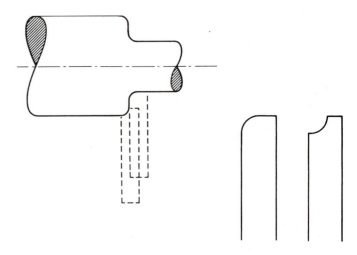

FIGURE E4.7 Form tools are used to produce the desired shape in the workpiece. External radius tools, for example, are used to make outside corners round; fillet radius tools are used on shafts to round the inside corners on shoulders.

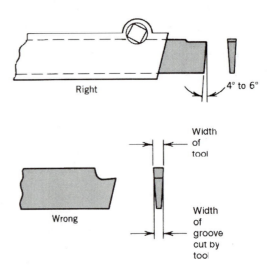

FIGURE E4.8 Correct and incorrect methods of grinding a cutoff tool for deep parting.

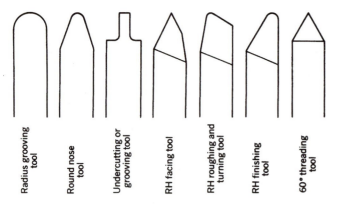

Radius grooving tool | Round nose tool | Undercutting or grooving tool | RH facing tool | RH roughing and turning tool | RH finishing tool | 60° threading tool

FIGURE E4.6 Some useful tool shapes most often used. The first tool shapes needed are the three on the right, which are the roughing or general turning tool, finishing tool, and threading tool.

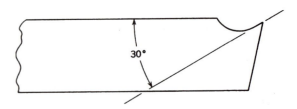

FIGURE E4.9 Cutoff tools are sometimes ground with large back rake angles for aluminum and other soft metals.

Tools that have been ground back for resharpening too many times often form a **chip trap,** causing the metal to be torn off or the tool to not cut at all (Figure E4.11). A good machinist will never allow tools to get in this condition, but will grind off the useless end and regrind a proper tool shape (Figure E4.12).

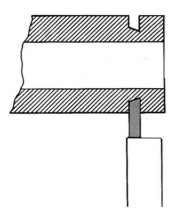

FIGURE E4.10 Parting tool ground on an angle to avoid burrs on the cutoff pieces.

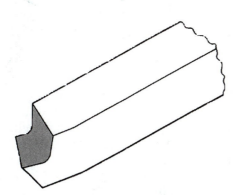

FIGURE E4.11 Deformed tool caused by many resharpenings. The chip trap should be ground off and a new point ground on the tool.

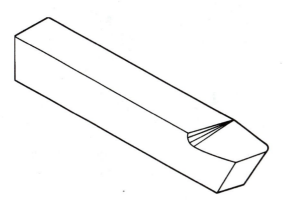

FIGURE E4.12 A properly ground right-hand roughing tool.

RAKE ANGLES AND CHIP BREAKERS

Modern lathes have toolholders that hold the tool horizontally, therefore it is sometimes necessary to grind a back rake into the tool when soft metals are being machined. The tool in Figure E4.13 has been ground with a back rake.

Threading tools should always have zero rake; any negative or positive rake would change the angle of the thread. The tool should also be checked for the 60-degree angle with a center gage while grinding. Relief should be ground on each side. A slight flat should be honed on the end with an oilstone.

Tools for brass or plastics should have zero to negative rake to keep the tools from "digging in" (Figure E4.14). Side rake, back rake, and relief angles are given for tools in Table E4.1 for machining various metals.

Figure E4.15 shows gages used for checking angles when grinding a tool. They are designed to check tool angles on any flat surface. The angles are for tools to be used in toolholders having $16\frac{1}{2}$-degree back rake. Tools for straight or horizontal toolholders should have an end relief of 10 degrees; in these cases, the end relief angle of the gage must be changed. A protractor or optical comparator can also be used to check tool angles.

Chip Breakers

For your safety, it is important to make tools that will produce chips that are not hazardous. Long, unbroken

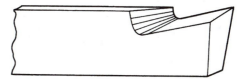

FIGURE E4.13 Right-hand roughing tool with back rake.

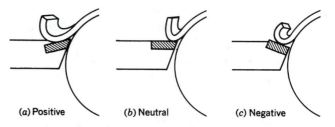

(a) Positive   (b) Neutral   (c) Negative

FIGURE E4.14 Side view of back rake angles. The tool point is thinner (and more subject to breakage) on the positive back rake than on the negative-rake tool.

TABLE E4.1 Angle degrees for high-speed steel tools

Material	End Relief	Side Relief	Side Rake	Back Rake
Aluminum	8 to 10	12 to 14	14 to 16	30 to 35
Brass, free-cutting	8 to 10	8 to 10	1 to 3	0
Bronze, free-cutting	8 to 10	8 to 10	2 to 4	0
Cast iron, gray	6 to 8	8 to 10	10 to 12	3 to 5
Copper	12 to 14	13 to 14	18 to 20	14 to 16
Nickel and monel	12 to 14	14 to 16	12 to 14	8 to 10
Steels, low-carbon	8 to 10	8 to 10	10 to 12	10 to 12
Steels, alloy	7 to 9	7 to 9	8 to 10	6 to 8

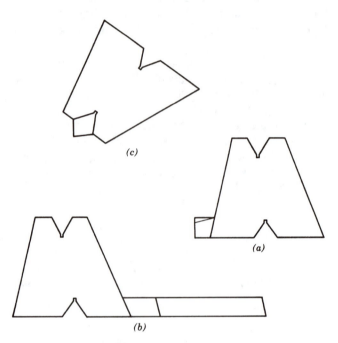

FIGURE E4.15 Tool gages for checking angles: *(a)* The side relief angle; *(b)* End relief; and *(c)* The wedge angle for steel.

chips are extremely dangerous. **Tool geometry,** especially side and back rakes, has a considerable effect on chip formation. Smaller side-rake angles tend to curl the chip more than larger angles, and the curled chips are more likely to break up. Coarse feeds for roughing and the maximum depth of cut also promote chip breaking. Feeds, speeds, and depth of cut will be further considered in Unit 8, "Turning Between Centers."

Chip breakers are extensively used on both carbide and high-speed tools to curl the chip as it flows across the face of the tool. Since the chip is curled back into the work, it can go no further and breaks (Figure E4.16). A C-shaped chip is often the result,

but a figure-9-shaped chip is considered ideal (Figure E4.17). This chip should drop safely into the chip pan without flying out.

Grinding the chip breaker too deep will form a chip trap that may cause binding of the chip and tool breakage (Figure E4.18). The correct depth to grind a chip breaker is approximately $\frac{1}{32}$ in. Chip breakers are typically of the parallel or angular types (Figure E4.19). More skill is needed to offhand grind a chip breaker on a high-speed tool than is required to grind the basic tool angles. Therefore, the basic tool should be ground and an effort made to produce safe chips through the use of correct feeds and depth of cut before a chip breaker is ground.

Care must be exercised while grinding on high-speed steel. A glazed wheel can generate heat up to 2000°F (1093°C) at the grinder-tool interface. Do not overheat the tool edge as this will cause small surface cracks that can result in the failure of the tool. Frequent cooling in water will keep the tool cool enough to handle. Do not quench in water, however, if you have overheated the tool. Let it cool in air.

Since the right-hand roughing tool is the one that is most commonly used by machinists and the first one that you will need, you should begin with it. A piece of **keystock** the same size as the tool bit should be used for practice until you are able to grind an acceptable tool.

Steps in Grinding a Right-Hand Roughing Lathe Tool

You will need a tool blank, a piece of keystock about 3 in. long, a tool gage, and a toolholder.

A. Grind one acceptable practice right-hand roughing tool. Have your instructor evaluate your progress until it is ground correctly.

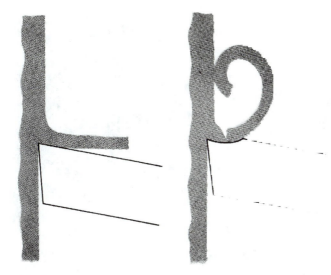

FIGURE E4.16 Chip flow with a plain tool and with chip breaker.

FIGURE E4.17 A figure-9 chip is considered the safest kind of chip to produce.

FIGURE E4.18 Crowding of the chip is caused by a chip breaker that is ground too deeply.

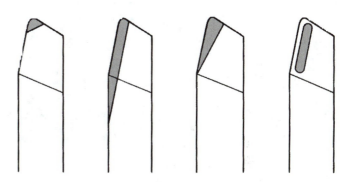

FIGURE E4.19 Four common types of chip breakers.

B. Grind one acceptable right-hand roughing tool from a high-speed tool blank.

C. Wear goggles and make certain the tool rest on the grinder is adjusted properly (about $\frac{1}{16}$ in. from the wheel). True up the wheels with a wheel dresser if they are grooved, glazed, or out-of-round.

1. Using the roughing wheel, grind the side relief angle and the side cutting angle about 10 degrees by holding the blank and supporting your hand on the tool rest (Figure E4.20).
2. Check the angle with a tool gage (Figure E4.21). Correct if needed.
3. Rough out the end relief angle about 14 degrees and the end cutting-edge angle (Figure E4.22).

4. Check the angle with the tool gage (Figure E4.23). Correct if needed.
5. Rough out the side rake. Stay clear of the side cutting edge by $\frac{1}{16}$ in. (Figure E4.24).
6. Check for wedge angle (Figure E4.25). Correct if needed.
7. Now change to the finer grit wheel and very gently finish grind the side and end relief angles. Try to avoid making several facets or grinds on one surface. A side-to-side oscillation will help to produce a good finish.
8. Grind the finish on the side rake as in Figure E4.24 and bring the ground surface just to the side cutting edge but avoid going deeper.

UNIT 4: CUTTING TOOLS FOR THE LATHE 185

FIGURE E4.20 Roughing the side relief angle and the side cutting-edge angle.

FIGURE E4.23 Checking the end relief angle with a tool gage.

FIGURE E4.21 Checking the side cutting-edge angle with a tool gage.

FIGURE E4.24 Roughing the side rake.

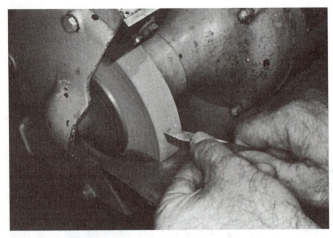

FIGURE E4.22 Roughing the end relief angle and the end cutting-edge angle.

FIGURE E4.25 Checking for wedge angle with a tool gage.

9. A slight radius on the point of the tool should be ground on the circumference of the wheel (Figure E4.26) and all the way from the nose to the heel of the tool.
10. A medium to fine oilstone is used to remove the burrs from the cutting edge (Figure E4.27). The finished tool is shown in Figure E4.28.

CARBIDES AND OTHER TOOL MATERIALS

The cutting tool materials such as carbon steels and high-speed steel that served the needs of machining in past years are not suitable in many applications today. Tougher and harder tools are required to machine the tough, hard, space-age metals and new alloys. The constant demand for higher productivity led to the need for faster stock removal and quick-changing tooling. You, as a machinist, must learn to achieve maximum productivity at minimum cost. Your knowledge of cutting tools and ability to select them for specific machining tasks will affect your productivity directly.

The various tool materials used in today's machining operations are high-speed steel, cemented carbides, ceramics, cubic boron nitride (CBN), and diamond.

High-Speed Steels

High-speed steels (HSS) may be used at higher speeds (100 FPM in mild steels) without losing their hardness. The relationship of cutting speeds to the approximate temperature of tool-chip interface is as follows:

100 FPM 1000°F (538°C)
200 FPM 1200°F (649°C)
300 FPM 1300°F (704°C)
400 FPM 1400°F (760°C)

Cemented Carbides

A carbide, generally, is a chemical compound of carbon and a metal. The term *carbide* is commonly used to refer to cemented carbides, the cutting tools composed of **tungsten carbide,** titanium carbide, or tantalum carbide and cobalt in various combinations. A typical composition of cemented carbide is 85 to 95 percent carbides of tungsten and the remainder a cobalt binder for the tungsten carbide powder.

Cemented carbides are made by compressing various metal powders and **sintering** (heating to weld

FIGURE E4.26 One method of grinding the nose radius is on the circumference of the wheel.

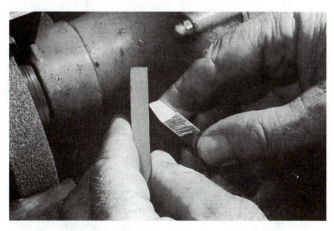

FIGURE E4.27 Using an oilstone to remove the burrs from the cutting edge.

FIGURE E4.28 The finished tool.

particles together without melting them) the briquettes. Cobalt powder is used as a binder for the carbide powder used, either tungsten, titanium, or tantalum carbide powder or a combination of these. Increasing the percentage of cobalt binder increases the toughness of the tool material and at the same time reduces its hardness or wear resistance. Carbides have greater hardness at both high and low temperatures than do high-speed steel or cast alloys. At temperatures of 1400°F (760°C) and higher, carbides maintain the hardness required for efficient machining. This makes possible machining speeds of approximately 400 FPM in steels. The addition of tantalum increases the red hardness of a tool material. Cemented carbides are extremely hard tool materials (above RA 90), have a high compressive strength, and resist wear and rupture.

Coated carbide inserts are often used to cut hard or difficult-to-machine workpieces. Titanium carbide (TiC) coating offers high wear resistance at moderate cutting speeds and temperatures. **Aluminum oxide** (Al_2O_3) coating resists chemical reactions and maintains its hardness at high temperatures. Titanium nitride (TiN) coating has high resistance to crater wear and reduces friction between the toolface and the chip, thereby reducing the tendency for a built-up edge.

Cemented carbides are the most widely used tool materials in the machining industry. They are particularly useful for cutting tough alloy steels.

Most cemented carbide inserts are provided with a chip breaker, whose object is to create the desirable chip formation (Figure E4.29). However, other factors, such as feed rate, depth of cut, and workpiece material, affect the chip curl and its final shape. A higher feed rate and deeper cut tend to curl the chip more, whereas a light cut and small feed rate will often produce a long, stringy chip even if the tool has a chip breaker. A tough or hard workpiece will curl the chip more than a soft material.

Ceramic Tools

Ceramic or "cemented oxide" tools are made primarily from aluminum oxide. Some manufacturers add titanium, magnesium, or chromium oxides in quantities of 10 percent or less. The tool materials are molded at pressures over 4000 PSI and sintered at temperatures of approximately 3000°F (1649°C). This process partly accounts for the high density of hardness of cemented oxide tools.

Cemented oxides are brittle and require that machines and setups be rigid and free of vibration. Some machines cannot obtain the spindle speeds

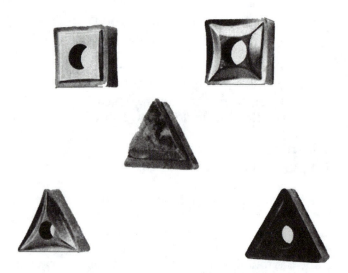

FIGURE E4.29 Chip breakers used are the adjustable chip deflator (center) with a straight insert and the type with the built-in chip control groove.

required in terms of SFPM to use cemented oxides at their peak capacity. When practicable, however, cemented oxides can be used to machine relatively hard materials at high speeds.

Ceramic tools are either cold pressed or hot pressed, that is, formed to shape. Some hot-pressed, high-strength ceramics are termed *cermets* because they are a combination of ceramics and metals. An insert tool composed of titanium carbide and titanium nitride is an example. A common combination of materials in a cermet is aluminum oxide and titanium carbide. These ceramics possess the high shear resistance of ceramics and the toughness and thermal shock resistance of metals. Cermets were originally designed for use at high temperatures such as those found in jet engines. These materials were subsequently adapted for cutting tools. In machining of steels, these tools have a low coefficient of friction and therefore less tendency to form a built-up edge. They are used at high cutting speeds. Another method of gaining the high-speed cutting characteristics of ceramics and the toughness of metal tools is to coat a carbide tool with a ceramic composite. Ceramic tools should be used as a replacement for carbide tools that are wearing rapidly but not to replace carbide tools that are breaking.

Cubic Boron Nitride (CBN) Tools

CBN is next to diamond in hardness and therefore can be used to machine plain carbon steels, alloy steels, and gray cast irons with hardnesses of 45 RC and above. Formerly, steels over 60 RC would have to be abrasive machined, but with the use of CBN they can often be cut with single-point tools.

CBN inserts consist of a cemented carbide substrate with an outside layer of CBN formed as an integral part of the tool. Tool life, finishes, and resistance to cracking and abrasion make CBN a tool material superior to both carbides and ceramics.

Diamond Tools

Industrial diamonds are sometimes used to machine extremely hard workpieces. Only relatively small removal rates are possible with diamond tools, but very high speeds are used and good finishes are obtained. Nonferrous metals are turned at 2000 to 2500 FPM, for example. Sintered polycrystalline diamond tools, available in shapes similar to those of ceramic tools, are used for materials that are abrasive and difficult to machine. Polycrystalline tools consist of a layer of randomly oriented synthetic diamond crystals brazed to a tungsten carbide insert.

Diamond tools are particularly effective for cutting abrasive materials that quickly wear out other tool materials. Nonferrous metals, plastics, and some nonmetallic materials are often cut with diamond tools. Diamond is not particularly effective with carbon steels or superalloys that contain cobalt or nickel. Ferrous alloys chemically attack single- or polycrystalline diamonds, causing rapid tool wear. Because of the high cost of diamond tool material, it is usually restricted to those applications where other tool materials do not cut very well or break down quickly.

Diamond or ceramic tools should never be used for interrupted cuts such as on splines or **keyseats** because they could chip or break. They must never be used at low speeds or on machines that are not capable of attaining the higher speeds at which these tools should operate.

Each of the cutting tool materials varies in hardness. The differences in hardness tend to become more pronounced at high temperatures, as can be seen in Table E4.2. Hardness is related to the wear resistance of a tool and its ability to machine materials that are softer than it is. Temperature change is important since the increase of temperature during machining results in the softening of the tool material.

SELF-TEST

1. Name the advantages of using high-speed steel for tools.
2. Other than hardness and toughness, what is the most important aspect of a lathe tool?

TABLE E4.2 Hardness of cutting materials at high and low temperatures

Tool Material	Hardness at Room Temperature	Hardness at 1400°F (760°C)
High-speed steel	RA 85	RA 60
Carbide	RA 92	RA 82
Cemented oxide (ceramic)	RA 93–94	RA 84
Cubic boron nitride		Near diamond hardness
Diamond		Hardest known substance

3. How do form tools work?
4. A tool that has been reground too many times on the same place can form a chip trap. Describe the problems that result from this condition.
5. Why is it not always necessary to grind a back rake into the tool?
6. When should a zero or negative rake be used?
7. Explain the purpose of the side and end relief angles.
8. What is the function of the side and back rakes?
9. How can these angles be checked while grinding?
10. Why should chips be broken up?
11. In what ways can chips be broken?
12. Overheating a high-speed tool bit can easily be done by using a glazed wheel that needs dressing or by exerting too much pressure. What does this cause in the tool?
13. What are ceramic tools made of?
14. When should ceramic tools be used?
15. What is the greatest advantage of carbide tools over high-speed steel?
16. Extremely hard or abrasive materials are machined with diamond tools. Is the material removal rate high or low? What kind of finishes are produced?
17. When turning a shaft with a keyseat in it, should ceramic or diamond tools be used? Explain your answer.
18. Polycrystalline diamond tools are similar in some ways to ceramic inserts. What are their major advantages and disadvantages?
19. In what way is a CBN tool superior to ceramic or diamond? Explain.
20. How are right-hand tools identified?

Lathe Spindle Tooling

Workholding and driving devices that are fastened to the spindle nose are very important to machining on lathes. Various types of these workholding devices, their uses, and proper care are detailed in this unit.

OBJECTIVES

After completing this unit, you should be able to:

1. Explain the uses and care of independent and universal chucks.
2. Explain the limitations and advantages of collets and describe a collet setup.
3. Explain the use of a face driver or drive center.
4. Explain the uses of and differences between drive plates and face plates.

LATHE SPINDLE NOSE

The lathe spindle nose is the carrier of a variety of workholding devices that are fastened to it in several ways. The spindle is hollow and has an internal Morse taper at the nose end, which makes possible the use of taper shank drills or drill chucks (Figure E5.1). This internal taper is also used to hold live centers, drive centers, or collet assemblies. The outside of the spindle nose can be threaded (Figure E5.2), have a long taper with key drive (Figure E5.3), or have a camlock (Figure E5.4).

Threaded spindle noses are mostly used on older lathes. The chuck or face plate is screwed on a coarse, right-hand thread until it is forced against a shoulder on the spindle, which aligns it. Two disadvantages of the threaded spindle nose are that the spindle cannot be rotated in reverse against a load and that it is sometimes very difficult to remove a chuck or face plate (Figure E5.5).

The long-taper key drive spindle nose relies on the principle that a tapered fit will always repeat its original position. The key gives additional driving power. A large nut having a right-hand thread is turned with a spanner wrench. It draws the chuck into position and holds it there.

Camlock spindle noses use a short taper for alignment. A number of studs arranged in a circle fit into holes in the spindle nose. Each stud has a notch into which a cam is turned to lock it in place.

All spindle noses and their mating parts must be carefully cleaned before assembly. Small chips or **grit** will cause a workholding device to run out of true and be damaged. A spring cleaner (Figure E5.6) is used on mating threads for threaded spindles. Brushes and cloths are used for cleaning. A thin film of light oil should be applied to threads and mating surfaces.

CHUCKS

Independent four-jaw and universal three-jaw chucks and, occasionally, drive or face plates are mounted on the spindle nose of engine lathes. Each of the four jaws of the independent chuck moves independently of the others, which makes it possible to set up oddly shaped pieces (Figure E5.7). The concentric rings on the chuck face help to set the work true before starting the machine. Very precise setups also can be made with the four-jaw chuck by using a dial indicator, especially on round material. Each jaw of the chuck can be removed and reversed to accommodate irregular shapes. Some types are fitted with top jaws that can be reversed after the removal of bolts on the jaw. Jaws in the reverse position can grip larger-diameter workpieces (Figure E5.8). The independent chuck will hold work more securely for heavy cutting than will the three-jaw universal chuck.

Universal chucks usually have three jaws, but some are made with two jaws (Figure E5.9) or six

FIGURE E5.1 Section view of the spindle. (Machine Tools and Machining Practices)

FIGURE E5.4 Camlock spindle nose.

FIGURE E5.2 Threaded spindle nose.

FIGURE E5.5 The threaded chuck can be removed by using a large monkey wrench on one of the chuck jaws while the spindle is locked in a low gear. A long steel bar may also be used between the jaws. The knockout bar should never be used to remove a chuck because it is too light and will bend.

FIGURE E5.3 Long taper with key drive spindle nose.

jaws (Figure E5.10). All the jaws are moved in or out equally in their slides by means of a scroll plate located at the back of the jaws. The scroll plate has a bevel gear on its reverse side that is driven by a pinion gear. This gear extends to the outside of the chuck body and is turned with the chuck wrench (Figure E5.11). Universal chucks provide quick and simple chucking and centering of round stock. Uneven or irregularly shaped material will damage these chucks.

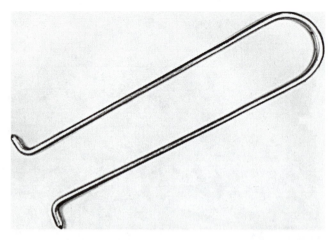

FIGURE E5.6 A spring cleaner is used for cleaning internal threads on chucks.

FIGURE E5.8 Four-jaw chuck in reverse position holding a large-diameter workpiece.

FIGURE E5.7 Four-jaw independent chuck holding an offset rectangular part.

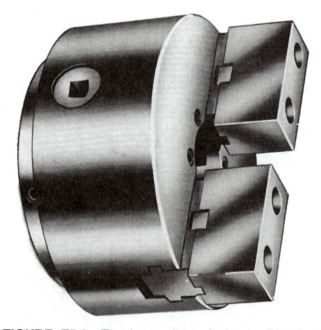

FIGURE E5.9 Two-jaw universal chuck. (Hardinge Brothers, Inc.)

The jaws of standard universal chucks will not reverse like independent chucks, so a separate set of reverse jaws is used (Figure E5.12) to hold pieces with larger diameters. The chuck and each of its jaws are stamped with identification numbers. Do not interchange any of these parts with another chuck or both will be inaccurate. Also each jaw is stamped 1, 2, or 3 to correspond to the same number stamped by the slot on the chuck. The jaws are removed from the chuck in the order 3, 2, 1 and should be returned in the reverse order, 1, 2, 3.

FIGURE E5.10 Six-jaw universal chuck. (Buck Tool Company)

A universal chuck with top jaws (Figure E5.13), in contrast to the standard type with two sets of jaws, is reversed by removing the bolts in the top jaws and reversing them. They must be carefully cleaned when this is done. Soft top jaws are frequently used when special gripping problems arise. Since the jaws are machined to fit the shape of the part (Figure E5.14), they can grip it securely for heavy cuts (Figure E5.15).

One disadvantage of most universal chucks is that they lose their accuracy when the scroll and jaws wear, and normally there is no compensation for wear other than regrinding the jaws. The three-jaw adjustable chuck in Figure E5.16 has a compensating adjustment for wear or misalignment.

FIGURE E5.11 Exploded view of a universal three-jaw chuck (Adjust-tru) showing the scroll plate and gear drive mechanism. (Buck Tool Company)

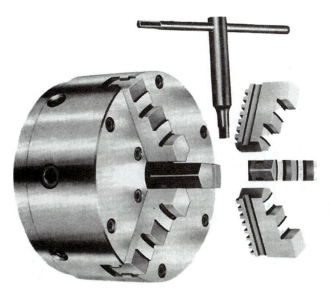

FIGURE E5.12 Universal three-jaw chuck (Adjust-tru) with a set of outside jaws. (Buck Tool Company)

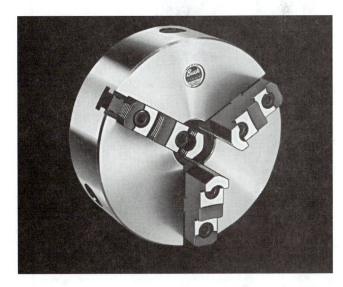

FIGURE E5.13 Universal chuck with top jaws. (Buck Tool Company)

FIGURE E5.14 Machining soft jaws to fit an odd-shaped workpiece on a jaw-turning fixture. (The Warner & Swasey Company)

FIGURE E5.15 Soft jaws have been machined to fit the shape of this cast steel workpiece in order to hold it securely for heavy cuts. (The Warner & Swasey Company)

Combination universal and independent chucks also provide for quick opening and closing and have the added advantage of independent adjustment on each jaw. These chucks are like the universal type since three or four jaws move in or out equally, but each jaw can be adjusted independently as well.

Magnetic chucks (Figure E5.17) are sometimes used for making light cuts on **ferromagnetic** material. They are useful for facing thin material that would be difficult to hold in conventional workholding devices. Magnetic chucks do not hold work very securely and so are not used much in lathe work.

FIGURE E5.16 Universal chuck (Adjust-tru) with special adjustment feature (G) makes it possible to compensate for wear. (Buck Tool Company)

1 Pinion (C) moves jaws

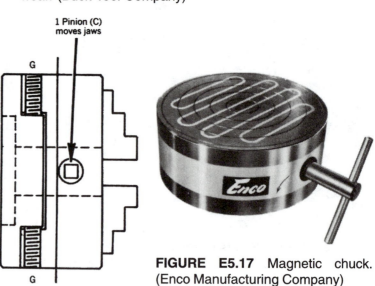

FIGURE E5.17 Magnetic chuck. (Enco Manufacturing Company)

All chucks need frequent cleaning of scrolls and jaws, which should be lightly oiled after cleaning. Chucks with grease fittings should be pressure lubricated. Chucks come in all diameters and are made for light-, medium-, and heavy-duty uses.

DRIVE PLATES AND FACE PLATES

Drive plates are used together with lathe dogs to drive work mounted between centers (Figure E5.18). The live center fits directly into the spindle taper and turns with the spindle. A sleeve is sometimes used if the spindle taper is too large in diameter to fit the center. The live center is usually made of soft steel, so the point can be machined as needed to keep it running true. Live centers are removed by means of a knock-out bar (Figure E5.19). Soft live centers should not be confused with hardened tailstock centers. A soft center in the tailstock will immediately fuse to the center hole in the work and damage it.

Often when a machinist wants to machine the entire length of work mounted between centers without the interference of a lathe dog, specially ground and hardened drive centers are used (Figure E5.20). These are serrated so they will turn the work, but only light cuts can be made. Modern drive centers or face drivers (Figure E5.21) can also be used to machine a part without interference from a lathe dog. Quite heavy cuts are possible with these drivers, which are used especially for manufacturing purposes.

Face plates are used for mounting workpieces or fixtures. Unlike drive plates that have only slots, face plates have T-slots and are more heavily built (Figure E5.22). Face plates are made of cast iron and so must be operated at relatively slow speeds. If the speed is too high, the face plate could fly apart.

COLLET CHUCKS

Collet chucks (Figure E5.23) are very accurate work-holding devices and are used in producing small

FIGURE E5.18 Drive plate for turning between centers.

FIGURE E5.19 A knockout bar is used to remove centers.

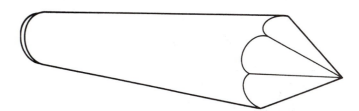

FIGURE E5.20 A hardened and serrated drive center is used for very light turning between centers.

FIGURE E5.21 Face driver is mounted in headstock spindle and work is driven by the drive pins that surround the center. (Copyright © 1976 Sandvik Madison, Inc. All rights reserved. Madison-Kosta is a registered trademark of Sandvik Madison, Inc. All dimensions and specifications subject to change without notice.)

FIGURE E5.22 T-slot face plate. Workpieces are clamped on the plate with T-bolts and strap clamps. (The Monarch Machine Tool Company, Sidney, Ohio)

FIGURE E5.23 Side and end views of a spring collet for round work.

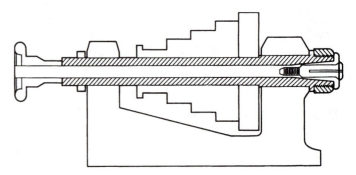

FIGURE E5.24 Cross section of spindle showing construction of a draw-in collet chuck attachment.

FIGURE E5.25 Rubber flex collet.

high-precision parts. Steel spring collets are available for holding and turning hexagonal, square, and round workpieces. They are made in specific sizes (which are stamped on them) with a range of only a few thousandths of an inch. Rough and inaccurate workpieces should not be held in the collet chuck since the gripping surfaces of the chuck would form an angle with the workpiece. The contact area would then be at one point on the jaws instead of along the entire length, and the piece would not be held firmly. If it is not held firmly, workpiece accuracy is impaired and the collet may be damaged. An adapter called a *collet sleeve* is fitted into the spindle taper and a draw bar is inserted into the spindle at the opposite end (Figure E5.24). The collet is placed in the adapter and the draw bar is rotated, which threads the collet into the taper and closes it. *Never tighten a collet without a workpiece in its jaws* because this will damage it. Before collets and adapters are installed, they should be cleaned to insure accuracy.

The rubber flex collet (Figure E5.25) has a set of tapered steel bars mounted in rubber. Each collet has

a range of about $\frac{1}{8}$ in., a much wider range than the spring collet. A large handwheel is used to open and close the collets instead of a draw bar (Figure E5.26).

The concentricity that you could expect from each type of workholding device is as follows:

Device	Centering Accuracy in Inches (Indicator Reading Difference)
Centers	Within .001
Four-jaw chuck	Within .001 (depending on the ability of machinist)
Collets	.0005 to .001
Three-jaw chuck	.001 to .003 (good condition) .005 or more (poor condition)

FIGURE E5.26 Collet handwheel attachment for rubber flex collets. (The Monarch Machine Tool Company, Sidney, Ohio)

SELF-TEST

1. Briefly describe the lathe spindle. How does the spindle support chucks and collets?
2. Name the spindle nose types.
3. What is an independent chuck and what is it used for?
4. What is a universal chuck and what is it used for?
5. What chuck types make possible the frequent adjusting of chucks so that they will hold stock with minimum runout?
6. Workpieces mounted between centers are driven with lathe dogs. Which type of plate is used on the spindle nose to turn the lathe dog?
7. What is a live center made of? How does it fit in the spindle nose?
8. Describe a drive center and a face driver.
9. On which type of plate are workpieces and fixtures mounted? How is it identified?
10. Name one advantage of using steel spring collets. Name one disadvantage.

Operating the Machine Controls

Before using any machine, you must be able to properly use the controls, know what they are for, and how they work. You must also be aware of the potential hazards that exist for you and the machine if it is mishandled. This unit prepares you to operate lathes.

OBJECTIVES

After completing this unit, you should be able to:

1. Explain drives and shifting procedures for changing speeds on lathes.
2. Describe the use of various feed control levers.
3. Explain the relationship between longitudinal feeds and cross feeds.
4. State the differences in types of crossfeed screw micrometer collars.

DRIVES

Most lathes have similar control mechanisms and operating handles for feeds and threading. Some machines, however, have entirely different driving mechanisms as well as different speed controls.

One drive system uses a variable speed drive (Figure E6.1) with a high and low range using a backgear. On this drive system, the motor must be running to change the speed on the vari-drive, but the motor must be turned off to shift the backgear lever. Geared head lathes are shifted with levers on the outside of the headstock (Figure E6.2). Several of these levers are used to set up the various speeds within the range of the machine. The gears will not mesh unless they are perfectly aligned, so it is sometimes necessary to rotate the spindle by hand. **Never try to shift gears with the motor running and the clutch lever engaged.**

FEED CONTROL LEVERS

The carriage is moved along the ways by means of the lead screw when threading, or by a separate feed rod when using feeds. On most small lathes, however, a lead screw-feed rod combination is used. In order to make left-hand threads and reverse the feed, the feed reverse lever is used. This lever re-verses the lead screw. It should never be moved when the machine is running.

The quick-change gearbox (Figures E6.3a and E6.3b) typically has two or more sliding gear shifter levers. These are used to select feeds or threads per inch. On those lathes also equipped with metric selections, the threads are expressed in pitches (measured in millimeters).

The carriage apron (Figure E6.4) contains the handwheel for hand feeding and a power feed lever that engages a clutch to a gear drive train in the apron.

Hand feeding should not be used for long cuts as there would be lack of uniformity and a poor finish. When using power feed and approaching a shoulder or the chuck jaws, disengage the power feed and hand feed the carriage for the last $\frac{1}{8}$ in. or so. The handwheel is used to quickly bring the tool close to the work before engaging the feed and for rapidly returning to the start of a cut after disengaging the feed. A feed change lever diverts the feed to either the carriage for longitudinal movement or to the crossfeed screw to move the cross slide. There is generally some slack or backlash in the crossfeed and compound screws. As long as the tool is being fed in one direction against the work load, there is no problem, but if the screw is *slightly* backed off, the readings will be in error. To correct this prob-

FIGURE E6.1 Variable speed control and speed selector. (Clausing Machine Tools)

FIGURE E6.3a Quick-change gearbox with index plate.

FIGURE E6.2 Speed change levers and feed selection levers on a geared head lathe.

FIGURE E6.3b Exposed quick-change gear mechanism for a large, heavy duty-lathe. (Lodge & Shipley Company)

lem, back off two turns and come back to the desired position.

Cross feeds are geared differently than longitudinal feeds. On most lathes, the cross feed is approximately one-third to one-half that of the longitudinal feed, so a facing job (Figure E6.5) with the quick-change gearbox set at about .012 in. feed would actually only be .004 in. for facing. The crossfeed ratio for each lathe is usually found on the quick-change gearbox index plate.

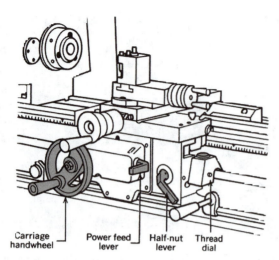

FIGURE E6.4 View of carriage apron with names of parts. (Clausing Machine Tools)

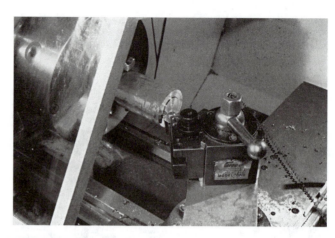

FIGURE E6.5 Facing on a lathe.

The half-nut or split-nut lever on the carriage engages the thread on the lead screw directly and is used **only** for threading. It cannot be engaged unless the feed change lever is in the neutral position.

Both the crossfeed screw handle and the compound restfeed screw handle are fitted with microm-

eter collars (Figure E6.6). These collars traditionally have been graduated in English units, but metric conversion collars (Figure E6.7) will help in the transition to the metric system.

Some micrometer collars are graduated to read single depth; that is, the tool moves as much as the read-

FIGURE E6.6 Micrometer collar on the crossfeed screw that is graduated in English units. Each division represents .001 in.

FIGURE E6.7 Crossfeed and compound screw handles with metric-English conversion collars. (The Monarch Machine Tool Company, Sidney, Ohio)

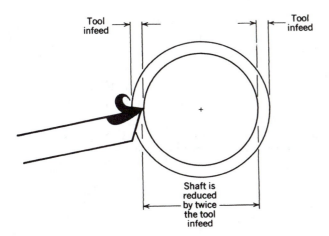

FIGURE E6.8 The diameter of the workpiece is reduced twice the amount in which the tool is moved. (Machine Tools and Machining Practices)

FIGURE E6.9 The clutch rod is actuated by moving the clutch level. This disengages the motor from the spindle. (Lane Community College)

ing shows. When turning a cylindrical object such as a shaft, dials that read single depth will remove twice as much from the diameter (Figure E6.8). For example, if the crossfeed screw is turned in .020 in. and a cut is taken, the diameter will have been reduced by .040 in. Sometimes only the compound dial is calibrated in this way and the cross slide will have graduations on its micrometer dial to compensate for double depth on turning. On this type of lathe, if the crossfeed screw is turned in until .020 in. shows on the dial and a cut is taken, the diameter will have been reduced .020 in. The tool would have actually moved into the work only .010 in. This is sometimes called *radius and diameter reduction.*

To determine which type of graduation you are using, set a fractional amount on the dial (such as .250 in. = $\frac{1}{4}$ in.) and measure on the cross slide with a rule. The actual slide movement you measure with the rule will be either the same as the amount set on the dial, for the single-depth collar, or one-half that amount, for the double-depth collar.

Some lathes have a brake and clutch rod that is the same length as the lead screw. A clutch lever connected to the carriage apron rides along the clutch rod (Figure E6.9). The spindle can be started and stopped without turning off the motor by using the clutch lever. Some types also have a spindle brake that quickly stops the spindle when the clutch lever is moved to the stop position. An adjustable automatic clutch kickoff is also a feature of the clutch rod.

When starting a lathe for the first time use the following checklist:

1. Move carriage and tailstock to the right to clear workholding device.
2. Locate feed clutches and half-nut lever and disengage before starting spindle.
3. Set up to operate at low speeds.
4. Read any machine information panels that may be located on the machine and observe precautions.
5. Note the feed direction; there are no built-in travel limits or warning devices to prevent feeding the carriage into the chuck or against the end of the slides.
6. When you are finished with a lathe, disengage all clutches, clean up chips, and remove any attachments or special setups.

SELF-TEST

1. On a geared-head drive, should the motor and drive clutch be turned on or off when shifting gears? Explain your answer.
2. Explain speed shifting procedure on the variable-speed drive.
3. In what way can speed changes be made on gearhead lathes?

4. What lever is shifted in order to reverse the lead screw?

5. The sliding gear shifter levers on the quick-change gearbox are used for just two purposes. What are they?

6. When is the proper time to use the carriage handwheel?

7. Why will you not get the same surface finish (tool marks per inch) on the face of a workpiece as you would get on the outside diameter when on the same power feed?

8. The half-nut lever is not used to move the carriage for turning. Name its only use.

9. Micrometer collars are attached on the cross-feed handle and compound handle. In what ways are they graduated?

10. How can you know if the lathe you are using is calibrated for single or double depth?

Facing and Center Drilling

Facing and center drilling the workpiece are often the first steps taken in a turning project to produce a stepped shaft or a sleeve from solid material. Much lathe work is done in a chuck and requires considerable facing and some center drilling. These important lathe practices will be covered in detail in this unit.

OBJECTIVES

After completing this unit, you should be able to:

1. Correctly set up a workpiece and face the ends.
2. Correctly center drill the ends of a workpiece.
3. Determine the proper feeds and speeds for a workpiece.
4. Explain how to set up to make facing cuts to a given depth and how to measure them.

SETTING UP FOR FACING

Facing is done to obtain a flat surface on the end of cylindrical workpieces or on the face of parts clamped in a chuck or face plate (Figures E7.1a and E7.1b). The work most often is held in a three- or four-jaw chuck. If the chuck is to be removed from the lathe spindle, a lathe board must first be placed on the ways. Figure E7.2 shows a camlock-mounted chuck being removed. The correct procedure for installing a chuck on a camlock spindle nose is shown in Figures E7.3a to E7.3f. The cams should be tightly snugged (Figure E7.3f) for one or two revolutions around the spindle.

Setting up work in an independent chuck is simple, but mastering the procedures takes some practice. Round stock can be set up by using a dial indicator (Figure E7.4). Square or rectangular stock can either be set up with a dial indicator or by using a toolholder turned backwards (Figures E7.5a and E7.5b).

Begin the setup by aligning two opposite jaws with the same concentric ring marked in the face of the chuck while the jaws are near the workpiece. This will roughly center the work. Set up the other two jaws with a concentric ring also when they are near the work. Next, bring all of the jaws firmly against the work. When using the dial indicator, zero the bezel at the lowest reading. Now rotate the chuck to the opposite jaw with the high reading and tighten it half the

amount of the runout. It might be necessary to loosen slightly the jaw on the low side. Always tighten the jaws at the position where the dial indicator contacts the work since any other location will give erroneous readings. When using the back of the toolholder, the micrometer dial on the cross slide will show the difference in runout. Chalk is sometimes used for setting up rough castings and other work too irregular to be measured with a dial indicator. Workpieces can either be chucked normally, internally, or externally (Figures E7.6a to E7.6c).

FACING

The material to be machined usually has been cut off in a power saw and so the piece is not square on the end or cut to the specified length. Facing from the center out (Figure E7.7) produces a better finish, but it is difficult to cut on a solid face in the center. Facing from the outside (Figure E7.8) is more convenient since heavier cuts may be taken and it is easier to work to the scribed lines on the circumference of the work. When facing from the center out, a right-hand turning tool in a left-hand toolholder is the best arrangement, but when facing from the outside to the center, a left-hand tool in a right-hand or straight toolholder can be used. Facing or other tool machining should not be done on workpieces extending more than five diameters from the chuck jaws.

FIGURE E7.1a Facing a workpiece in a chuck. (Lane Community College)

FIGURE E7.1b Facing the end of a shaft. (Lane Community College)

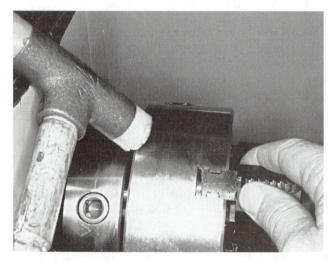

FIGURE E7.2 Removing a camlock chuck that is mounted on a lathe spindle.

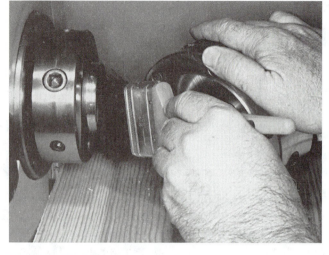

FIGURE E7.3a Chips are cleaned from spindle nose with a brush.

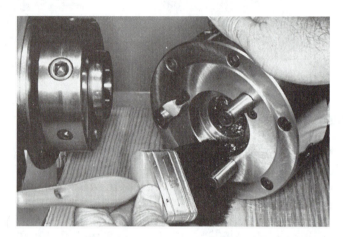

FIGURE E7.3b Cleaning the chips from the chuck with a brush.

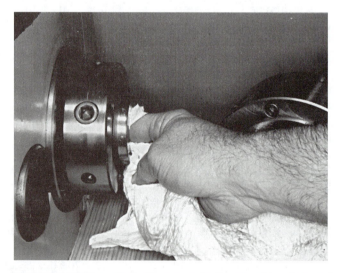

FIGURE E7.3c Spindle nose is thoroughly cleaned with a soft cloth.

FIGURE E7.3*d* Chuck is thoroughly cleaned with a soft cloth.

FIGURE E7.3*e* Chuck is mounted on spindle nose.

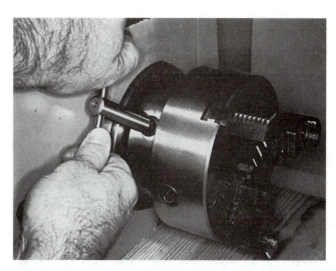

FIGURE E7.3*f* All cams are turned clockwise until locked securely.

FIGURE E7.4 Setting up round stock in an independent chuck with a dial indicator.

The point of the tool should be set to the center of the work (Figure E7.9). This is done by setting the tool to the tailstock center point or by making a trial cut to the center of the work. If the tool is off center, a small uncut stub will be left. The tool can then be reset to the center of the stub.

The carriage must be locked when taking facing cuts as the cutting pressure can cause the tool and carriage to move away (Figure E7.10), which would make the faced surface curved rather than flat. Finer feeds should be used for finishing than for roughing. Remember, the cross feed is one-half to one-third that of the longitudinal feed. The ratio is usually listed on the index plate of the quick-change gearbox. A roughing feed could be from .005 to .015 in. and a finishing feed from .003 to .005 in. Use of cutting oils will help produce better finishes on finish facing cuts.

Facing to length may be accomplished by trying a cut and measuring with a hook rule (Figure E7.11) or by facing to a previously made layout line. A more precise method is to use the graduations on

FIGURE E7.5a Rectangular stock being set up by using a toolholder turned backwards. The micrometer dial is used to center the workpiece.

FIGURE E7.6a Normal chucking position.

FIGURE E7.5b Adjusting the rectangular stock from Figure E7.5a at 90 degrees.

the micrometer collar of the compound. The compound is set so its slide is parallel to the ways (Figure E7.12). The carriage is locked in place and a trial cut is taken with the micrometer collar set on zero index. The workpiece is measured with a micrometer and the desired length is subtracted from the measurement; the remainder is the amount you should remove by facing. If more than .015 to .030 in. (depth left for finish cut) has to be removed, it should be taken off in two or more cuts by moving the compound micrometer dial the desired amount. A short trial cut (about $\frac{1}{8}$ in.) should again be taken on the finish cut and adjustment made if necessary.

FIGURE E7.6b Internal chucking position.

FIGURE E7.6c External chucking position.

FIGURE E7.7 Facing from the center to the outside of a workpiece.

FIGURE E7.8 Facing from the outside toward the center of the workpiece. (Clausing Machine Tools)

FIGURE E7.9 Setting the tool to the center of the work using the tailstock center.

FIGURE E7.10 The carriage must be locked before taking a facing cut.

Roughing cuts should be approximately .060 in. in depth.

Quite often the compound is kept at 30 degrees for threading purposes (Figure E7.13). It is convenient to know that at this angle, the tool feeds into the face of the work .001 in. for every .002 in. that the slide is moved. For example, if you wanted to remove .015 in. from the workpiece, you would turn

FIGURE E7.11 Facing to length using a hook rule for measuring.

FIGURE E7.12 The compound set at 90 degrees for facing operations.

FIGURE E7.13 Setting the compound at 30 degrees.

FIGURE E7.14 Half centers make facing shaft ends easier.

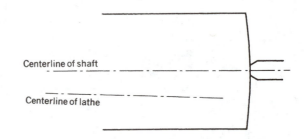

FIGURE E7.15a Convex shaft ends caused by the tailstock being moved off center away from the operator.

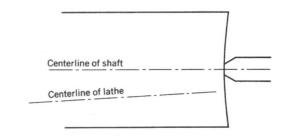

FIGURE E7.15b Concave shaft ends caused by the tailstock being moved toward the operator.

in .030 in. on the micrometer dial (assuming it reads single depth). It should be noted that the angle graduations on lathes are not standardized. On some lathes, Figure E7.12 would show the compound set at 0 degrees, and Figure E7.13 could show the compound at 60 degrees.

A specially ground tool is used to face the end of a workpiece that is mounted between centers. The right-hand facing tool is shaped to fit in the angle between the center and the face of the workpiece. Half centers (Figure E7.14) are made to make the job easier, but they should be used only for facing and not for general turning. If the tailstock is moved off center away from the operator, the shaft end will be **convex** (Figure E7.15a) and, if it is moved toward the operator, it will be **concave** (Figure E7.15b). Both right-hand and left-hand facing tools are used for facing work held on mandrels (Figure E7.16).

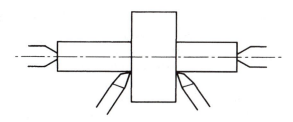

FIGURE E7.16 Work that is held between centers on a mandrel can be faced on both sides with right-hand or left-hand facing tools.

SPEEDS

Speeds (RPM) for lathe turning a workpiece are determined in essentially the same way as speeds for drilling tools. The only difference is that the diameter of the work is used instead of the diameter of the drill. For facing work, the outside diameter is always used to determine RPM. Thus:

$$RPM = \frac{CS \times 4}{D}, \text{ where}$$

$$D = \text{diameter of workpiece}$$
$$\text{(where machining is done)}$$
$$RPM = \text{revolutions per minute}$$
$$CS = \text{cutting speed (surface feet per minute)}$$

Cutting speeds for various metals are given in Table E8.1 of Unit 8, "Turning Between Centers."

EXAMPLE 1

The cutting speed for low-carbon steel is 90 SFM (surface feet per minute) and the workpiece diameter to be faced is 6 in. Find the correct RPM.

$$RPM = \frac{90 \times 4}{6} = 60$$

EXAMPLE 2

A center drill has a $\frac{1}{8}$-in. drill point. Find the correct RPM to use on low-carbon steel (CS 90).

$$RPM = \frac{90 \times 4}{\frac{1}{8}} = \frac{360}{1} \times \frac{8}{1} = 2880$$

These are only approximate speeds and will vary according to the conditions. If chatter marks (vibration marks) appear on the workpiece, the RPM should be reduced. If this does not help, ask your instructor for assistance. For more information on speeds and feeds, see Unit 8, "Turning Between Centers."

CENTER DRILLS AND DRILLING

When work is held and turned between centers, a center hole is required on each end of the workpiece. The center hole must have a 60-degree angle to conform to the center and have a smaller drilled hole to clear the center's point. This center hole is made with a combination drill and countersink, sometimes referred to as a center drill. These drills are available in a range of sizes from $\frac{1}{8}$- to $\frac{3}{4}$-in. body diameter and are classified by numbers from 00 to 8, which are normally stamped on the drill body. For example, a number 3 center drill has a $\frac{1}{4}$-in. body diameter and a $\frac{7}{64}$-in. drill diameter. Full listings can be found in the *Machinery's Handbook*.

Center drills are usually held in a drill chuck in the tailstock, while the workpieces are most often supported and turned in a lathe chuck for center drilling (Figure E7.17). A workpiece could also be laid out and supported in a vertical position for center drilling in a drill press. This method, however, is not used very often.

As a rule, center holes are drilled by rotating the work in a lathe chuck and feeding the center drill into the work by means of the tailstock spindle. Long workpieces, however, are generally faced by chucking one end and supporting the other in a steady rest (Figure E7.18). Since the end of stock is never sawed square, it should be center drilled only after spotting a small hole with the lathe tool. A slow feed is needed to protect the small, delicate drill end. Cutting oil should be used and the drill should be backed out frequently to remove chips.

FIGURE E7.17 Center drilling a workpiece held in a chuck.

FIGURE E7.18 Center drilling long material that is supported in a steady rest.

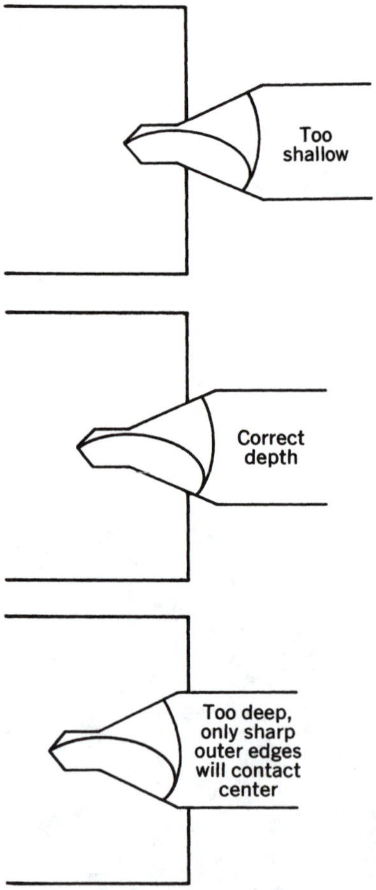

The greater the work diameter and the heavier the cut, the larger the center hole should be.

The size of the center hole can be selected by the center drill size and then regulated to some extent by the depth of drilling. You must be careful not to drill too deeply (Figure E7.19) as this causes the center to contact only the sharp outer edge of the hole, which is a poor bearing surface. It soon becomes loose and out-of-round and causes such machining problems as chatter and roughness. Center drills are often broken from feeding the drill too fast with the lathe speed too slow or with the tailstock off center.

FIGURE E7.20 Center drill is brought up to work and lightly fed into material.

Too shallow

Correct depth

Too deep, only sharp outer edges will contact center

FIGURE E7.19 Correct and incorrect depth for center drilling. Remember, the speed of the lathe usually has to be made faster when center drilling to avoid breakage.

FIGURE E7.21 Center drill is fed into work with a slow, even feed.

Center drills are often used as starting or spotting drills when a drilling sequence is to be performed (Figures E7.20 and E7.21). This keeps the drill from "wandering" off the center and making the hole run **eccentric.** Spot drilling is done when work is chucked or is supported in a steady rest. Care must be taken that the workpiece is centered properly in the steady rest or the center drill will be broken.

SELF-TEST

1. You have a rectangular workpiece that needs a facing operation plus center drilling, and a universal chuck is mounted on the lathe spindle. What is your procedure to prepare for machining?
2. Should the point of the tool be set above, below, or at the center of the spindle axis when taking a facing cut?
3. If you set the quick-change gearbox to .012 in., would that be considered a roughing feed for facing?
4. An alignment step must be machined on a cover plate .125 in., plus or minus .003 in., in depth (Figure E7.22). What procedure should be taken to face to this depth? How can you check your final finish cut?

5. What tool is used for facing shaft ends when they are mounted between centers? In what way is this tool different from a turning tool?
6. If the cutting speed of aluminum is 300 SFM and the workpiece diameter is 4 in., what is the RPM? The formula is

$$RPM = \frac{CS \times 4}{D}$$

7. Name two reasons for center drilling a workpiece in a lathe.
8. How is laying out and drilling center holes in a drill press accomplished?
9. Name two causes for center drill breakage.
10. What happens when you drill too deeply with a center drill?

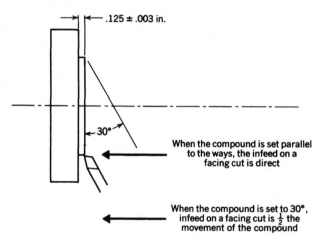

.125 ± .003 in.

30°

When the compound is set parallel to the ways, the infeed on a facing cut is direct

When the compound is set to 30°, infeed on a facing cut is ½ the movement of the compound

FIGURE E7.22

UNIT 8   Turning Between Centers

Since for a large percentage of lathe work the workpiece is held between centers or between a chuck and a center, turning between centers is a good way for you to learn the basic principles of lathe operation. The economics of machining time and quality will be detailed in this unit, as heavy roughing cuts are compared to light cuts, and speeds and feeds for turning operations are presented.

OBJECTIVES

After completing this unit, you should be able to:

1. Describe the correct setup procedure for turning between centers.
2. Select correct feeds and speeds for a turning operation.
3. Detail the steps necessary for turning to size predictably.

SETUP FOR TURNING BETWEEN CENTERS

To turn a workpiece between centers, it is supported between the dead center (tailstock center) and the live center in the spindle nose. A lathe dog (Figure E8.1) clamped to the workpiece is driven by a drive or dog plate (Figure E8.2) mounted on the spindle nose. Machining with a single-point tool can be done anywhere on the workpiece except near or at the location of the lathe dog.

Turning between centers has some disadvantages. A workpiece cannot be cut off with a parting tool while being supported between centers as this will bind and break the parting tool and ruin the workpiece. For drilling, boring, or machining the end of a long shaft, a steady rest is normally used to support the work. But these operations cannot very well be done when the shaft is supported only by centers.

The advantages of turning between centers are many. A shaft between centers can be turned end for end to continue machining without eccentricity if the centers are in line (Figure E8.3). This is why shafts that are to be subsequently finish-ground between centers should be machined between centers on a lathe. If a partially threaded part is removed from between centers for checking and everything is left the same on the lathe, the part can be returned to the lathe and the threading resumed where it was left off.

A considerable amount of straight turning on shafts is done with the work held between a chuck and the tailstock center (Figure E8.4). The advantages of this method are quick setup and a positive drive. One disadvantage is that eccentricities in the shaft are caused by inaccuracies in the chuck jaws. Another is the tendency for the workpiece to slip endwise into the chuck jaws under a heavy cut, thus allowing the workpiece to loosen or to come out of the tailstock center. This is a hazard that can damage the machine and possibly throw the workpiece out of the lathe toward the operator. A chalk mark near the jaws can alert the operator that the workpiece is sliding out of the center and into the chuck jaws. An even better means of preventing this from occurring is to first machine a step onto the shaft where it will be held in the chuck, as seen in Figure E8.4. This will keep the shaft from sliding into the chuck jaws even with heavy cuts. Of course, this solution is not always possible if the chucked end of the shaft has to remain at its full diameter.

As in other lathe operations, chip formation and handling are important to safety. Coarser feeds, deeper cuts, and smaller rake angles all tend to increase chip curl, which breaks up the chip into small, safe pieces. Fine feeds and shallow cuts, on the other hand, produce a tangle of wiry, sharp, hazardous chips (Figures E8.5a and E8.5b) even with a chip breaker on the tool. Long strings may come off the tool, suddenly wrap in the work, and be drawn

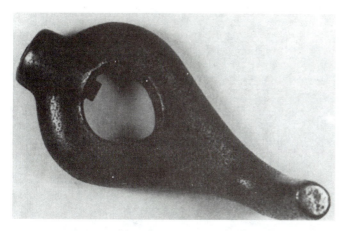

FIGURE E8.1 Lathe dog.

FIGURE E8.2 Dog plate or drive plate on spindle nose of the lathe. (The Monarch Machine Tool Company, Sidney, Ohio)

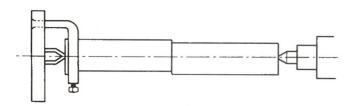

FIGURE E8.3 Eccentricity in the center of the part because of the live center being off center.

FIGURE E8.4 Work being machined between the chuck and tailstock center. Note chip formation. The chip guard has been removed for clarity.

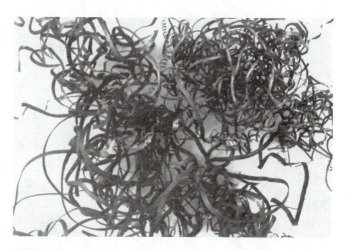

FIGURE E8.5a A tangle of wiry chips. These chips can be hazardous to the operator.

FIGURE E8.5b A better formation of chips. This type of chip will fall into the chip pan and is more easily handled.

FIGURE E8.6 Inserting the tapered spindle nose sleeve.

FIGURE E8.8 Installing the center.

FIGURE E8.7 Make sure the bushing is firmly seated in the taper.

FIGURE E8.9 Checking the live center for runout with a dial indicator.

back rapidly to the machine. The edges are like saws and can cause very severe cuts.

The center for the headstock spindle is called a *live center,* and it is usually not hardened since its point frequently needs machining to keep it true. Thoroughly clean the inside of the spindle with a soft cloth and wipe off the live center. If the live center is too small for the lathe spindle taper, use a tapered bushing that fits the lathe (Figure E8.6). Seat the bushing firmly in the taper (Figure E8.7) and install the center (Figure E8.8). Set up a dial indicator on the end of the center (Figure E8.9) to check for runout. If there is runout, remove the center by using a knockout bar through the spindle. Be sure to catch the center with one hand. Check the

outside of the center for nicks or burrs. These can be removed with a file. Check the inside of the spindle taper with your finger for nicks or grit. If nicks are found, *do not* use a file but check with your instructor. After removing nicks, if the center still runs out more than the acceptable tolerances (usually .0001 to .0005 in.) a light cut by tool or grinding can be taken with the compound set at 30 degrees off the center axis of the lathe.

A live center is often machined from a short piece of soft steel mounted in a chuck (Figure E8.10). It is then left in place and the workpiece is mounted between it and the tailstock center. A lathe dog with the bent tail placed against a chuck jaw is used to drive the workpiece. This procedure sometimes

FIGURE E8.10 Live center being machined in a four-jaw chuck. The lathe dog on the workpiece is driven by one of the chuck jaws.

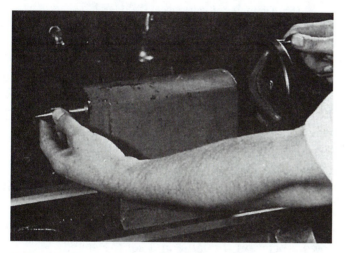

FIGURE E8.11 The dead center is hardened to resist wear. It is made of high-speed steel or steel with a carbide insert.

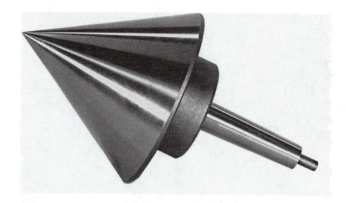

FIGURE E8.12a Pipe center used for turning. (The Monarch Machine Tool Company, Sidney, Ohio)

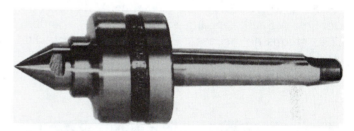

FIGURE E8.12b Antifriction ball-bearing center. (The Monarch Machine Tool Company, Sidney, Ohio)

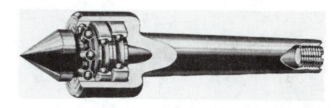

FIGURE E8.12c Cutaway view of a ball-bearing tailstock center. (The DoAll Company)

saves time on large lathes where changing from the chuck to a drive plate is cumbersome and the amount of work to be done between centers is small.

The tailstock center (Figure E8.11) is hardened to withstand machining pressures and friction. Clean inside the taper and on the center before installing. Ball-bearing antifriction centers are often used in the tailstock as they will withstand high-speed turning without the overheating problems of dead centers. Pipe centers are used for turning tubular material (Figures E8.12a to E8.12c).

To set up a workpiece that has been previously center drilled, slip a lathe dog on one end with the bent tail toward the drive plate. Do not tighten the dog yet. Put antifriction compound into the center hole toward the tailstock and then place the workpiece between centers (Figure E8.13). The tailstock spindle should not extend out too far as some rigidity in the machine would be lost and chatter or vibration may result. Set the dog in place and avoid any binding of the bent tail (Figure E8.14). Tighten the dog and then adjust the tailstock so there is no end play, but so the bent tail of the dog moves freely in its slot. Tighten the tailstock binding lever. The heat of machining will expand the workpiece and cause the dead center to heat from friction. If overheated, the center may be ruined and may even be welded into your workpiece. You should check the adjustment of the centers periodically, or at the end of each heavy cut, and reset if necessary.

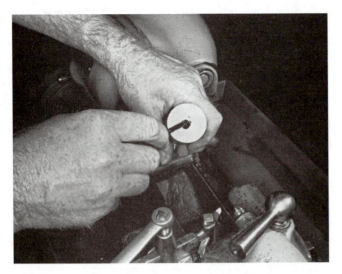

FIGURE E8.13 Antifriction compound put in the center hole before setting the workpiece between centers. This step is not necessary when using an antifriction ball-bearing center.

FIGURE E8.14 Lathe dog in position.

In roughing operations, the tool should not be extended any farther than necessary. Excessive overhang can cause the tool to turn into the work, gouging it as well as causing chatter.

SPEEDS AND FEEDS FOR TURNING

Since machining time is an important factor in lathe operations, it is necessary for you to understand fully the principles of speeds and feeds in order to make the most economical use of your machine. Speeds are determined for turning between centers by using the same formula as given for facing operations in the last unit.

$$RPM = \frac{CS \times 4}{D}, \text{ where}$$

$$RPM = \text{Revolutions per minute}$$

$$D = \text{Diameter of workpiece}$$

$$CS = \text{Cutting speed in surface feet per minute (SFM)}$$

Cutting speeds for various materials are given in Table E8.1.

EXAMPLE

If the cutting speed is 40 for a certain alloy steel and the workpiece is 2 in. in diameter, find the RPM.

$$RPM = \frac{40 \times 4}{2} = 80$$

After calculating the RPM, use the nearest or next lower speed on the lathe and set the speed.

Feeds are expressed in inches per revolution (IPR) of the spindle. A .010-in. feed will move the carriage and tool .010 in. for one full turn of the headstock spindle. If the spindle speed is changed, the feed ratio still remains the same. Feeds are selected by means of an index chart (Figure E8.15) found either on the quick-change gearbox or on the

TABLE E8.1 Cutting speeds and feeds for high-speed steel tools

	Low-Carbon Steel	High-Carbon Steel Annealed	Alloy Steel Normalized	Aluminum Alloys	Cast Iron	Bronze
Roughing speed SFM	90	50	45	200	70	100
Finishing speed SFM	120	65	60	300	80	130
Feed IPR roughing	.010–.020	.010–.020	.010–.020	.015–.030	.010–.020	.010–.020
Feed IPR finishing	.003–.005	.003–.005	.003–.005	.005–.010	.003–.010	.003–.010

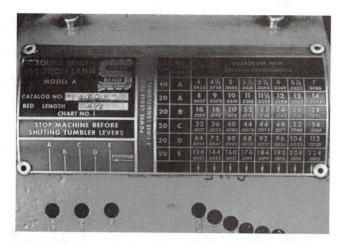

FIGURE E8.15 Index chart on the quick-change gear-box.

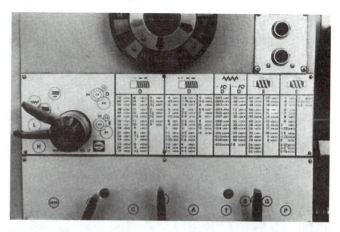

FIGURE E8.16 Index chart for feed mechanism on a modern geared head lathe with both metric and inch thread and feed selections.

side of the headstock housing (Figure E8.16). The sliding gear levers are shifted to different positions to obtain the feeds indicated on the index plate. The lower decimal numbers on the plate are feeds and the upper numbers are threads per inch.

Feeds and depth of cut should be as much as the tool, workpiece, or machine can stand without undue stress. A small 10- or 12-in. swing lathe should handle $\frac{1}{8}$-in. depth of cut in soft steel, but in some cases this may have to be reduced to $\frac{1}{16}$ in. If .100 in. were selected as a trial depth of cut, then the feed could be anywhere from .010 to .020 in. If the machine seems to be overloaded, reduce the feed. Finishing feeds can be from .003 to .005 in. for steel. Feeds smaller than .003 in. often produce poor finishes because of low tool pressure from the small chip. Use a tool with a larger nose radius for finishing.

TURNING TO SIZE

The cut-and-try method of turning a workpiece to size or making a cut and measuring how close you came to the desired result was used in the past when calipers and a rule were used for measuring work diameters. A more modern method of turning to size predictably uses the compound and crossfeed micrometer collars and micrometer calipers for measurement. If the micrometer collar on the crossfeed screw reads in single depth, it will remove twice the amount from the diameter of the work as the reading shows. A micrometer collar that reads directly or double depth will remove the same amount from the diameter that the reading shows, though the tool will actually move in only half that amount.

After taking one or several roughing cuts (depending on the diameter of the workpiece), .015 to .030 in. should be left for finishing. This can be taken in one cut if the tolerance is large, such as plus or minus .003 in. If the tolerance is small (plus or minus .0005 in.), two finish cuts should be taken, but enough stock must be left for the second cut to make a chip. If insufficient material is left for machining, .001 in. for example, the tool will rub and will not cut. Between .005 and .010 in. should be left for the last finish cut.

The position of the tool is set in relation to the micrometer dial reading, and the first of the two finish cuts is made (Figure E8.17). *The tool is then returned to the start of the cut without moving the crossfeed screw.* The diameter of the workpiece is checked with a micrometer (Figure E8.18) and the remaining amount to be cut is dialed on the crossfeed micrometer dial. A short trial cut is taken (about $\frac{1}{8}$ in. long) and the lathe stopped. A final check with a micrometer is made to validate the tool setting, and then the cut is completed. If the lathe makes a slight taper, see the next unit on "Alignment of the Lathe Centers" to correct this problem.

Finishing of machined parts with a file and abrasive cloth should not be necessary if the tools are sharp and honed and if the feeds, speeds, and depth of cut are correct. A machine-finished part looks better than a part finished with a file and abrasive cloth. In the past, filing and polishing the precision surfaces of lathe workpieces were necessary because of lack of rigidity and repeatability of machines. In the same way, worn lathes are not dependable for close tolerances and so an extra allowance must be made

FIGURE E8.17 A trial cut is made to establish a setting of a micrometer dial in relation to the diameter of the workpiece.

FIGURE E8.18 Measuring the workpiece with a micrometer.

for filing. The amount of surface material left for filing ranges from .0005 to .005 in., depending on the diameter, finish, and finish diameter. If more than the tips of the tool marks are removed with a file and abrasive cloth, a wavy surface will result. For most purposes, .002 in. is sufficient material to leave for finishing.

When filing on a lathe, use a low speed, long strokes, and file left-handed (Figure E8.19). For polishing with abrasive cloth, set the lathe for a high speed and move the cloth back and forth across the work. Hold an end of the cloth strip in each hand (Figures E8.20a and E8.20b). Abrasive cloth leaves grit on the ways of the lathe, so a thorough cleaning of the ways should be done after polishing.

Machining shoulders to specific lengths can be done in several ways. Using a machinist's rule (Figure E8.21) to measure workpiece length to a shoulder is a very common but semiprecision method. Preset carriage stops (Figure E8.22) can be used to limit carriage movement and establish a shoulder. This method can be very accurate if it is set up correctly. Another very accurate means of machining shoulders

FIGURE E8.19 Filing in the lathe, left-handed.

is by using special dial indicators with long travel plunger rods. These indicators show the longitudinal position of the carriage. Whichever method is used, *the power feed should be turned off $\frac{1}{8}$ in. short of the workpiece shoulder* and the tool hand fed to the de-

FIGURE E8.20a Using abrasive cloth for polishing.

FIGURE E8.20b Using a file for backing abrasive cloth for more uniform polishing.

FIGURE E8.21 Measuring the workpiece length to a shoulder with a machinist's rule.

FIGURE E8.22 Micrometer carriage stop set to limit tool travel in order to establish a shoulder.

sired length. If the tool should be accidentally fed into an existing shoulder, the feed mechanism may jam and be very difficult to release. A broken tool, toolholder, or lathe part may be the result.

Mandrels, sometimes called *lathe arbors,* are used to hold work that is turned between centers (Figures E8.23a and E8.23b). Tapered mandrels are made in standard sizes and have a taper of only .006 in./ft. A flat is milled on the big end of the mandrel for the lathe dog set screw. High pressure lubricant is applied to the bore of the workpiece and the mandrel is pressed into the workpiece with an arbor press. The assembly is mounted between centers and the workpiece is turned or faced on either side.

Expanding mandrels (Figure E8.24) have the advantage of providing a uniform gripping surface for the length of the bore (Figure E8.25). A tapered mandrel grips tighter on one end than the other. Gang mandrels (Figure E8.26) grip several pieces of similar size, such as discs, to turn their circumference. These are made with collars, thread, and a nut for clamping.

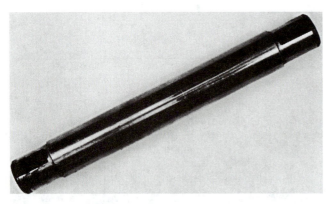

FIGURE E8.23*a* Tapered mandrel or arbor. *(Machine Tools and Machining Practices)*

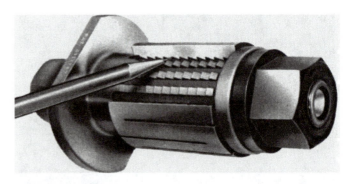

FIGURE E8.24 Operating principle of a special type of an expanding mandrel. The "saber-tooth" design provides a uniform gripping action in the bore. (Buck Tool Company)

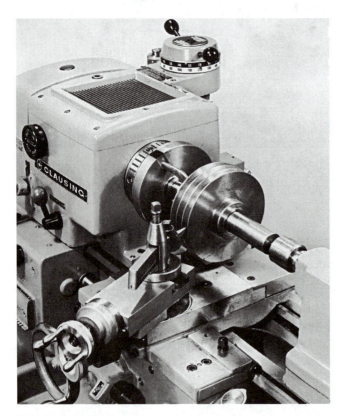

FIGURE E8.23*b* Tapered mandrel and workpiece set up between centers. (Clausing Machine Tools)

FIGURE E8.25 Expanding mandrel and workpiece set up between centers. (Lane Community College)

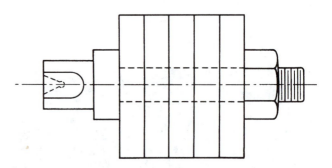

FIGURE E8.26 Many similar parts are machined at the same time with a gang mandrel.

Stub mandrels (Figures E8.27*a*–E8.27*c*) are used in chucking operations. These are often quickly machined for a single job and then discarded. Expanding stub mandrels (Figures E8.28*a* and E8.28*b*) are used when production of many similar parts is carried out.

Threaded stub mandrels are used for machining the outside surfaces of parts that are threaded in the bore.

Coolants are used for heavy-duty and production turning. Oil-water emulsions and synthetic coolants are the most commonly used, while sulfurized oils

FIGURE E8.27a Stub mandrel being machined to size with a slight taper, about .006 in./ft.

FIGURE E8.27b Part to be machined being affixed to the mandrel. The mandrel is oiled so that the part can be easily removed.

FIGURE E8.27c Part being machined after assembly on mandrel.

FIGURE E8.28a Special stub mandrel with adjust-tru feature. (Buck Tool Company)

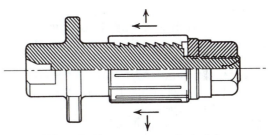

MC standard between centers nut actuated

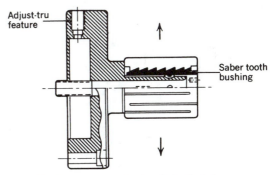

Adjust-tru feature

Saber tooth bushing

MLDR — flange mount — draw bolt actuated stationary sleeve

FIGURE E8.28b Precision adjustment can be maintained on the expanding stub mandrels with the adjust-tru feature on the flange mount. (Buck Tool Company)

usually are not used for turning operations except for threading. Most job work or single-piece work is done dry. Many shop lathes do not have a coolant pump and tank, so, if any coolant or cutting oil is used, it is applied with a pump oilcan. Coolants and cutting oils for various materials are given in Table E8.2.

TABLE E8.2 Coolants and cutting oils used for turning

Material	Dry	Water-Soluble Oil	Synthetic Coolants	Kerosene	Sulfurized Oil	Mineral Oil
Aluminum		x	x	x		
Brass	x	x	x			
Bronze	x	x	x			x
Cast iron	x					
Steel						
Low-carbon		x	x		x	
Alloy		x	x		x	
Stainless		x	x		x	

SELF-TEST

1. Name two advantages and two disadvantages of turning between centers.
2. What other method besides turning between centers is extensively used for turning shafts and long workpieces that are supported by the tailstock center?
3. What factors tend to promote or increase chip curl so that safer chips are formed?
4. Name three kinds of centers used in the tailstock and explain their uses.
5. How is the dead center correctly adjusted?
6. Why should the dead center be frequently adjusted when turning between centers?
7. Why should you avoid excess overhang with the tool and toolholder when roughing?
8. Calculate the RPM for roughing a $1\frac{1}{2}$-in. diameter shaft of machine steel.
9. What would the spacing or distance between tool marks on the workpiece be with a .010-in. feed?
10. How much should the feed rate be for roughing?
11. How much should be left for finishing?
12. Describe the procedure in turning to size predictably.

Alignment of the Lathe Centers

As a lathe operator, you must be able to check a workpiece for taper and properly set the tailstock of a lathe. Without these skills, you will lose much time in futile attempts to restore precision turning between centers when the workpiece has an unintentional taper. This unit will show you several ways to align the centers of a lathe.

OBJECTIVES

After completing this unit, you should be able to:

1. Check for taper with a test bar and restore alignment by adjusting the tailstock.
2. Check for taper by taking a cut with a tool and measuring the workpiece, and restore alignment by adjusting the tailstock.

The tailstock will normally stay in good alignment on a lathe that is not badly worn. If a lathe has been used for taper turning with the tailstock offset, however, the tailstock may not have been realigned properly (Figure E9.1). The tailstock also could be slightly out of alignment if an improper method of adjustment was used. It is therefore a good practice to occasionally check the center alignment of the lathe you usually use and to always check the alignment before using a different lathe.

It is often too late to save the workpiece by realigning centers if a taper is discovered while making a finish cut. A check for taper should be made on the workpiece while it is still in the roughing stage. You can do this by taking a light cut for some distance along the workpiece or on each end *without resetting* the crossfeed dial. Then check the diameter on each end with a micrometer; the difference between the two readings is the amount of taper in that distance.

Four methods are used for aligning centers on a lathe. In one method, the center points are brought together and visually checked for alignment (Figure E9.2). This is, of course, not a precision method for checking alignment.

Another method of aligning centers is by using the tailstock witness marks. Adjusting the tailstock to the witness marks (Figure E9.3), however, is only an approximate means of eliminating taper. The tailstock is moved by means of a screw or screws. A typ-

ical arrangement is shown in Figure E9.4, where one set screw is released and the opposite one is tightened to move the tailstock on its slide (Figure E9.5). The tailstock clamp bolt must be released before the tailstock is offset.

Two more-accurate means of aligning centers are by using a test bar and by machining and measuring. A **test bar** is simply a shaft that has true centers (is not off center) and has no taper. Some test bars are made with two diameters for convenience. When checking alignment with a test bar, no dog is necessary because the bar is not rotated. A dial indicator is mounted, preferably in the tool post, so it will travel with the carriage (Figure E9.6). Its contact point should be on the center of the test bar.

Begin with the indicator at the headstock end, and set the indicator bezel to zero (Figure E9.7). Now move the setup to the tailstock end of the test bar (Figure E9.8), and check the dial indicator reading. If no movement of the needle has occurred, the centers are in line. If the needle has moved clockwise, the tailstock is misaligned toward the operator. This will cause the workpiece to taper with the smaller end near the tailstock. If the needle has moved counterclockwise, the tailstock is away from the operator too far, and the workpiece will taper with the smaller end at the headstock. In either case, move the tailstock until both diameters have the same reading.

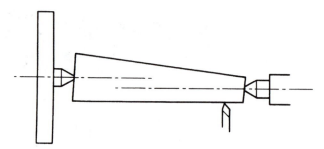

FIGURE E9.1 Tailstock out of line causing a tapered workpiece.

FIGURE E9.2 Checking alignment by matching center points.

Witness mark

FIGURE E9.3 Adjusting the tailstock to the witness marks for alignment.

Set screw

FIGURE E9.4 Hexagonal-socket set screw that, when turned, moves the tailstock provided the opposite screw is loosened.

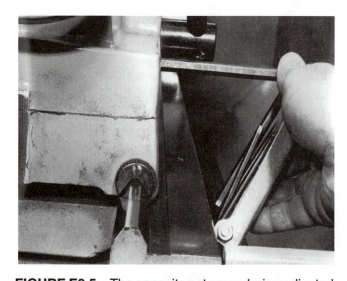

FIGURE E9.5 The opposite set screw being adjusted.

FIGURE E9.6 Test bar setup between centers with a dial indicator mounted in the tool post.

FIGURE E9.7 Indicator is moved to measuring surface at headstock end and the bezel is set on zero.

FIGURE E9.8 The carriage with the dial indicator is moved to the measuring surface near the tailstock. In this case the dial indicator did not move, so the tailstock is on center.

FIGURE E9.9 Checking for taper by taking a cut on a workpiece. After the cut is made for the length of the workpiece, a micrometer reading is taken at each end to determine any difference in diameter.

FIGURE E9.10 Using a dial indicator to check the amount of movement of the tailstock when it is being realigned.

Since usually only a minor adjustment is needed while a job is in progress, cutting and measuring is the method most used for aligning centers on a lathe. It is also the most accurate. This method, unlike the test bar method, usually uses the workpiece while it is in the roughing stage (Figure E9.9). A light cut is taken along the length of the test piece and both ends are measured with a micrometer. If the diameter at the tailstock end is smaller, the tailstock is toward the operator, and if the diameter at the headstock end is smaller, the tailstock is away from the operator. Set up a dial indicator (Figure E9.10) and move the tailstock half the difference of the two micrometer readings. Make another light cut and check for taper.

SELF-TEST

1. What is the result on the workpiece when the centers are out of line?

2. What happens to the workpiece when the tailstock is offset toward the operator?
3. Name three methods of aligning the centers.
4. Which measuring instrument is used when using a test bar?
5. By what means is the measuring done when checking taper by taking a cut?

Lathe Operations

Much of the versatility of the lathe as a machine tool is due to the variety of tools and workholding devices used. This equipment makes possible the many special operations that you will begin to do in this unit.

OBJECTIVES

After completing this unit, you should be able to:

1. Explain the procedures for drilling, boring, reaming, knurling, recessing, parting, and tapping in the lathe.
2. Set up to drill, ream, bore, and tap on the lathe and complete each of these operations.
3. Set up for knurling, recessing, die threading, and parting on the lathe and complete each of these operations.

DRILLING

The lathe operations of boring, tapping, and reaming usually begin with spotting and drilling a hole. The workpiece, often a solid material that requires a bore, is mounted in a chuck, collet, or face plate, and the drill is typically mounted in the tailstock spindle that has a Morse taper. If there is a slot in the tailstock spindle, the drill tang must be aligned with it when inserting the drill.

Drill chucks with Morse taper shanks are used to hold straight shank drills and center drills (Figure E10.1a). Center drills are used for spotting or making a start for drilling (Figure E10.1b). When drilling with large-size drills, a pilot drill should be used first. If a drill wobbles when started, place the heel of a toolholder against it near the point to steady the drill while it is starting in the hole. Taper shank drills (Figure E10.2) are inserted directly into the tailstock spindle. The friction of the taper is usually all that is needed to keep the drill from turning while a hole is being drilled (Figures E10.3 and E10.4), but when using larger drills, the friction is not enough. A lathe dog is sometimes clamped to the drill just above the flutes (Figure E10.5) with the bent tail resting on the compound. Hole depth can be measured with a rule or by means of the graduations on top of the tailstock spindle. The alignment of the tailstock with the lathe center line should be checked before drilling or reaming.

Drilled holes are not accurate enough for many applications, such as for gear or pulley bores, which should not exceed the nominal size by more than .001 to .002 in. Drilling typically produces holes that are oversize and run eccentric to the center axis of the lathe (Figure E10.6). This is not true in the case of some manufacturing procedures such as gun drilling. However, truer axial alignment of holes is possible when the work is turned and the drill remains stationary, as in a lathe operation, in comparison to operations where the drill is turned and the work is stationary, as on a drill press. Drilling also produces holes with rough finishes, which like size errors can be corrected by boring or reaming. The hole must first be drilled slightly smaller than the finish diameter in order to leave material for finishing by either of these methods.

BORING

Boring is the process of enlarging and truing an existing or drilled hole. A drilled hole for boring can be from $\frac{1}{32}$- to $\frac{1}{16}$-in. undersize, depending on the situation. Speeds and feeds for boring are determined in the same way as they are for external turning. Boring to size predictably is also done in the same way as in external turning except that the crossfeed screw is turned counterclockwise to move the tool into the work.

FIGURE E10.1*a* Mounting a straight shank drill in a drill chuck in the tailstock spindle.

FIGURE E10.1*b* Center drilling is the first step prior to drilling, reaming, or boring.

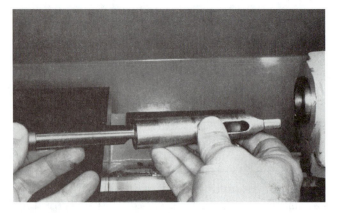

FIGURE E10.2 A drill sleeve is placed on the drill so that it will fit the taper in the tailstock spindle.

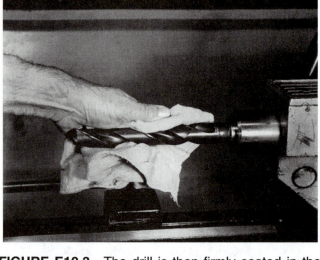

FIGURE E10.3 The drill is then firmly seated in the tailstock spindle.

FIGURE E10.4 The feed pressure when drilling is usually sufficient to keep the drill seated in the tailstock spindle, thus keeping the drill from turning.

An inside spring caliper and a rule are sometimes useful for rough measurement. Vernier calipers are also used by machinists for internal measuring, though the telescoping gage and outside micrometer are most commonly used for the precision measurement of small bores. Inside micrometers are used for larger bores. Other means of measurement are an inside spring caliper used with an outside micrometer and, on large bores, an inside micrometer used with an outside caliper. Precision bore gages are used where many bores are checked for similar size, such as for acceptable tolerance.

FIGURE E10.5 A lathe dog is used when the drill has a tendency to turn in the tailstock spindle.

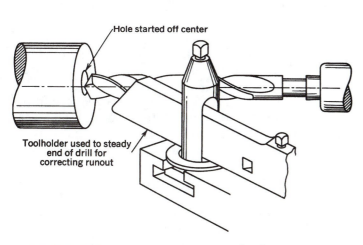

Hole started off center

Toolholder used to steady end of drill for correcting runout

FIGURE E10.6 An exaggerated view of the runout and eccentricity that is typical of drilled holes.

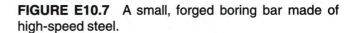

FIGURE E10.7 A small, forged boring bar made of high-speed steel.

FIGURE E10.8 Two boring bars with inserted tools set at different angles.

FIGURE E10.9 A boring bar with carbide insert. When one edge is dull, a new one is selected. (Kennametal Inc., Latrobe, Pa.)

A boring bar is clamped in a holder mounted on the carriage compound. Several types of boring bars and holders are used. Boring bars designed for small holes ($\frac{1}{2}$ in. and smaller) are usually the forged type (Figure E10.7). The forged end is sharpened by grinding. When the bar gets ground too far back, it must be reshaped or discarded. Boring bars for holes with diameters over $\frac{1}{2}$ in. (Figure E10.8) use high-speed tool inserts, which are typically hand ground in the form of a left-hand turning tool. These tools can be removed from the bar for resharpening when needed. The cutting tool can be held at various angles to obtain different results, which makes the boring bar useful for many applications. Standard bars generally come with a tool angle of 30, 45, or 90 degrees. Some boring bars are made for carbide inserts (Figure E10.9).

Chatter is the rattle or vibration between a workpiece and a tool because of the lack of rigid support for the tool. Chatter is a great problem in boring operations since the bar must extend away from the support of the compound (Figure E10.10). For this reason, boring bars should be inserted into their holders as far as practicable. This machining problem is also called **harmonic chatter** (see Glossary). Tuned boring bars can be adjusted so that their vibration is dampened (Figure E10.11). The stiffness of boring bars is increased by making them of solid

FIGURE E10.10 A boring bar setup with a large over-hang for making a deep bore. It is difficult to avoid chatter with this arrangement.

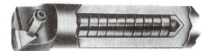 

FIGURE E10.11 Tuned boring bars contain dampening slugs of heavy material that can be adjusted by applying pressure with a screw. (Kennametal Inc., Latrobe, Pa.)

tungsten carbide. If chatter occurs when boring, one or more of the following may help to eliminate the vibration of the boring tool.

1. Shorten the boring bar overhang, if possible.
2. Make sure the tool is on center.
3. Reduce the spindle speed.
4. Use a boring bar as large in diameter as possible without it binding in the hole.
5. Reduce the nose radius on the tool.
6. Apply cutting oil to the bore.

Boring bars sometimes spring away from the cut and cause bell mouth, a slight taper at the front edge of a bore. One or two extra (free) cuts taken without moving the cross feed will usually eliminate this problem.

A large variety of boring bar holders are used. Some types are designed for small, forged bars while others of more rigid construction are used for larger, heavier work.

Boring tools are made with side relief and end relief, but usually with zero back rake (Figure E10.12). Insufficient end relief will allow the heel of the tool

FIGURE E10.12 Boring tools must have sufficient side relief and side rake to be efficient cutting tools. Back rake is not normally used.

to rub on the workpiece (Figure E10.13). The end relief should be between 10 and 20 degrees. The machinist must use judgment when grinding the end relief because the larger the bore, the less end relief is required (Figure E10.14). If the end of the tool is relieved too much, the cutting edge will be weak and break down.

The point of the cutting tool should be positioned exactly on the centerline of the workpiece (Figures E10.15a to E10.15c). There must be a space to allow the chips to pass between the bar and the surface being machined, or the chips will wedge and bind on the back side of the bar, forcing the cutting tool deeper into the work (Figure E10.16).

Through boring is the boring of a workpiece from one end to the other or all the way through. For through boring, the tool is held in a bar that is perpendicular to the axis of the workpiece. A slight side

FIGURE E10.13 A tool with insufficient end relief will rub on the heel of the tool and will not cut.

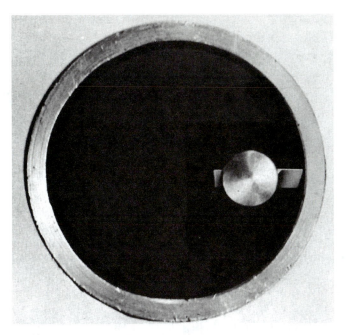

FIGURE E10.14 End relief angle varies depending on the diameter of the workpiece bore. These views are looking outward from inside the chuck.

FIGURE E10.15b If the boring tool is too low, the heel of the tool will rub and the tool will not cut, even if the tool has the correct relief angle.

FIGURE E10.15a The point of the boring tool must be positioned on the centerline of the workpiece.

FIGURE E10.15c If the tool is too high, the back rake becomes excessively negative and the tool point is likely to be broken off. A poor quality finish is the result of this position.

FIGURE E10.16 Allowance must be made for the chips to clear the space between the bar and the surface being machined. This setup has insufficient chip clearance (where the arrow is pointing).

cutting-edge angle is often used for through boring (Figure E10.17). Back facing is sometimes done to true up a surface on the back side of a through bore. This is done with a straight-ground right-hand tool also held in a bar perpendicular to the workpiece. The amount of facing that can be done in this way is limited to the movement of the bar in the bore.

FIGURE E10.17 Bar and tool arrangement for through boring.

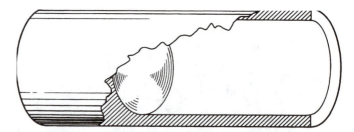

FIGURE E10.18 A blind hole machined flat in the bottom.

A blind hole is one that does not go all the way through the part to be machined (Figure E10.18). Machining the bottom or end of a blind hole to a flat is easier when the drilled center does not need to be cleaned up. A bar with the tool set at an angle, usually 30 or 45 degrees, is used to square the bottom of a hole with a drilled center (Figure E10.19).

Most boring is performed on workpieces mounted in a chuck. But it is also done in the end of workpieces supported by a steady rest. Boring and other operations are infrequently done on workpieces set up on a face plate (Figure E10.20).

A thread relief is a bore that is a little larger at the blind end of a hole. The purpose of a thread relief is to allow a threading tool to disengage the work at the end of a pass (Figure E10.21). When the work will allow it, a hole can be drilled deeper than necessary. This will give the end of the boring bar enough space so that the tool can reach into the area to be relieved and still be held at a 90-degree angle (Figure E10.22). When the work will not allow for deeper drilling, a special tool must be ground (Figure E10.23).

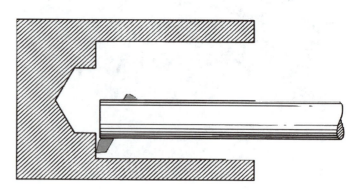

FIGURE E10.19 A bar with an angled tool used to square the bottom of a hole with a drilled center.

FIGURE E10.20 Workpiece clamped on face plate has been located, drilled, and bored.

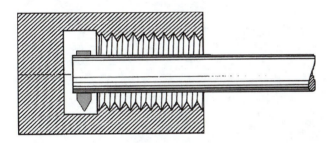

FIGURE E10.21 Ample thread relief is necessary when making internal threads.

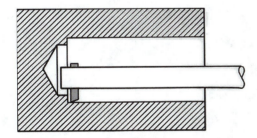

FIGURE E10.22 The hole is drilled deeper than necessary to allow room for the boring bar.

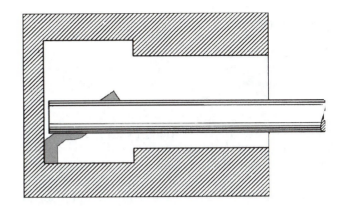

FIGURE E10.23 A special tool is needed when the thread relief must be next to a flat bottom.

FIGURE E10.24 A tool that is ground to the exact width of the desired groove can be moved directly into the work to the correct depth.

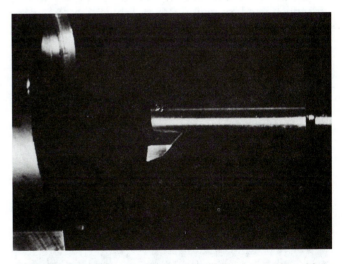

FIGURE E10.25 A square shoulder is made with a counterboring tool.

Grooves in bores are made by feeding a form tool (Figure E10.24) straight into the work. Snap ring, O-ring, and oil grooves are made in this way. Cutting oil should always be used in these operations.

Counterboring in a lathe is the process of enlarging a bore for definite length (Figure E10.25). The shoulder that is produced in the end of the counterbore is usually made square (90 degrees) to the lathe axis. Boring and counterboring are also done on long workpieces that are supported in a steady rest. All boring work should have the edges and corners broken or chamfered.

REAMING

Reaming is done in the lathe to quickly and accurately finish drilled or bored holes to size. Machine reamers, like drills, are held in the tailstock spindle of the lathe. Floating reamer holders are sometimes used to assure alignment of the reamer, since the reamer follows the eccentricity of drilled holes. This helps eliminate bell-mouth bores that result from reamer wobble but does not eliminate the hole eccentricity. Only boring will remove the bore runout.

Roughing reamers (rose reamers) are often used in drilled or cored holes followed by machine or finish reamers. When drilled or cored holes have excessive eccentricity, they are bored .010- to .015-in. undersize and machine reamed. If a greater degree of accuracy is required, the hole is bored to within .003 to .005 in. of finish size and hand reamed (Figure E10.26). For hand reaming, the machine is shut off and the hand reamer is turned with a tap wrench. Types of machine reamers are shown in Section D, Unit 6, "Reaming in the Drill Press."

Cutting oils used in reaming are similar to those used for drilling holes. (See Table D6.3, "Coolants Used for Reaming," in Section D, Unit 6, "Reaming in the Drill Press.") Cutting speeds are dependent on machine and workpiece material finish requirements. A rule of thumb for reaming speeds is to use one-half the speed used for drilling. (See Table D6.1, "Reaming Speed," in Section D, Unit 6, "Reaming in the Drill Press.")

Feeds for reaming are about twice those used for drilling. The cutting edge should not rub without cutting as it causes glazing, work hardening, and dulling of the reamer.

A simple machine reaming sequence would be as follows:

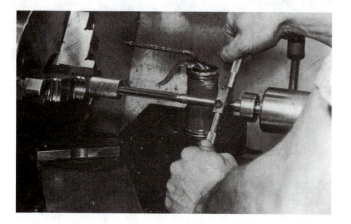

FIGURE E10.26 Hand reaming in the lathe.

Step 1. Assuming that the hole has been drilled $\frac{1}{64}$-in. undersize, a taper shank machine reamer is seated in the taper by hand pressure (Figure E10.27).

Step 2. Cutting oil is applied to the hole and the reamer is started into the hole by turning the tailstock handwheel (Figure E10.28). The reamer should be backed out with the machine running forward. Never reverse the machine when reaming.

Step 3. The reamer is removed from the tailstock spindle and cleaned with a cloth (Figure E10.29). The reamer is then returned to the storage rack.

Step 4. The lathe is cleaned with a brush (Figure E10.30).

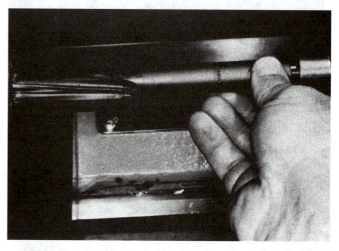

FIGURE E10.27 The reamer must be seated in the taper.

FIGURE E10.28 Starting the reamer in the hole. Kerosene is being used as a cutting fluid since the material is aluminum.

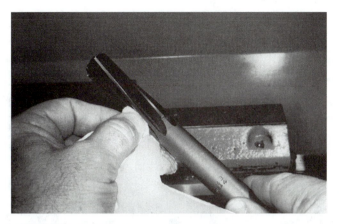

FIGURE E10.29 The reamer should be cleaned and put away after using it.

FIGURE E10.30 The chips are brushed into the chip pan.

FIGURE E10.31 Plug tap. (TRW Inc., 1980)

FIGURE E10.32 Spiral-point tap or gun tap. (TRW Inc., 1980)

FIGURE E10.33 Bottoming tap. (TRW Inc., 1980)

TAPPING

The tapping of work mounted in a chuck is a quick and accurate means of producing internal threads. Tapping in the lathe is similar to tapping in the drill press, but it is generally reserved for small-size holes, as tapping is the only way they can be internally threaded. Large internal threads are made in the lathe with a single-point threading tool. A large tap requires considerable force or torque to turn, more than can be provided by hand turning. A tap that is aligned by the dead center will make a straight tapped hole that is in line with the lathe axis.

A plug tap (Figure E10.31) or spiral-point tap (Figure E10.32) may be used for tapping through holes. When tapping blind holes, a plug tap could be followed by a bottoming tap (Figure E10.33) if threads are needed to the bottom of the hole. A good practice is to drill a blind hole deeper than the required depth of threads.

Two approaches may be taken for hand tapping. Power is not used in either case. One method is to turn the tap by means of a tap wrench or adjustable wrench with the spindle engaged in a low gear so that it will not turn (Figure E10.34). The other method is to disengage the spindle and turn the chuck by hand while the tap wrench handle rests on the compound (Figure E10.35). In both cases the tailstock is clamped to the ways, and the dead center is kept in the center of the tap by slowly turning the tailstock handwheel. The tailstock on small lathes need not be clamped to the ways for small taps but held firmly with one hand.

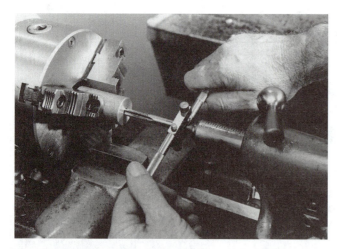

FIGURE E10.34 Hand tapping in the lathe by turning the tap wrench.

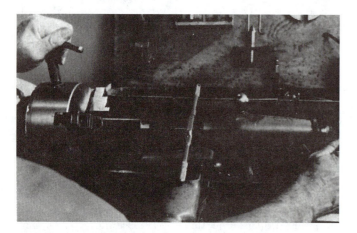

FIGURE E10.35 Tapping by turning the chuck.

FIGURE E10.36 Starting die on rod to be threaded in the lathe. The tailstock spindle (without a center) is used to start the die squarely onto the work.

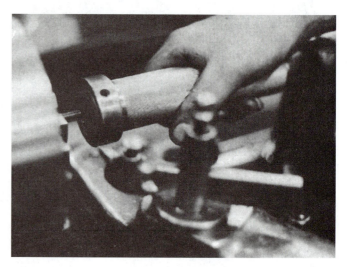

FIGURE E10.37 A button die holder used to guide the die onto the work. This insures a more uniform thread than by using a die stock.

Cutting oil should be used and the tap backed off every one or two turns to break chips unless it is a spiral-point tap, sometimes called a *gun tap*.

The correct tap drill size should be obtained from a tap drill chart. Drills tend to drill slightly oversize, and tapping the oversize hole can produce poor internal thread with only a small percentage of thread cut. Make sure the drill produces a correctly sized hole by drilling first with a slightly smaller drill, then using the tap drill as a reamer.

Tapping can be done on the lathe with power, but it is recommended that it be done only if the spindle rotation can be reversed, if a spiral-point tap is used, and if the hole is clear through the work. The tailstock is left free to move on the ways. Insert the tap in a drill chuck in the tailstock and set the lathe on a low speed. Use cutting oil and slide the tailstock so the tap engages the work. Reverse the tap and remove it from the work every

$\frac{3}{8}$ to $\frac{1}{2}$ in. When reversing, apply light hand pressure on the tailstock to move it to the right until the tap is all of the way out.

External threads cut with a **die** should only be used for nonprecision purposes, since the die may wobble and the pitch (the distance from a point on one thread to the same point on the next) may not be uniform. The rod to be threaded extends a short distance from the chuck and a die and diestock are started on the end (Figure E10.36). Cutting oil is used. The handle is rested against the compound. The chuck may be turned by hand, but if power is used, the machine is set for low speed and reversed every $\frac{3}{8}$ to $\frac{1}{2}$ in. to clear the chips. Finish the last $\frac{1}{4}$ in. by hand if approaching a shoulder. Reverse the lathe to remove the die. An alternative method is to use a button die holder to guide the die onto the work (Figure E10.37). Thread cutting only takes place when the holder is gripped by the operator, a method that provides more control over the cutting action. This method provides a much truer, more-aligned thread than does the method shown in Figure E10.36.

RECESSING, GROOVING, AND PARTING

Recessing and grooving on external diameters (Figures E10.38*a* to E10.38*c*) is done to provide grooves for thread relief, snap rings, and O-rings. Special tools (Figure E10.39) are ground for both external and internal grooves and recesses. Parting tools are sometimes used for external grooving and thread relief.

Parting or cutoff tools (Figure E10.40) are designed to withstand high cutting forces, but if chips are not sufficiently cleared or cutting oil is not used, these tools can quickly jam and break. Parting tools must be set on center and square with the work (Figure E10.41). Lathe tools are often specially ground as parting tools for small or delicate parting jobs (Figure E10.42). Diagonally ground parting tools leave no burr.

Parting alloy steels and other metals is sometimes difficult, and step parting (Figure E10.43) may help in these cases. When deep parting difficult material, extend the cutting tool from the holder a short distance and part to that depth. Then back off the

FIGURE E10.38*b* The tool is fed to the single depth of the thread or the required depth of the groove. If a wider groove is necessary, the tool is moved over and a second cut is taken as shown.

FIGURE E10.38*c* The finished groove.

FIGURE E10.38*a* The undercutting tool is brought to the workpiece and the micrometer dial is zeroed. Cutting oil is applied to the work.

cross feed and extend the tool a bit farther; part to that depth. Repeat the process until the center is reached. Sulfurized cutting oil works best for parting unless the lathe is equipped with a coolant pump and a steady flow of soluble oil is available. Parting tools are made in either straight or offset types. A right-hand offset cutoff tool is necessary when parting very near the chuck.

All parting and grooving tools have a tendency to chatter; therefore, any setup must be as rigid as possible. A low speed should be used for parting; if the tool chatters, reduce the speed. Work should not extend very far from the chuck when parting or grooving, and no parting should be done in the middle of

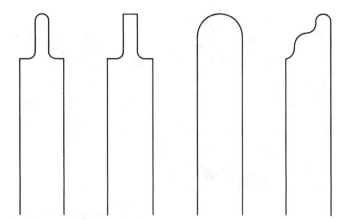

FIGURE E10.39 Recessing or grooving tools for internal and external use.

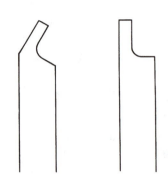

FIGURE E10.42 Special parting tools that have been ground from lathe tools for small or delicate parting jobs.

FIGURE E10.40 Parting tool making a cut.

FIGURE E10.43 Step parting.

FIGURE E10.41 Parting tools must be set to the center of the work.

a workpiece or at the end near the dead center. A feed that is too light can cause a chatter, but a feed that is too heavy can jam the tool. The tool should always be making a chip. Hand feeding the tool is best at first.

KNURLING

A **knurl** is a raised impression on the surface of a workpiece produced by two hardened rolls and is usually of two patterns, diamond or straight (Figure E10.44). The diamond pattern is formed by a right-hand and a left-hand helix mounted in a self-centering head. The straight pattern is formed by two straight rolls. These common knurl patterns can be either fine, medium, or coarse.

Diamond knurling is used to improve the appearance of a part and to provide a good gripping surface for levers and tool handles. Straight knurling is used to increase the size of a part for press fits in

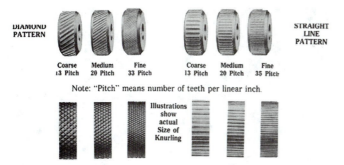

DIAMOND PATTERN ... STRAIGHT LINE PATTERN

Coarse 13 Pitch   Medium 20 Pitch   Fine 33 Pitch   Coarse 13 Pitch   Medium 20 Pitch   Fine 35 Pitch

Note: "Pitch" means number of teeth per linear inch.

Illustrations show actual Size of Knurling

FIGURE E10.44 Set of straight knurls and diagonal knurls. (J. H. Williams Division of TRW Inc.)

FIGURE E10.45a Knuckle-joint knurling toolholder.

light-duty applications. A disadvantage to this use of knurls is that the fit has less contact area than a standard fit.

Three basic types of knurling toolholders are used: the knuckle-joint holder (Figure E10.45a), the revolving head holder (Figure E10.45b), and the straddle holder (Figure E10.45c). The straddle holder allows small diameters to be knurled with less distortion. This principle is used for knurling on production machines.

Knurling works best on workpieces mounted between centers. When held in a chuck and supported by a center, the workpiece tends to crawl back into the chuck and out of the supporting center with the high pressure of the knurl. This is especially true when the knurl is started at the tailstock end and the feed is toward the chuck. Long slender pieces push away from the knurl and will stay bent if the knurl is left in the work after the lathe is stopped.

Knurls do not cut but displace the metal with high pressures. Lubrication is more important than cooling, so a lard oil or lubricating oil is satisfactory. Low speeds (about the same as for threading) and a feed of about .015 to .030 in. are used for knurling.

The knurls should be centered vertically on the workpiece (Figure E10.46) and the knurl toolholder should be square with the work, unless the knurl pattern is difficult to establish, as it often is in tough materials. In that case, the toolholder should be angled about 5 degrees to the work so the knurl can penetrate deeper (Figure E10.47).

A knurl should be started in soft metal about half depth and the pattern checked. An even diamond pattern should develop (Figure E10.48). But, if one roll is dull or placed too high or too low, a double impression will develop (Figure E10.49) because the rolls are not tracking evenly. If this happens, move the knurls to a new position along the workpiece, readjust up or down, and try again. The knurls

FIGURE E10.45b Revolving head knurling toolholder.

FIGURE E10.45c Straddle knurling toolholder. (Ralmike's Tool-A-Rama)

FIGURE E10.46 Knurls are centered on the work-piece.

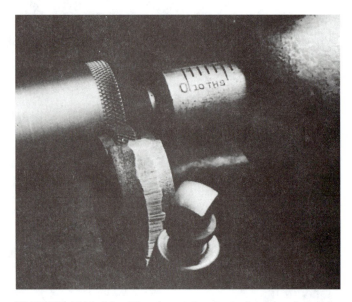

FIGURE E10.48 The knurl is started approximately half depth. Notice the diamond pattern is not fully developed.

FIGURE E10.47 Angling the toolholder 5 degrees often helps establish the diamond pattern.

FIGURE E10.49 Double impression on the left is the result of the rolls not tracking evenly.

should be cleaned with a wire brush between passes.

Material that hardens as it is worked, such as high-carbon or spring steel, should be knurled in one pass if at all possible, but in not more than two passes. Even in ordinary steel, the surface will work harden after a diamond pattern has developed to

points. It is best to stop knurling just before the points are sharp (Figure E10.50). Metal flaking off the knurled surface is evidence that work hardening has occurred. Avoid knurling too deeply as it produces an inferior knurled finish.

Knurls are also produced with a type of cutting tool (Figure E10.51) that is similar in appearance to

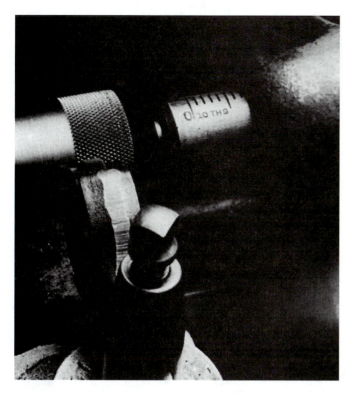

FIGURE E10.50 More than one pass is usually required to bring the knurl to full depth.

FIGURE E10.51 A knurling tool that cuts a knurl rather than forming it by pressure.

a knurling tool. The serrated rolls form a chip on this edge (Figure E10.52). Material that is difficult to knurl by pressure rolling, such as tubing and work-hardening metals, can be knurled by this cutting tool. Sulfurized cutting oil should be used when knurling steel with this kind of knurling tool.

FIGURE E10.52 A knurl being cut showing formation of chip.

SELF-TEST

1. Why are drilled holes not used for bores in machine parts such as pulleys, gears, and bearing fits?
2. Describe the procedure used to produce drilled holes on workpieces in the lathe with minimum oversize and runout.
3. What is the chief advantage of boring over reaming in the lathe?
4. List five ways to eliminate chatter in a boring bar.
5. Explain the differences between through boring, counterboring, and boring blind holes.
6. By what means are grooves and thread relief made in a bore?
7. Reamers will follow an eccentric drilled hole, thus producing a bell-mouth bore with runout. What device can be used to help eliminate bell mouth? Does it help remove the runout?
8. Machine reamers produce a better finish than is obtained by boring. How can you get an even better finish with a reamer?
9. Cutting speeds for reaming are (twice, half) that used for drilling; feeds used for reaming are (twice, half) that used for drilling.
10. Are large internal threads produced with a tap or a boring tool? Explain the reason for your answer.
11. How can you avoid drilling oversize with a tap drill?
12. Standard plug or bottoming taps can be used when hand tapping in the lathe. If power is used, what kind of tap works best?
13. Why would threads cut with a hand die in a lathe not be acceptable for using on a feedscrew with a micrometer collar?

14. By what means are thread relief on external grooves produced?
15. If cutting oil is not used on parting tools or chips do not clear out of the groove because of a heavy feed, what is generally the immediate result?
16. How can you avoid chatter when cutting off stock with a parting tool?
17. State three reasons for knurling.
18. Ordinary knurls do not cut. In what way do they make the diamond or straight pattern on the workpiece?
19. If a knurl is producing a double impression, what can you do to make it develop a diamond pattern?
20. How can you avoid producing a knurled surface on which the metal is flaking off?

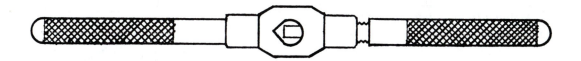

Prior to starting each procedure for this project, study and complete Post-Tests for:

Section E, Units 7, 8, and 10

Note: If you have already completed Post-Tests for any of these units, do not take the test again, but simply refer to the units to refresh your memory of the procedures.

Objectives

1. Learn to set up and turn rectangular stock.
2. Learn to make external threads with a die and to tap threads on the lathe.
3. Learn to knurl slender stock.

Materials

A $5\frac{1}{2}$ in. of HRMS flat bar $\frac{1}{2} \times 1$ in., and $4\frac{3}{4}$ in. of $\frac{3}{8}$ in. dia. CR round stock.

Procedure for Making Part No. 1

1. Cut off a piece of hot rolled mild steel (HRMS) flat bar $\frac{1}{2} \times 1 \times 5\frac{1}{2}$ in. long.
2. Clean the lathe spindle nose and mount a four-jaw chuck on it.
3. Fasten your workpiece in the chuck and let approximately 1 in. extend out from it.
4. Center your work to within .003 in., using methods shown in Section E, Unit 7.
5. Use a right hand turning tool and face the end with a good finish. Do not remove more material than necessary and do not leave a stub in the center.
6. Mount a No. 3 centerdrill in the tailstock drill chuck.
7. Select and set the speed for center drilling. If the speed is too slow the centerdrill will break.
8. Center drill the end. Use cutting oil.
9. Take the workpiece out of the chuck, reverse it, and repeat steps 4 through 8.
10. Remove the workpiece from the chuck.
11. Remove the chuck and substitute a drive plate and center in the headstock.

12. Fasten the dog on your workpiece and set it up for turning between centers.
13. Select the *speed* for roughing.
14. Select the *feed* for roughing.
15. Mark your workpiece 2 in. from the tailstock end and take roughing cuts to this mark until the diameter is approximately .800 in.
16. Now mark your workpiece $\frac{7}{16}$ in. from the tailstock end and rough turn that portion to approximately .560 in.
17. Reverse your workpiece end for end, clamping the dog on the end you have just turned.
18. Mark your work piece $1\frac{9}{16}$ in. from the *headstock* end and rough turn to that mark until the diameter is approximately .560 in.
19. Now make a mark $2\frac{9}{16}$ in. from the headstock end and turn your piece to approximately .430 in. dia. to that mark. Part No. 1 is now all machined to oversize dimensions and is ready for the finishing operation.
20. Resharpen your tool so it can be used for finishing, or make a finishing tool.
21. Select a *speed* for finishing. This will be slightly higher than for roughing.
22. Select a *feed* for finishing. This will be much lower than that used for roughing.
23. Mark the work at 2 in. from the headstock end.
24. Finish turn the $\frac{3}{8}$ in. diameter with a tolerance of .377 to .373 in. Machine to the mark.
25. Finish turn the $\frac{1}{2}$ in. diameter with a tolerance of .500 to .494 in. It should end exactly $1\frac{1}{2}$ in. from the headstock end.
26. Set up a medium diamond knurling tool in the toolholder.
27. Use a relatively low speed and a feed of about .010 to .020 in.
28. Make a mark $2\frac{13}{16}$ in. from the headstock *end;* this is how far your knurl will go.
29. Bring the knurls to the work near the dead center and feed in until you can see a definite diamond pattern develop. If there is one solid line one direction and short broken lines in the other, either raise the tool up, lower it, or turn it slightly sideways so it will bite in better. If that does not work, find a different knurling tool and try that. Use lubricating or cutting oil

on the knurls and on the work. When a full diamond pattern develops, stop knurling, since further working will ruin the job. Two passes should be sufficient; however, this workpiece will spring in the middle so more passes may be necessary at the center of the workpiece. Remove the knurls while the work rotates to avoid bending the work.

30. Reverse your workpiece; protect the knurled section with some soft material such as aluminum or copper when you put the dog on.

31. Change speed and feed for a finishing cut and set up your right-hand turning tool again.

32. Turn the large diameter with a tolerance of .750 to .748 in.

33. Turn the $\frac{1}{2}$ in. diameter nearest the tailstock end with a tolerance of .500 to .494 in. and $\frac{1}{2}$ in. long.

34. Set the compound rest so it will cut the 30 degree angle; also set up the tool for the cut (Figure E10.53).

35. Lock the carriage in place and cut the taper. Make sure you get a smooth transition between the $\frac{1}{2}$ in. diameter and the taper.

36. Swivel the compound rest to cut the other 30 degree bevel (Figure E10.54).

37. Turn this taper, being careful also to make a smooth transition between the taper and $\frac{1}{2}$ in. diameter.

38. Set the cutting edge of the tool at 45 degrees to the work and make the two required chamfers.

39. Remove the dog from the workpiece and change to a four-jaw chuck on the spindle.

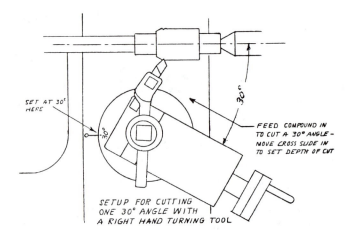

FIGURE E10.54

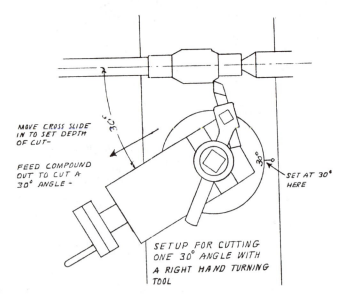

MOVE CROSS SLIDE IN TO SET DEPTH OF CUT —

FEED COMPOUND OUT TO CUT A 30° ANGLE —

SET AT 30° HERE

SETUP FOR CUTTING ONE 30° ANGLE WITH A RIGHT HAND TURNING TOOL

FIGURE E10.53

40. Mark your workpiece 5 in. long. Use a protective aluminum sleeve and chuck the workpiece with the small end out. Allow approximately 1 in. to extend from the chuck.

41. Adjust the chuck jaws until the workpiece runs true.

42. Face off the excess material (to your mark).

43. Rough turn the $\frac{3}{16}$ in. radius and use a file to complete the shape and to finish it. Check the radius with a gage.

44. Remove your workpiece from the chuck.

45. Chuck the workpiece on the $\frac{3}{8}$ in. diameter using the protective sleeve; this time let the large part extend out from the chuck.

46. Adjust the jaws until the work runs concentric.

47. Find the size of the tap drill for $\frac{5}{16}$-24 NF threads and secure a drill.

48. Hold this drill in the tailstock drill chuck and drill the part to the depth of one inch. The center drilled hole will act as a pilot to start the tap drill.

49. Remove the drill and fasten a $\frac{5}{16}$-24 NF tap in the drill chuck.

50. Loosen the tailstock clamp so it can slide freely on the ways.

51. Put cutting oil on the tap and turn the chuck with the workpiece in it by hand; after you bring the tap in contact with the work, it will pull itself into the work. Back up the tap frequently to break up the chips.

52. When the tap has reached full depth, turn the chuck in reverse by hand. This will push the tap out.

53. Lay out for the $\frac{3}{8}$-in. drilled hole on the flat side of the large diameter. Center punch.

54. Clamp the large diameter in a drill press vise, using soft protective material on the jaws. Use a level on the part to make sure it is set up square

to the spindle *after the vise is tightened.* Check to see if the drill press table is level; if not, use the same bubble position for setting up the part.

55. Spot drill with a small center drill.
56. Drill the $\frac{3}{8}$ in. hole and chamfer lightly on both sides with a countersink.

Procedure for Making Part No. 2

1. Cut off a piece of $\frac{3}{8}$ in. diameter cold rolled round stock 4-in. long.
2. Set up your workpiece in a three-jaw chuck leaving 1 in. extending from the chuck. Check it for runout. If it is excessive (over .005 in.), use a four-jaw chuck.
3. Center drill the workpiece. Then extend it 3 in. from the chuck.
4. Support this end with a center in the tailstock. Use center lube.
5. Put a mark on your workpiece $\frac{29}{16}$ in. from the tailstock end. This is how far the knurl will extend.
6. Repeat steps 26 to 29 in the procedure for making part No. 1.
7. Use a protective sleeve to hold the knurled piece so that the end with the center hole extends 1 in. from the chuck.
8. Use your right-hand tool and cut $\frac{3}{8}$ in. off the end of the workpiece. This should remove the center hole.
9. Rough turn and use a file to make the $\frac{3}{16}$ in. radius.
10. Remove your workpiece from the chuck and mark it $2\frac{11}{16}$ in. from the rounded end.
11. Chuck the workpiece with protective material over the knurl so your mark is $\frac{1}{2}$ in. from the chuck.
12. Center drill the end and support it with a center.
13. Turn the $\frac{5}{16}$ in. diameter with a tolerance of .312 to .310 in. diameter to your mark.

14. Turn the $\frac{1}{4}$ in. diameter at the end with a tolerance of .250 to .248 in. diameter so that 1 in. remains of the $\frac{5}{16}$ in. diameter.
15. Make a small undercut $\frac{3}{32}$ in. wide to the minor diameter of your thread.
16. Use a button die holder in the tailstock or a standard die holder to make the thread. (If you wish to make the thread with a single-point tool, you must first study Unit 11.)
17. Use your right hand tool to cut the $\frac{1}{4}$ in. diameter length so it will be $\frac{5}{16}$ in. long. If an aluminum protective sleeve is used on the knurl, the end may be turned off while set up as a chucking operation (without the dead center).
18. Chamfer the $\frac{1}{4}$ in. diameter to the specifications on the drawing.
19. This completes the turning of Part 2.

Assembling the Tap Handle

1. Use a small square file and make the vee in Part 1. It should be symmetrical to the centerline.
2. The short $\frac{1}{4}$ in. end may be case hardened if desired in order to prevent it from splaying out with repeated use. This can be done with an acetylene torch. Only the small end is heated to a bright cherry red and then placed in a hardening compound. It is then reheated and quenched in water.
3. Assemble both pieces. Use oil as a rust preventative.
4. Check to see if it will hold a tap square to the handle; if not, correct by filing.

Evaluation

Show your finished tap handle to your instructor for grading.

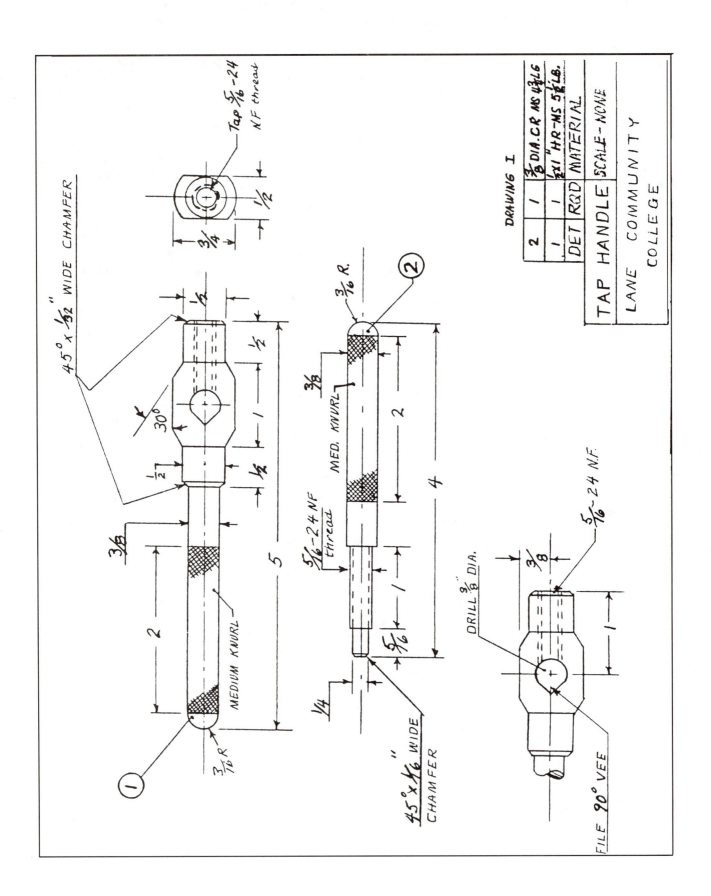

Tap 5/16 - 24
N.F. thread

45° × 1/32" WIDE CHAMFER

30°

MEDIUM KNURL

3/16 R

3/8

1/2

1/2

1

1/2

5

2

1

DRAWING I

2	1	7/8 DIA. CR MS 4 3/16
1	1	1/2 × 1" HR-MS 5 3/16 LB.
DET	R&D	MATERIAL

TAP HANDLE   SCALE - NONE

LANE COMMUNITY
COLLEGE

3/8 R.

2

MED. KNURL

5/16 - 24 N.F.
thread

4

1

5/16

45° × 1/16"
WIDE
CHAMFER

1/4

DRILL 3/8 DIA.

3/8

1

5/16 - 24 N.F.

FILE 90° VEE

Cutting External and Internal Threads

A manual or CNC lathe operator is frequently called upon to cut threads of various forms on a lathe. The threads most commonly made are the vee-form, that is, American National, Unified, or Metric. Although the Unified thread form is the basic standard in the United States, the American National, which is interchangeable with Unified except for systems of tolerances, will be used in this unit.

As a CNC operator, you will need a clear understanding of the process of "chasing a thread," which is making a series of cuts in the same groove. Small, internal threads are typically tapped, but larger sizes, from one inch and up, are usually cut in a lathe with a single-point tool. The problems and calculations involved with cutting internal threads differ in some ways from those of cutting external threads. Measuring and testing of both external and internal threads is of utmost importance in order to insure interchangeability.

These concepts will all be covered in this unit.

OBJECTIVES

After completing this unit, you should be able to:

1. Detail the steps and procedures necessary to cut an external thread to the correct depth.
2. Set up a lathe for threading and cut several different thread pitches and diameters.
3. Cut one or more internal threads on the lathe to fit a plug gage.
4. Select taps and tap drill sizes for tapping threads on the lathe.
5. Identify several methods of testing and inspecting external and internal threads for correct size.

HOW THREADING IS DONE ON A LATHE

Thread cutting on a lathe with a single-point tool is done by taking a series of cuts in the same helix of the thread. This is sometimes called *chasing a thread*. A direct ratio exists between the headstock spindle rotation, the lead screw rotation, and the number of threads on the lead screw. This ratio can be altered by the quick-change gearbox to make a variety of threads. When the half-nuts are clamped on the thread of the lead screw, the carriage will move a given distance for each revolution of the spindle. This distance is the **lead** of the thread.

If the infeed of a thread is made with the cross slide (Figure E11.1), equal-size chips will be formed on both cutting edges of the tool. This causes higher tool pressures, which can result in tool breakdown and sometimes causes tearing of the threads because of insufficient chip clearance. A more accepted practice is to feed in with the compound, which is set at 29 degrees (Figure E11.2) toward the right of the operator, for cutting right-hand threads. This insures a cleaner cutting action than setting the compound at 30 degrees when most of the chip will be taken from the leading edge and a scraping cut will be made from the following edge of the tool (Figure E11.3).

The dimensions and measurements of screw threads are given in Figure E11.4. Note that pitch is the distance from one thread crest to the next one, while threads per inch are the number of thread crests in one inch. The American National form thread (Figure E11.5) is the one used in this unit.

Begin the setup by obtaining or grinding a 60-degree threading tool. The flat on the end of the

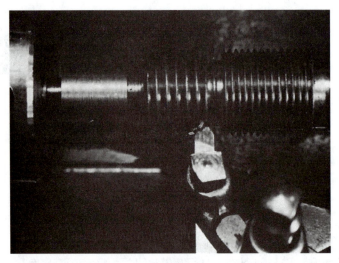

FIGURE E11.1 An equal chip is formed on each side of the threading tool when the infeed is made with the cross slide. (Lane Community College)

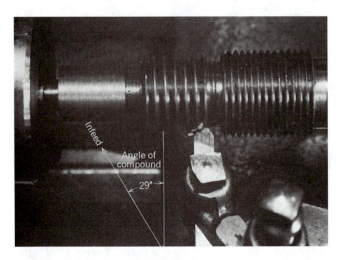

FIGURE E11.2 A chip is formed on the leading edge of the tool when the infeed is made with the compound set at 29 degrees to the right of the operator to make right-hand threads. (Lane Community College)

tool should be .125P. A center gage (Figure E11.6) may be used to check the tool angle. An adequate allowance for the helix angle on the leading edge will insure sufficient side relief.

The part to be threaded is set up between centers, in a chuck or in a collet. The tool is clamped in the holder and set on the centerline of the workpiece (Figure E11.7). A center gage is used to align the tool to the workpiece (Figure E11.8). The toolholder is clamped tightly after the tool is properly aligned.

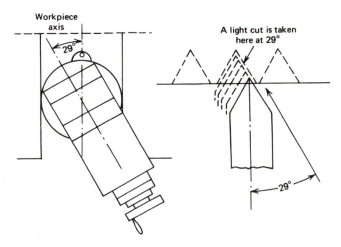

FIGURE E11.3 The compound at 29 degrees for cutting 60-degree threads. The reason for setting the compound at 29 degrees instead of 30 degrees for 60-degree threads is to provide a finish on the trailing side of the thread.

Setting Dials on the Compound and Cross Feed

The point of the tool is brought into contact with the work by moving the cross feed handle, and the micrometer collar is set on the zero mark (Figure E11.9). The compound micrometer collar should also be set on zero (Figure E11.10), but first be sure all slack or backlash is removed by turning the compound feed handle clockwise.

Setting Apron Controls

On some lathes a feed change lever, which selects either cross or longitudinal feeds, must be moved to a neutral position for threading. This action locks out the feed mechanism so that no mechanical interference is possible. All lathes have some interlock mechanism to prevent interference when the half-nut lever is used. The half-nut lever causes two halves of a nut to clamp over the leadscrew. The carriage will move the distance of the lead of the thread on the leadscrew for each revolution of the leadscrew.

Threading dials operate off the leadscrew and continue to turn when the leadscrew is rotating and the carriage is not moving. When the half-nut lever is engaged, the threading dial stops turning and the carriage moves. The marks on the dial indicate when it is safe to engage the half-nut lever. If the half-nuts are engaged at the wrong place, the threading tool will not track in the same groove as

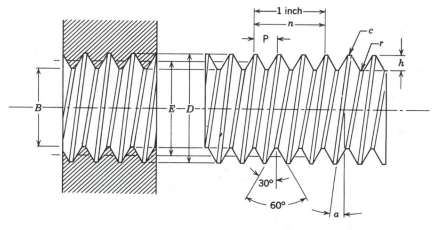

FIGURE E11.4 General dimensions of screw threads.

D	Major Diameter	a	Helix (or Lead) Angle
E	Pitch Diameter	c	Crest of Thread
B	Minor Diameter	r	Root of Thread
n	Number of threads per inch (TPI)	h	Basic Thread Height (or depth)
P	Pitch		

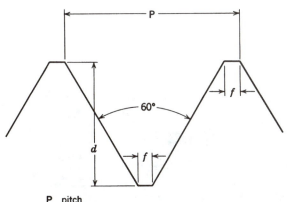

P pitch
d depth of thread = .6495P
f flat at crest and root of thread = $\frac{P}{8}$

FIGURE E11.5 The American National form thread.

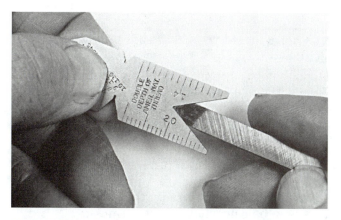

FIGURE E11.6 Checking the tool angle with a center gage. *(Machine Tools and Machining Practices)*

FIGURE E11.7 The tool is adjusted to the dead center for height. A tool that is set too high or too low will not produce a true 60-degree angle in the cut thread.

before but may cut into the center of the thread and ruin it. With any even number of threads such as 4, 6, 12, and 20, the half-nut may be engaged at any line. Odd-numbered threads such as 5, 7, 13 may be engaged at any numbered line. With fractional threads it is safest to engage the half-nut at the same line every time.

The Quick-Change Gearbox

The settings for the gear shift levers on the quick-change gearbox are selected according to the threads per inch desired (Figure E11.11). If the lathe

FIGURE E11.8 The tool is properly aligned by using a center gage. The toolholder is adjusted until the tool is aligned. The toolholder is then tightened.

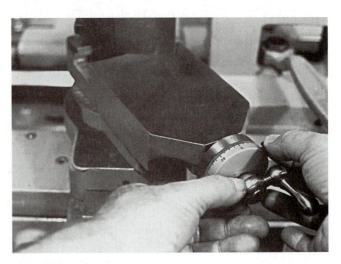

FIGURE E11.10 The operator then sets the compound micrometer collar to the zero index.

FIGURE E11.9 After the tool is brought into contact with the work, the cross feed micrometer collar is set to the zero index.

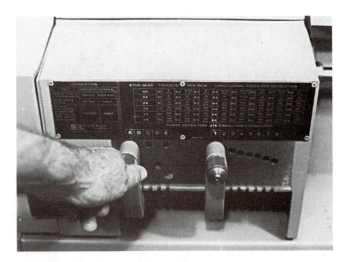

FIGURE E11.11 The threads per inch selection is made on the quick-change gearbox.

has an interchangeable stud gear, be sure the correct one is in place.

To cut 60-degree form threads, the tool is fed into the work with the compound (Figure E11.7), which is set at 29 degrees. The infeed depth along the flank of the thread at 30 degrees is greater than the depth at 90 degrees from the work axis. This depth may be calculated for American National threads by dividing the number of threads per inch (n) into .75,

$$\text{Infeed} = \frac{.75}{n} \quad \text{or} \quad P \times .75$$

Thus, for a thread with 10 threads per inch (.100 in. pitch):

$$\text{Infeed} = \frac{.75}{10} = .075 \text{ in.}$$

or

$$\text{Infeed} = .75 \times .100 = .075 \text{ in.}$$

For external Unified threads, the infeed at 29 degrees may be calculated by the formula:

$$\text{Infeed} = \frac{.708}{n} \quad \text{or} \quad .708P$$

FIGURE E11.12 Setting the correct speed for threading.

FIGURE E11.13 The half-nut lever is engaged at the correct line or numbered line depending on whether the thread is an odd-, even-, or fractional-numbered thread.

Thus, for a thread with 10 threads per inch (.100 in. pitch):

$$\text{Infeed} = \frac{.708}{10} = .0708 \text{ in.}$$

or

$$\text{Infeed} = .708 \times .100 = .0708 \text{ in.}$$

Spindle Speeds

Spindle speeds for thread cutting are approximately one-fourth turning speeds. The speed should be slow enough so you will have complete control of the thread cutting operation (Figure E11.12).

Choose a thread diameter and pitch from a thread table. It should be a standard thread so that a nut or ring gage may be used to check your thread for fit. Your instructor may have a process sheet with a thread project that already has thread and pitch sizes.

CUTTING THE EXTERNAL THREAD

The following is the procedure for cutting right-hand threads:

Step 1. Move the tool off the work and turn the crossfeed micrometer dial back to zero.

Step 2. Feed it in .002 in. on the compound dial.

Step 3. Turn on the lathe and engage the half-nut lever (Figure E11.13).

Step 4. Take a scratch cut without using cutting fluid (Figure E11.14). Disengage the half-nut at the

FIGURE E11.14 A light scratch cut is taken for the purpose of checking the pitch.

end of the cut. Stop the lathe and back out the tool using the cross feed. Return the carriage to the starting position.

Step 5. Check the thread pitch with a screw pitch gage or a rule (Figure E11.15). If the pitch is wrong, it can still be corrected.

Step 6. Apply appropriate cutting fluid to the work (Figure E11.16).

Step 7. Feed the compound in .005 to .020 in. for the first pass, depending on the pitch of the thread. For a coarse thread, heavy cuts can be taken on the first few cuts. The depth of cut should be reduced

FIGURE E11.15 The pitch of the thread is being checked with a screw pitch gage.

FIGURE E11.17 The second cut is taken after feeding in the compound .010 in.

FIGURE E11.16 Cutting fluid is applied before taking the first cut.

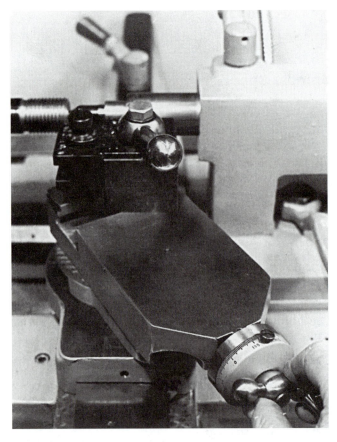

FIGURE E11.18 The finish cut is taken with infeed of .001 to .002 in. (Lane Community College)

for each pass until it is about .002 in. at the final passes. Bring the crossfeed dial to zero.

Step 8. Make the second cut (Figure E11.17).

Step 9. Continue this process until the tool is within .010 in. of the finished depth (Figure E11.18).

Step 10. Brush the threads to remove the chips. Check the thread fit with a ring gage (Figure E11.19), standard nut or mating part (Figure E11.20), or screw thread micrometer (Figure E11.21). The work may be removed from between centers and returned without disturbing the threading setup, provided that the tail of the dog is returned to the same slot.

Step 11. Continue to take cuts of .001 or .002 in. (as shown in Figure E11.18) and check the fit be-

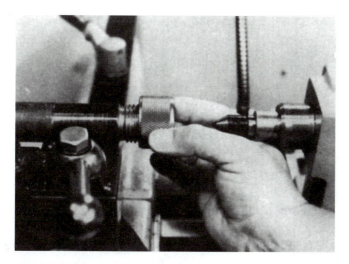

FIGURE E11.19 The thread is checked with a ring gage.

FIGURE E11.21 The screw thread micrometer may be used to check the pitch diameter of the thread.

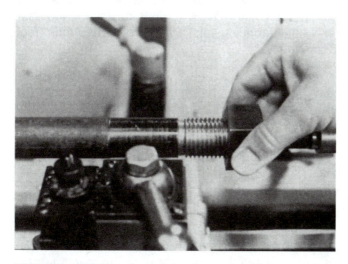

FIGURE E11.20 A standard nut is often used to check a thread.

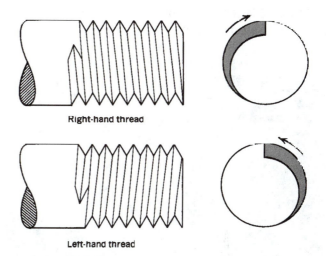

Right-hand thread

Left-hand thread

FIGURE E11.22 The difference between right-hand and left-hand threads as seen from the side and end.

tween each cut. Thread the nut with your fingers; it should go on easily but without end play. A class 2 fit is desirable for most purposes.

Step 12. Chamfer the end of the thread to protect it from damage.

Left-Hand Threads

The procedure for cutting left-hand threads (Figure E11.22) is the same as that used for cutting right-hand threads with two exceptions: The compound is set at 29 degrees to the left of the operator (Figure E11.23), and the leadscrew rotation is reversed so that the cut is made from the left to the right. The feed reverse lever is moved to reverse the leadscrew. Sufficient undercut must be made for a start-

ing place for the tool. Also, sufficient relief must be provided on the *right* side of the tool.

Methods of Terminating Threads

Undercuts are often used for terminating threads. They should be made the single depth of the thread plus .005 in. The undercut should have a radius to lessen the possibility of fatigue failure resulting from stress concentration in the sharp corners.

Machinists sometimes simply remove the tool quickly at the end of the thread while disengaging the half-nuts. If a machinist misjudges and waits too long, the point of the threading tool will be broken off. A dial indicator is sometimes used to locate the exact position for removing the tool. When this

FIGURE E11.23 Compound set for cutting a left-hand thread. (Lane Community College)

FIGURE E11.25 By placing the blade of the threading tool in the upper position, this tool can be made to thread on the bottom side. The lathe is reversed and the thread is cut from left to right making the job easier when threading next to a shoulder. (Aloris Tool Company, Inc.)

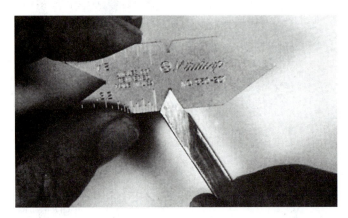

FIGURE E11.24 A threading tool that is used for threading to a shoulder.

tool-withdrawal method is used, an undercut is not necessary.

Terminating threads close to a shoulder requires specially ground tools (Figure E11.24). Sometimes, when cutting right-hand threads to a shoulder, it is convenient to turn the tool upside down and reverse the lathe spindle, which then moves the carriage from left to right. Some commercial threading tools are made for this purpose (Figure E11.25).

Picking Up a Thread

It sometimes becomes necessary to reset the tool when its position against the work has been changed during a threading operation. This position change may be caused by removing the threading tool for grinding, by the work slipping in the chuck or lathe dog, or by the tool moving from the pressure of the cut.

To reposition the tool the following steps may be taken.

Step 1. Check the tool position with reference to the work by using a center gage. If necessary, re-align the tool.

Step 2. With the tool backed away from the threads, engage the half-nuts with the machine running. Turn off the machine with the half-nuts still engaged and the tool located over the partially cut threads.

Step 3. Position the tool in its original location in the threads by moving both the crossfeed and compound handles (Figure E11.26).

Step 4. Set the micrometer dial to zero on the crossfeed collar and set the dial on the compound to the last setting used.

Step 5. Back off the cross feed and disengage the half-nuts. Resume threading where you left off.

Tapping Threads in the Lathe

If the thread is made by tapping in the lathe, the hole size of the workpiece can be varied to obtain the desired percentage of thread. See Table 11 in the Appendix for determining percentage of thread for tapping and the correct tap drill size.

FIGURE E11.26 Repositioning the tool.

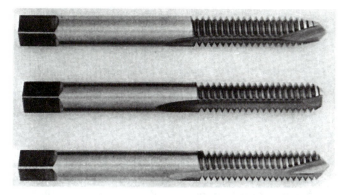

FIGURE E11.27 Set of spiral-pointed (or "gun") taps.

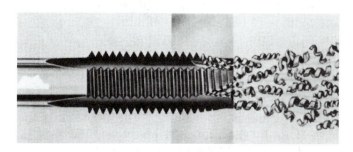

FIGURE E11.28 Cutting action of spiral-pointed taps. (© TRW Inc., 1980)

Full-depth threads are very difficult to tap in soft metals and impossible in tough materials. Tests have proven that above 60 percent of thread, very little additional strength is gained. Lower percentages, however, provide less flank surface for wear. Most commercial internal threads in steel are about 75 percent. Therefore, the purpose of making an internal thread with less than 100-percent thread depth is only for easier tapping since the hole size is larger and the internal thread depth is smaller and less material needs to be removed.

When a thread is to be made in a through hole, the spiral-point tap (Figure E11.27) cuts easier and cleaner and usually does not require frequent stopping to back it up and break the chips. The spiral-point tap cannot only be operated at higher speeds but all of the chips are pushed ahead of it and do not clog the hole and the tap (Figure E11.28). General tap terms are given in Figure E11.29.

Standard or fluteless taps (Figure E11.30) are recommended for tapping blind holes (those that do not go clear through). Fluteless, thread-forming taps do not cut threads in the same manner as conventional taps. They are forming tools and work best on ductile material such as aluminum, brass, copper, die castings, and leaded steels. The problem of

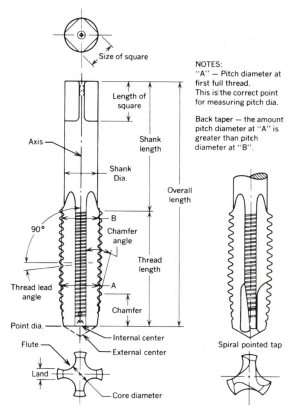

FIGURE E11.29 General tap terms. (Bendix Industrial Tools Division)

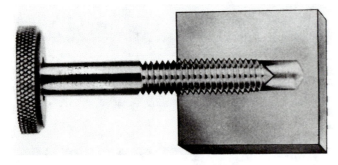

FIGURE E11.30 The thread-forming action of a fluteless thread-forming tap. (© TRW Inc., 1980)

chip congestion and removal often associated with the tapping of blind holes is eliminated. Thread formers also provide a stronger thread.

INTERNAL THREADS

Many of the same rules used for external threading apply to internal threading. The tool must be shaped to the exact form of the thread, and the tool must be set on the center of the workpiece. When cutting an internal thread with a single-point tool, the inside diameter (hole size) of the workpiece should be the minor diameter of the internal thread (Figure E11.31).

Single-Point Tool Threading

The advantages of making internal threads with a single-point tool are that large threads of various forms can be made and that the threads are concentric to the axis of the work. The threads may not be concentric when they are tapped. There are some difficulties encountered when making internal

threads. The tool is often hidden from view and tool spring must be taken into account.

The hole to be threaded is first drilled to $\frac{1}{16}$-in. diameter less than the minor diameter. Then a boring bar is set up, and the hole is bored to the minor diameter of the thread. If the thread is to go completely through the work, no recess is necessary, but if threading is done in a blind hole, a recess must be made. The compound rest should be swiveled 29 degrees to the left of the operator for cutting right-hand threads (Figures E11.32 and E11.33).

An alternative method sometimes used by machinists is to simply turn the boring bar 180 degrees so that the tool bit is opposite and the cutting edge faces downward (Figure E11.34). With this method, the compound is swiveled to the right for right-

FIGURE E11.32 For right-hand internal threads, the compound is swiveled to the left.

FIGURE E11.33 The compound rest is swiveled to the right for left-hand internal threads.

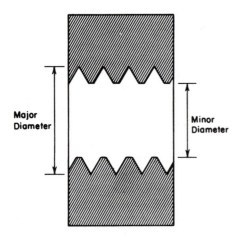

FIGURE E11.31 View of internal threads showing major and minor diameters.

hand internal threads. The advantages of this method are that the cutting operations can be more easily observed and the compound can remain in the same position as for external right-hand threads. For both methods, the threading tool is clamped in the bar in the same way and aligned by means of a center gage (Figure E11.35).

The compound micrometer collar is moved to the zero index after the slack has been removed by turning the screw outwards or counterclockwise. The tool is brought to the work with the cross-slide handle, and its collar is set on zero. Threading may now proceed in the same manner as it is done with external threads. The compound is advanced outwards a few thousandths of an inch, a scratch cut is made, and the thread pitch is checked with a screw pitch gage.

FIGURE E11.34 An alternative method of internal threading in which the tool is turned 180 degrees away from the operator is shown here.

FIGURE E11.35 Aligning the threading tool with a center gage. (Lane Community College)

The cross slide is backed out of the cut and reset to zero before the next pass. Cutting oil is used. The compound is advanced a few thousandths (.001 to .010 in.).

Often it is necessary to realign an internal threading tool with the thread when the tool has been moved for sharpening or when the setup has moved during the cut. The tool is realigned in the same way as it is done for external threads: by engaging the half-nut and positioning the tool in a convenient place over the threads, then moving both the compound and cross slides to adjust the tool position.

The exact amount of infeed depends on how rigid the boring bar and holder are and how deep the cut has progressed. Too much infeed will cause the bar to spring away and produce a bell-mouth internal thread.

If a slender boring bar is necessary or there is more than usual overhang, lighter cuts must be used to avoid chatter. The bar may spring away from the cut, causing the major diameter to be less than the calculated amount, or that amount fed in on the compound. If several passes are taken through the thread with the same setting on the compound, this problem can often be corrected.

The single depth of the American National thread equals $P \times .6495$, and the infeed on the compound set at 29 degrees is $P \times .75$. These figures may be substituted in the previous calculations to determine the single depth and infeed for a $1\frac{1}{2}$-6 NC internal thread.

The minor diameter is found by subtracting the double depth of the thread from the major diameter. Thus, if

$$D = \text{major diameter}$$
$$d = \text{minor diameter}$$
$$P = \text{thread pitch}$$
$$P \times .541 = \text{single depth}$$

the formula is

$$d = D - (P \times .541 \times 2)$$

EXAMPLE

A $1\frac{1}{2}$–6 UNC nut must be bored and threaded to fit a stud. Find the dimension of the bore.

$$P = \tfrac{1}{6} = .1666$$
$$d = 1.500 - (.1666 \times .541 \times 2)$$
$$d = 1.500 - .180$$
$$d = 1.320$$

Thus the bore should be made 1.320 in.

Cutting the Internal Thread

The following is the procedure for cutting right-hand internal threads.

Step 1. After correctly setting up the boring bar and tool, touch the threading tool to the bore and set both crossfeed and compound micrometer dials to zero.

Step 2. Feed in counterclockwise .002 in. on the compound dial.

Step 3. Turn on the lathe and engage the half-nut lever.

Step 4. Take a scratch cut. Disengage the half-nut when the tool is through the workpiece.

Step 5. Check the thread pitch with a screw pitch gage.

Step 6. Apply cutting fluid to the work.

Step 7. Feed the compound in an appropriate amount. (See Step 7 of "Cutting the Thread" for external threads for an explanation of depth of cut when threading.) Slightly less depth for cutting internal threads than for similar external threads may be necessary because of the spring of the boring bar.

Step 8. When nearing the calculated depth, test the thread with a plug gage between each pass. When the plug gage turns completely into the thread without being loose, the thread is finished.

Step 9. Several free cuts (passes without infeed) should be taken when nearing the finish depth to compensate for boring bar spring. The test plug should be used between each free pass.

Step 10. The inside and outside edges of the internal thread should be chamfered.

THREAD MEASUREMENT

Basic Internal Thread Measurement

Since small internal threads are most often made by tapping, the pitch diameter and fit are determined by the tap used. However, internal threads cut with a single-point tool need to be checked. A precision thread plug gage (Figure E11.36) is generally sufficient for most purposes. These gages are available in various sizes, which are stamped on the handle. An external screw thread is on each end of the handle. The longer threaded end is called the "go" gage, while the shorter end is called the "no go" gage. The no go end is made to a slightly larger dimension than the pitch diameter for the class of fit that the gage tests. To test an internal thread, both

FIGURE E11.36 Thread plug gage. (PMC Industries)

the go and no go gages should be tried in the hole. If the part is within the range or tolerance of the gage, the go end should turn in flush to the bottom of the internal thread, but the no go end should just start into the hole and become snug with no more than three turns. The gage should never be forced into the hole. If no gage is available, a shop gage can be made by cutting the required external thread to very precise dimensions. If only one threaded part of a kind is to be made and no interchangeability is required, the mating part may be used as a gage.

Basic External Thread Measurement

The simplest method for checking a thread is to try the mating part for fit. The fit is determined solely by feel with no measurement involved. While a loose, medium, or close fit can be determined by this method, the threads cannot be depended upon for interchangeability with others of the same size and pitch.

Thread ring gages are used to check the accuracy of external threads (Figure E11.37). The outside of the ring gage is knurled and the no go gage can be easily identified by a groove on the knurled surface. When these gages are used, the go ring gage should enter the thread fully. The no go gage should not exceed more than $1\frac{1}{2}$ turns on the thread being checked.

Thread roll snap gages are used to check the accuracy of external screw threads (Figure E11.38). These common measuring tools are easier and faster to use than thread micrometers or ring gages. The part size is compared to a preset dimension on the roll gage. The first set of rolls are the go and the second the no go rolls.

Thread roll snap gages, ring gages, and plug gages are used in production manufacturing where quick gaging methods are needed. These gaging methods depend on the operator's "feel," and the

FIGURE E11.37 Thread ring gage. (PMC Industries)

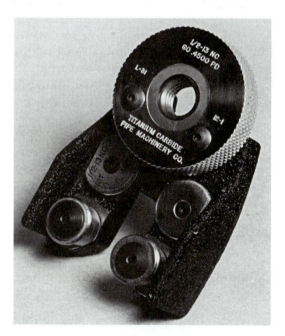

FIGURE E11.38 Thread roll snap gage. (PMC Industries)

level of precision is only as good as the accuracy of the gage. The thread sizes are not measurable in any definite way.

The thread comparator micrometer (Figure E11.39) has two conical points. This micrometer

FIGURE E11.39 Thread comparator micrometer.

does not measure the **pitch diameter** of a thread but is used only to make a comparison with a known standard. The micrometer is first set to the threaded part and then compared to the reading obtained from a thread plug gage.

Advanced Methods of Thread Measurement

The most accurate place to measure a screw thread is on the flank or angular surface of the thread at the pitch diameter. The outside diameter measured at the **crest** or the minor diameter measured at the root could vary considerably. Threads may be measured with standard micrometers and specially designated wires (Figure E11.40) or with a screw thread micrometer. The pitch diameter is measured directly by these methods. The kit in Figure E11.40 has wire sizes for most thread pitches. Most kits provide a number to be added to the nominal OD of the threads, which is the reading that should be seen on

FIGURE E11.40 Set of "best" three-wires for various thread sizes.

the micrometer if the pitch diameter of the thread is correct.

The Three-Wire Method

The three-wire method of measuring threads is considered one of the best and most accurate. Figure E11.41 shows three wires placed in the threads with the micrometer measuring over them. Different sizes and pitches of threads require different-sized wires. For greatest accuracy, a wire size that will contact the thread at the pitch diameter should be used. This is called the "best" wire size.

The pitch diameter of a thread can be calculated by subtracting the wire constant (which is the single depth of a sharp V thread, or $.866 \times P$) from the measurement over the three wires when the best wire size is used. The wires used for three-wire measurement of threads are hardened and lapped steel and are available in sets that cover a large range of threaded pitches.

A formula by which the best wire size may be found is

$$\text{Wire size} = \frac{.57735}{n}$$

where n = the number of threads per inch.

EXAMPLE

To find the best wire size for measuring a $1\frac{1}{4}$ in.-12 UNC screw thread,

$$\text{Wire size} = \frac{.57735}{12} = .048$$

The best wire size to use for measuring a 12-pitch thread would be .048 in.

If the best wire sizes are not available, smaller or larger wires may be used within limits. They should

FIGURE E11.41 Measuring threads with the three-wire and micrometer.

not be so small that they are below the major diameter of the thread, or so large that they do not contact the flank of the thread. Subtract the constant $(.866 \times P)$ for best wire size from the pitch diameter and add three times the diameter of the available wire when the best wire size is not available.

EXAMPLE

The best wire size for $1\frac{1}{4}$–12 is .048 in., but only a $\frac{3}{64}$-in.-diameter drill rod is available, which has a diameter of .0469 in.

1.1959	Pitch diameter of $1\frac{1}{4}$–12
− .0722	Constant for best wire size
1.1237	
+ .1407	$3 \times .0469$ available wire size
= 1.2644	Measurement over wires

The measurement over the wires will be slightly different than that of the best wire size because of the difference in wire size. After the best size wire is found, the wires are positioned in the threaded grooves as shown in Figure E11.41. The anvil and spindle of a standard outside micrometer are then placed against the three wires and the measurement is taken.

To calculate what the reading of the micrometer should be if a thread is the correct finished size, use the following formula when measuring Unified coarse threads or American National threads:

$$M = D + 3W - \frac{1.5155}{n}, \text{ where}$$

M = micrometer measurement over wires
D = diameter of the thread
n = number of threads per inch
W = diameter of wire used

EXAMPLE

To find M for a $1\frac{1}{4}$–12 UNC thread proceed as follows. Where

$$W = .048$$
$$D = 1.250$$
$$n = 12$$

then

$$M = 1.250 + (3 \times .048) - \frac{1.5155}{12}$$

$$= 1.250 + 1.44 - .126$$
$$= 1.268 \text{ (micrometer measurement)}$$

When measuring a Unified fine thread, the same method and formula are used, except that the constant should be 1.732 instead of 1.5155.

The wire method of thread measurement is also used for other thread forms such as Acme and Buttress. Information and tables may be found in *Machinery's Handbook*. Another method of measuring threads with a standard micrometer is with precision-ground triangular bars (Figure E11.42). These also come in kits.

The screw thread micrometer (Figure E11.43) may be used to measure sharp V, Unified, and American National threads. The spindle is pointed to a 60-degree included angle. The anvil, which

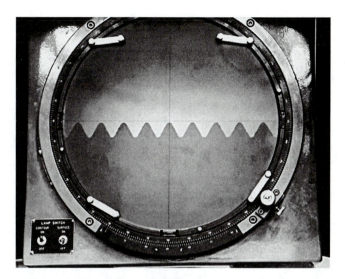

FIGURE E11.44 Profile of thread as shown on the screen of the optical comparator.

swivels, has a double-V shape to contact the pitch diameter. The thread micrometer measures the pitch diameter directly from the screw thread. This reading may be compared with pitch diameters given in handbook tables. Thread micrometers have interchangeable anvils that will fit a wide range of thread pitches. Some are made in sets of four micrometers that have a capacity up to 1 in., and each covers a range of threads. The range of these micrometers depends on the manufacturer. Typical ranges are as follows:

No. 1   8 to 14 threads per inch
No. 2   14 to 20 threads per inch
No. 3   22 to 30 threads per inch
No. 4   32 to 40 threads per inch

The optical comparator is sometimes used to check thread form, helix angle, and depth of thread on external threads (Figure E11.44). The part is mounted in a screw thread accessory that is adjusted to the helix angle of the thread so that the light beam will show a true profile of the thread.

SELF-TEST

1. By what method are threads cut or chased with a single-point tool in a lathe? How can a given helix or lead be produced?
2. The better practice is to feed the tool in with the compound set at 29 degrees rather than with the cross slide when cutting threads. Why is this so?
3. By what means should a threading tool be checked for the 60-degree angle?

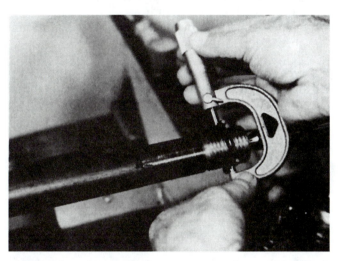

FIGURE E11.42 Using precision-ground triangular bars with a standard micrometer to measure threads.

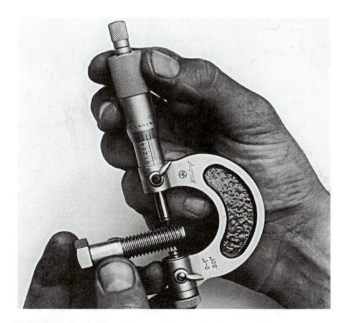

FIGURE E11.43 Screw thread micrometer.

4. How can the number of threads per inch be checked?

5. How is the tool aligned with the work?

6. Is the carriage moved along the ways by means of gears when the half-nut lever is engaged? Explain.

7. Explain which positions on the threading dial are used for engaging the half-nuts for even-, odd-, and fractional-numbered threads.

8. How fast should the spindle be turning for threading?

9. What is the procedure for cutting left-hand threads?

10. If for some reason it becomes necessary for you to temporarily remove the tool or the entire threading setup before a thread is completed, what procedure is needed when you are ready to finish the thread?

11. Explain how you would find the correct tap drill size for tapping a thread.

12. What would be the best kind of tap to use in a blind hole in soft aluminum?

13. When internal threads are made with a tool, what should the bore size be?

14. In what way is percentage of thread obtained? Why is this done?

15. What percentage of thread are tap drill charts usually based on?

16. What are two advantages in turning a single-point threading tool 180 degrees opposite to the operator?

17. Name two advantages of making internal threads with a single-point tool on the lathe.

18. When making internal right-hand threads, which direction should the compound be swiveled?

19. After a scratch cut is made, what would be the most convenient method to measure the pitch of the internal thread?

20. What does deflection or spring of the boring bar cause when cutting internal threads?

21. Using $P \times .65$ as a constant for single depth internal threads, what would the minor diameter be for a ANC 1–8 thread?

22. Name two methods of checking an internal thread for size.

Exercise

Your instructor may assign you one or several thread pitches and diameters to cut. Manual practice of thread cutting should give you a better understanding of how the CNC lathe operates in thread-cutting operations.

WORKSHEET: TURNING AND THREADING PROJECT

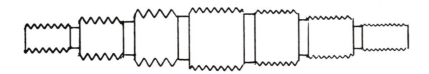

Objective

Learn to use a small engine lathe with sufficient skill to be able to do light, external turning and threading for the purpose of making mechanical repairs and for thread restoration.

Note: The instructor may choose to require only one side of the project to be completed and thread pickup practice to be done on the other side.

Materials

One $1\frac{1}{4}$-in. dia. HR mild steel round bar cut off $7\frac{1}{8}$ in. long, one high-speed tool blank, a Number 3 center drill, a small lathe and tooling, and appropriate measuring tools.

Procedure for Precision Measurement

The units on safety, measuring, and lathe work *must* be studied before proceeding with this project. You will need to master the use of micrometers, dial indicators, and other precision tools for any branch of the mechanical field.

Procedure for Introduction to the Lathe

After studying Units 1 through 11, Section E in Turning Machines, go to the lathe you will be using and familiarize yourself with the controls without turning on the power. When you are confident, turn on the motor and operate the feed mechanism and half-nut lever without doing any cutting.

Procedure for Grinding a Lathe Tool

Follow the procedure given in Section E, Unit 4, for grinding and sharpening a right-hand turning tool.

Procedure for Facing and Center Drilling

1. Set up the $1\frac{1}{4}$-in. diameter workpiece in a chuck on a lathe and face one end without leaving a center stub.

2. Set up a Number 3 center drill in the drill chuck in the tailstock and set the correct speed. Using cutting oil, drill the center hole to the desired depth (about three quarters of the way into the countersink section would be a good depth in this case).
3. Break the corner with a file; that is, deburr the sharp outer edge left by machining.
4. Turn the part around in the chuck and repeat steps 1, 2, and 3.

Procedure for Turning between Centers

The specifications and tolerances as given in Figure E11.45 must be observed and held unless your instructor has given you a different set of allowances.

1. Set up the lathe for turning between centers. Make sure the centers and sockets are clean and the centers are well seated.
2. Apply high pressure lubricant to the center hole at the tailstock end. Ordinary oil will not work as it will run out and the center will heat up and be damaged.
3. Slip on the dog with the bent tail pointed toward the drive plate.
4. Place the workpiece between the centers.
5. Put the dog in place and tighten the set screw. Make sure the bent tail is clear of the bottom of the slot.
6. Adjust the dead center so there is no end play, yet with sufficient clearance so the bent tail of the dog is free to move in the slot. Tighten the tailstock binding lever. *Note:* Further adjustment of the dead center will be necessary when the work heats up from machining (unless you are using a live center in the tailstock).
7. Fasten a right-hand turning tool in a straight or left-hand toolholder or in any appropriate holder provided on your lathe. Adjust the holder so the tool will not gouge the work if it should slip and turn from heavy cutting. Set the tool for height.
8. Set the lathe to the correct RPM for the work.
9. Adjust the feed for roughing; 0.008 to 0.012 in. would be appropriate for a small ten to twelve in. swing lathe.

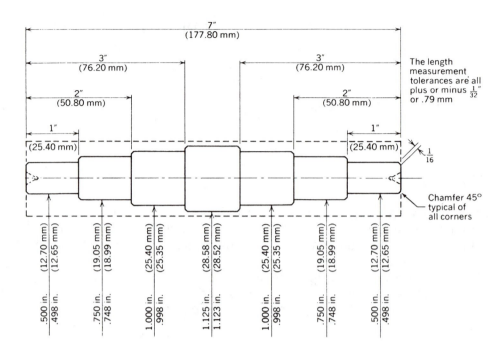

FIGURE E11.45

Length dimensions (left to right): 7" (177.80 mm); 3" (76.20 mm); 2" (50.80 mm); 1" (25.40 mm); 3" (76.20 mm); 2" (50.80 mm); 1" (25.40 mm).

The length measurement tolerances are all plus or minus $\frac{1}{32}$" or .79 mm

Chamfer 45° typical of all corners

$\frac{1}{16}$

Diameters (left to right): .500 in. (12.70 mm) / .498 in. (12.65 mm); .750 in. (19.05 mm) / .748 in. (18.99 mm); 1.000 in. (25.40 mm) / .998 in. (25.35 mm); 1.125 in. (28.58 mm) / 1.123 in. (28.52 mm); 1.000 in. (25.40 mm) / .998 in. (25.35 mm); .750 in. (19.05 mm) / .748 in. (18.99 mm); .500 in. (12.70 mm) / .498 in. (12.65 mm).

10. *Always check to see that the workpiece or dog is clear by turning the spindle by hand before using the power. Wear safety glasses.* Turn on the machine.

11. The first roughing cuts should extend past the center section of the workpiece so that a distance of roughly $4\frac{1}{2}$ in. should be marked from the tailstock end.

12. Touch the tool to the work, zero the crossfeed dial, and take a skim or clean-up cut of about 0.005 in. depth for about $\frac{1}{2}$ in. Turn off the machine. *Leave the dial setting as it is.*

13. Check this turned size with your micrometer.

14. In most turning operations, it is standard practice to leave all diameters of a turned part slightly oversize to allow for any warping before taking the final finish cut. This operation is called "roughing." About 0.030 in. is sufficient to leave for finish. Most small lathes should have sufficient power to make the first diameter with one roughing cut and one or two finishing cuts. For example, let us assume the $1\frac{1}{4}$ in. diameter is now 1.245 in. and the finish diameter will be 1.125 in. Thus, 0.120 in. is the total amount of stock to be removed from the diameter of the middle section. However about 0.030 in. should be left for finishing, leaving 0.090 in. for the roughing cut.

15. In this case, you would now set your depth of cut for 0.045- or 0.090-in. diameter depending on the type of graduation on the crossfeed collar of your lathe. Complete the roughing cut for a length of about $4\frac{1}{2}$ in. Check for taper and, if it is more than 0.001 in., reset the tailstock. (See Unit 9, "Alignment of Lathe Centers.")

16. The next size is a 1 in. diameter that has a shoulder. All diameters are roughed on both ends before taking any finish cuts.

17. Scribe or mark a line on the workpiece 3 in. from the tailstock end.

18. Make a roughing cut to within $\frac{1}{16}$ in. of the line. This allowance is left for later clean up of the shoulder. Increase lathe speed to compensate for smaller diameters.

19. Scribe a line 2 in. from the tailstock end and repeat step 18. Scribe a line 1 in. from the tailstock and repeat step 18.

20. Turn the workpiece end for end and rough turn each step as before.

21. Set up a right-hand finishing tool or a modified right-hand turning tool (by grinding a larger nose radius on it) in the toolholder.

22. Set the lathe for a finishing feed and speed.

23. Run the tool in until it touches the workpiece and move it to the starting position. Add 0.003 in. more and zero the micrometer collar. Take a light cut for about $\frac{1}{4}$ in.

24. Check the diameter with the micrometer caliper. It should now be between 0.020 and 0.030 in. oversize. Set the crossfeed micrometer collar for the appropriate amount and take a trial cut $\frac{1}{8}$ in. long. Check for size with a micrometer and adjust, if necessary, so that the finish size will be within the tolerances given on the drawing. Finish the cut.

25. When all of the diameters are finished on one end, turn the tool so that a square shoulder can be made. Pick up the cut (just touch the finish diameter) near the shoulder and hand feed to the

line. Back out the tool with the cross feed handle to square the shoulder. Turn the tool SCEA about 45 degrees and chamfer the edges.

26. Turn the work end for end and repeat.

Procedure for Undercutting

Before the threading operation is begun, a means of clearing the threading tool should be arranged. Undercutting to the depth of the thread is one such means.

1. Using either a parting tool or a specially ground tool, make an undercut for each thread equal to its single depth plus 0.005 in. Use cutting oil. Make the undercuts $\frac{3}{16}$ in. wide. If the tool chatters, check for too much toolholder or tool overhang; slow down the machine.

2. The following formula will give you the single depth for undercutting American National threads:

$$d = P \times 0.65 \quad \text{or} \quad d = 0.65/n$$

where:

> d = single depth
> P = pitch
> n = number of threads per inch (TPI)

Since there is no undercut for the $1\frac{1}{8}$-12 thread, consider only the other six diameters and threads. Fill out the following list for the single depth of undercutting on each:

14 TPI _____   8 TPI _____
16 TPI _____   10 TPI _____
20 TPI _____   13 TPI _____

Use this list as a guide for depth of undercutting as you undercut each step on the project shown below (Figure E11.46).

3. Set up a parting tool and touch it to the finish diameter you wish to undercut. Zero the micrometer collar and back off the tool. Now move the tool to

the shoulder and make the cut. You may need to double the calculated depth on some lathes that read directly for diameters. For single depth reading crossfeed collars, use the figures as they are. You may have to take two cuts side-by-side to get the $\frac{3}{16}$-in. width.

Procedure for Threading

1. For this operation, you will need to grind the opposite end of your lathe tool bit to make a 60-degree threading tool as shown in Section H, Unit 4. Use a center gage to check it. Provide for a zero or neutral back rake on the tool, depending on the type of toolholder you are using.

2. Set the compound at 29 degrees to the right, off the cross slide index line.

3. Determine the depth the tool is fed into the work when the compound slide is set at 29 degrees for the following threads. The formula for UNS (American Standard threads) is:

$$d = P \times 0.75 \quad \text{or} \quad d = 0.75/n$$

Fill out the following table for the infeed when threading each step:

12 TPI _____   8 TPI _____
14 TPI _____   10 TPI _____
16 TPI _____   13 TPI _____
20 TPI _____

4. Set up the threading tool, making sure it is on center, and square to the work axis. Use the center gage for aligning the tool to the work.

5. Select a slow RPM and set the quick-change gears to give you 12 threads per inch so you can do the $1\frac{1}{8}$ in. diameter thread first.

6. Put the feed change lever or its equivalent in neutral.

7. Set the compound micrometer collar on zero.

8. Bring the tool in contact with the work on the largest diameter using the cross slide. Then set its dial on zero.

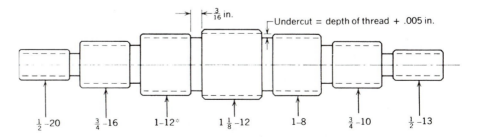

FIGURE E11.46

$\frac{1}{2}$-20 $\frac{3}{4}$-16   1-12* $1\frac{1}{8}$-12   1-8 $\frac{3}{4}$-10 $\frac{1}{2}$-13

9. Move the carriage to the right until the tool is clear of the work; feed in the compound 0.003 in. Apply cutting oil to the work. *Always* use cutting oil when threading.
10. When the index mark on the thread dial is in line with the proper position on the dial, engage the half-nut lever. Keep your hand close to it so you can disengage the lever as soon as it has cut the full length of the thread.
11. With the cross slide back the tool away from the work one or two turns. Return the carriage by hand to the right, where you started.
12. Return the cross slide dial to its zero position.
13. Feed the compound another 0.005 in. (single depth); 0.010 in. (double depth).
14. Repeat steps 10, 11, 12, and 13 until 0.010 in. remains of the infeed dimension. Then feed only 0.002 in. per pass. The last pass should be made without additional feed in order to clean up the thread. Check the thread for size. When you are nearing the last few passes, use a standard nut for testing the thread. It should fit snugly, but you should be able to turn it on with your fingers. When turning between centers using a dog, the workpiece may be removed for testing without disturbing the tracking of the thread if the dog is returned to the same slot.

15. Repeat steps 5 through 14 *with the changes made for each specific thread.* When reversing the workpiece between centers, protect the thread with two nuts locked on each other, or use a threaded dog.

Measuring Threads

Materials

Screw thread comparison micrometer, precision plug gage, or three-wire set and micrometer, or screw thread micrometer. Finished threading project.

Procedure

1. Measure all of the thread diameters and pitches.
2. Record the type of measuring system you are using and the measurement of each thread.
3. Record the correct measurement for each thread and note the difference.

Type of measuring instrument: _____

Evaluation

Clean up your project and turn it in for grading.

TPI	13	10	8	12	14	16	20
Your measurement							
Correct measurement							
Total error oversize							
Total error undersize							

WORKSHEET: INTERNAL THREAD

Prior to starting this project, study and complete the Post-Test for cutting an internal thread with a single point tool on the lathe, Section E, Unit 11.

Objective

Learn to bore and cut an internal thread in a nut blank.

Materials

One inch length of $1\frac{1}{2}$ in. dia. hexagonal or round stock, a precision 1–8 thread plug gage, a shop-made plug gage, or a standard 1-8 capscrew.

Procedure for Cutting an Internal Thread on the Lathe

1. Chuck the nut blank and make a facing cut. Chamfer.
2. Calculate the minor diameter for the 1-8 ANC thread. The formula is:

$$d = D - (P \times .541 \times 2)$$

when:

$$P = \text{pitch} = \tfrac{1}{8} \text{ in.} = .125 \text{ in.}$$
$$d = 1 \text{ in.} - .125 \text{ in.} \times .541 \times 2$$
$$d = .86475 \text{ in.} = .865 \text{ in.}$$

3. Spot drill the nut blank and drill through with a $\frac{3}{4}$- to $\frac{13}{16}$-in. drill.
4. Set up a boring bar and tool and bore the hole to .865 in., the minor diameter of the thread.
5. Set up your lathe for threading 8 TPI and place a threading tool in the boring bar.
6. Swivel the compound rest to 29 degrees toward the chuck.
7. Align the tool with the center gage. Check to see that the chuck jaws will clear the compound when the tool is moved through the bore.

8. Take the slack out of the compound screw by turning it outward and set the tool to the work. Zero both micrometer collars.
9. Take a scratch cut and check the threads with a screw pitch gage.
10. Select a slow speed for threading. Use cutting oil.
11. Determine the infeed on the compound for a 1-8 NC thread. The formula is:

$$\begin{aligned} \text{Infeed} &= P \times .75 \\ &= .125 \text{ in.} \times .75 \\ &= .094 \text{ in.} \end{aligned}$$

12. Advance the compound before each cut .005 to .010 in. depending on the rigidity of your setup. Reduce the infeed to .002 in. or less near the last few cuts.
13. When the total infeed depth has been reached, clean out all the chips from the threads and try the thread plug gage for fit. If it will not go in, run the tool through one or two times without advancing the compound slide and try the gage again. If the gage still does not fit, it indicates that the tool form, flat, or alignment is not correct, or that something on your setup has slipped out of place, or that tool spring has not allowed a full depth of cut. Sometimes several free cuts can be taken to enlarge the diameter. If you are not getting results, continue advancing the tool .002 in. per pass and check with the thread plug gage between each cut.
14. Chamfer the thread and reverse the workpiece in the chuck.
15. Face this side of the nut to length and chamfer both the thread and the outside diameter.

Evaluation

Clean up your project in solvent and bring both the test plug and your project to your instructor for grading.

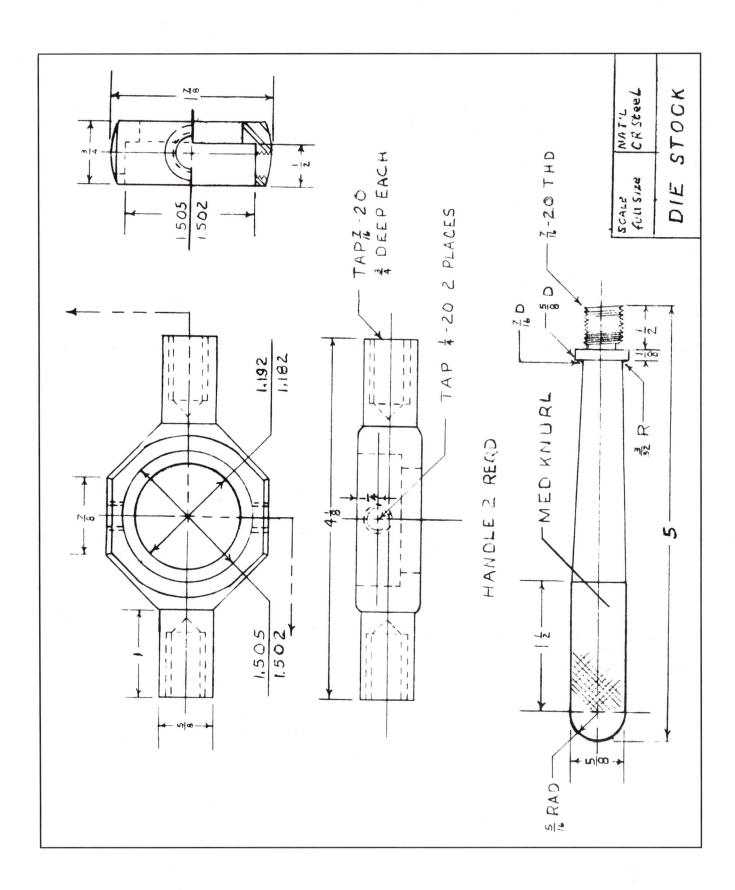

1 7/8

3/4

1.505
1.502

1/2

SCALE full size | NAT'L CR Steel

DIE STOCK

TAP 1/2-20 3/4 DEEP EACH

TAP 1/4-20 2 PLACES

7/16 D

5/8 D

1/2-20 THD

1 1/8 — 1/2

3/32 R

MED KNURL

HANDLE 2 REQD

1.192
1.182

7/8

1.505
1.502

4 1/8

1/4

1

5/8

1 1/2

5

5/8

5/16 RAD

WORKSHEET: DIE STOCK

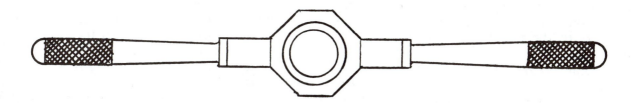

Prior to starting each procedure for this project, study and complete the Post-Tests for:

1. Making the body: Section E, Units 8 and 10.
2. Making the handles: Section E, Units 11, 12, and 13.

Note: If you have already completed Post-Tests for any of these units, do not take the test again, but simply refer to the units to refresh your memory of the procedures.

Objectives

1. Learn how to set up rectangular stock in a four-jaw chuck.
2. Bore and counterbore precision internal diameters.
3. Learn to use the taper attachment and perform taper calculations.

Materials

$4\frac{3}{16}$ inches of $\frac{3}{4} \times 2$ in. CR flat bar and 10 inches of $\frac{5}{8}$ CR shaft. Also two $1\frac{1}{4}$-20 socket head set screws $\frac{1}{4}$ in. long will be needed.

Procedure for Making the Body

1. Place a bar of $\frac{3}{4}$ 3 2 in. cold rolled mild steel (CRMS) in the power saw and cut off $4\frac{1}{8}$ in. plus $\frac{1}{16}$ in. for finishing, a total of $4\frac{3}{16}$ in.
2. Deburr the sharp edges and clean the piece in solvent.
3. Find the center of the large flat side and center punch.
4. Set up in a four-jaw chuck with the center punched side out. Two of the jaws may have to be reversed. The punch mark should be adjusted to run true with the machine axis. A wiggler and a dial indicator may be used for this or a sharp dead center in the tailstock could be used to locate the punch mark. True up the face using a dial indicator.
5. Drill through the material, using first a center drill followed by a pilot drill and a $\frac{3}{4}$ in. diameter drill.
6. Set up a boring holder with a $\frac{1}{2}$ in. bar. Set the bar to hold the toolbit in an angular position. Grind the tool to cut properly at that angle so that an internal square shoulder can be made. The bar should extend no further from its holder than necessary.
7. Using the correct RPM and feed, bore to $1\frac{3}{16}$ in. dia. with a tolerance of plus or minus .005 in.
8. Counterbore to a rough size of about .025 in. under finish diameter. Leave approximately $\frac{1}{32}$ in. in the bottom of the counterbore for finish; that is, make the depth of the counterbore $\frac{15}{32}$ in.
9. Check your toolbit for built-up edge or dullness and hone if necessary.
10. Set the feed and RPM for finishing.
11. One method of turning a counterbore to depth is to touch the toolbit to the outside surface of the workpiece near the bore. Using a steel rule laid flat on the ways and touching the carriage, mark at an inch line or at the end of the rule on the lathe way with a grease or graphite pencil. (*Note:* These markers will not harm a precision way, but do not use a scriber or any hard or sharp marking device.) This will provide an index position for moving the carriage. Position the boring bar with the cross slide so the bar and tool can enter the bore. Another method is to use the compound micrometer collar to set the depth of the tool with the compound parallel to the center axis of the lathe.
12. Bring the tool to the inner edge of the bottom of the counterbore and set it to $\frac{1}{2}$ in. depth with the rule and mark. Lock the carriage.
13. Start the lathe and slowly hand feed the cross slide to finish the face to depth. Watch the tool in the bore, and when the cut is at the diameter of the large bore, stop feeding and turn off the lathe.
14. Unlock the carriage and carefully clear the tool and hack it out from the work.

15. Take a light trial cut on the $1\frac{1}{2}$ in. bore (about .005 in.) for a distance of $\frac{1}{8}$ in.
16. Measure the trial cut and subtract the measurement from 1.505 in. The difference will be the depth of the finish cut as measured on the diameter.
17. If the lathe you are using is graduated on the micrometer collar to read on the diameter of the work, then simply set the number you have found in step 16. If, on the other hand, the dial is graduated for single depth, then set the depth of cut at half the value found in step 16.
18. Take another trial cut, this time for about $\frac{1}{4}$ in. Measure and correct any error caused by tool spring. The error will generally be undersize.
19. Finish bore to full depth. Stop the power feed before the tool comes to the bottom and hand feed to make a sharp corner.
20. Chamfer all sharp edges with a tool set at about 45 degrees. Remove the workpiece from the lathe.
21. Lay out for center on both ends of the material. Center punch.
22. Center drill in a drill press on both ends, or use an alternate method: center drill in the lathe and face to length.
23. Tap drill both holes for set screws.
24. Set up for turning between centers. Alternately, use a four-jaw chuck and dead center.
25. Rough out both ends to $\frac{21}{32}$ in. dia. and $\frac{31}{32}$ in. long.
26. Turn center section to $1\frac{7}{8}$ in. diameter using a finishing tool and feed for the last cut.
27. Set the compound for 45 degrees and rough turn both angular surfaces to within $\frac{1}{32}$ in. of finish size.
28. Face both ends so that the total length is $4\frac{1}{8}$ inches.
29. Finish turn the $\frac{5}{8}$ in. dia. ends to finish diameter and to the 1 inch length.
30. Finish turn the angular surface with the compound set at 45 degrees. Turn the compound hand crank slowly and evenly to provide a good finish.
31. Remove the part from the lathe and install a three-jaw chuck.
32. Place the $\frac{5}{8}$ in. end in the chuck and tighten.
33. Tap drill $\frac{3}{4}$ in. deep for a $\frac{7}{16}$-20 tap. Chamfer hole.
34. Place tap in the Jacobs chuck in the tailstock and start by turning the three-jaw chuck by hand.

Leave the tailstock loose on the ways. Use cutting oil on the tap. Continue to alternately turn the chuck forward and reverse to break the chips until the tap offers more than usual resistance, showing that it has reached the bottom of the hole. Back out the tap and repeat the process on the other end.
35. Remove the part from the lathe and tap the two holes for the set screws. Clean up the part and deburr where necessary. The body is now completed.

Procedure for Making the Handles

1. Place some $\frac{5}{8}$ in. CR round stock in the saw. Cut off $5\frac{1}{2}$ inches.
2. Set up material in a three-jaw chuck. Face off and center drill both ends.
3. With one end in the chuck and the other supported with a dead center, knurl from the tailstock end to a mark $3\frac{3}{16}$ in. from the headstock end.
4. Machine the step and shoulder for the $\frac{7}{16}$-20 thread.
5. Thread the end by single-point tool method. Do not use a die. Check with a $\frac{7}{16}$-20 nut or with the body of the die stock.
6. Machine the taper using the taper attachment. You will find the formula for the various methods of producing a taper in your textbook. Determine the taper per foot and set that amount on the taper attachment.
7. Chuck the knurled end extending the end opposite the thread, using soft aluminum or copper to protect the knurl. Face the part to length.
8. Free hand the $\frac{5}{16}$ in. radius. Check frequently with a radius gage. Finish with a single cut file.
9. Repeat steps 1 through 8 to make the other handle.
10. Clean all parts in solvent and wipe dry. The die stock is now ready for assembly.

Evaluation

Assemble your die stock and show it to your instructor for grading.

Taper Turning, Taper Boring, and Forming

Tapers are very useful machine elements that are used for many purposes. A manual machinist should be able to calculate a specific taper quickly and set up a lathe to produce it. The machinist should also be able to accurately measure tapers to determine precision fits.

On CNC lathes, precision tapers are easily programmed into the operation. However, CNC machine operators need to know the process and principles involved in making tapers. Measuring tapers for accuracy may be part of a CNC machine operator's responsibility.

This unit should help you to understand the various methods of producing and measuring tapers.

OBJECTIVE

After completing this unit, you should be able to:

Describe different types of tapers and the methods used to produce and measure them.

USES OF TAPERS

Tapers are used on machines because of their capacity to align and hold machine parts and to realign when they are repeatedly assembled and disassembled. This repeatability insures that tools such as centers in lathes, taper shank drills in drill presses, and arbors in milling machines will run in perfect alignment when placed in the machine. When a taper is slight, such as a Morse taper that is about $\frac{5}{8}$ in. taper/ft, it is called a *self-holding taper* since it is held in and driven by friction (Figure E12.1). A steep taper, such as a quick-release taper of $3\frac{1}{2}$ in./ft used on most milling machines, must be held in place with a draw bolt (Figure E12.2).

A taper may be defined as a uniform increase in diameter on a workpiece for a given length measured parallel to the axis. Internal or external tapers are expressed in taper per foot (TPF), taper per inch (TPI), or in degrees. The TPF or TPI refers to the difference in diameters in the length of one foot or one inch, respectively (Figure E12.3). This difference is measured in inches. Angles of taper, on the other hand, may refer to the included angles or the angles with the centerline (Figure E12.4).

Some machine parts that are measured in taper per foot are mandrels (.006 in./ft), taper pins and reamers ($\frac{1}{4}$ in./ft), the Jarno taper series (.600 in./ft), the Brown and Sharpe taper series ($\frac{1}{2}$ in./ft), and the Morse taper series (about $\frac{5}{8}$ in./ft). Morse tapers include eight sizes that range from size 0 to size 7. Tapers and dimensions vary slightly from size to size in both the Brown and Sharpe and the Morse series. For instance, a No. 2 Morse taper has .5944 in./ft taper and a No. 4 has .6233 in./ft taper. See Table E12.1 for more information on Morse tapers.

METHODS OF MAKING A TAPER

There are four methods of turning a taper on a lathe. They are the compound slide method, the offset tailstock method, the taper attachment method, and the form tool method. Each method has its advantages and disadvantages, so the kind of taper needed on a workpiece should be the deciding factor in the selection of the method that will be used.

The Compound Slide Method

Both internal and external short, steep tapers can be turned on a lathe by hand feeding the compound slide (Figure E12.5). The swivel base of the compound is divided in degrees. When the compound slide is in

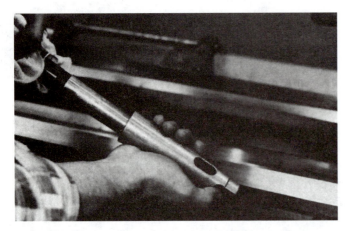

FIGURE E12.1 The Morse taper shank on this drill keeps the drill from turning when the hole is being drilled.

FIGURE E12.2 The milling machine taper is driven by lugs and held in by a draw bolt. (Lane Community College)

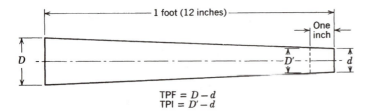

$$TPF = D - d$$
$$TPI = D' - d$$

FIGURE E12.3 The difference between taper per foot (TPF) and taper per inch (TPI).

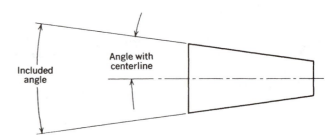

FIGURE E12.4 Included angles and angles with centerline.

FIGURE E12.5 Making a taper using the compound slide.

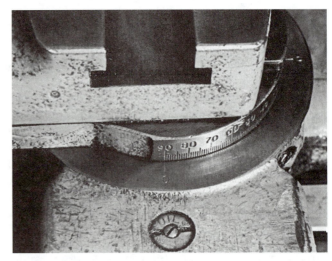

FIGURE E12.6 Alignment of the compound parallel with the ways.

line with the ways of the lathe, the 0 degree line will align with the index line on the cross slide (Figure E12.6). When the compound is swiveled off the index, which is parallel to the centerline of the lathe, a direct reading may be taken for the half angle or

angle to centerline of the machined part (Figure E12.7). When a taper is machined off the lathe centerline, its included angle will be twice the angle that is set on the compound. Not all lathes are indexed in this manner.

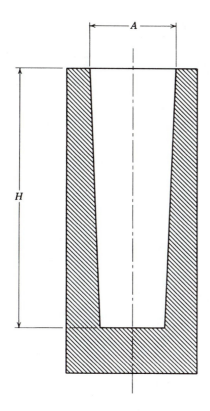 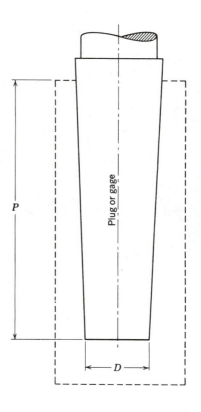

TABLE E12.1 Morse tapers information

Number of Taper	Taper per Foot	Taper per Inch	P Standard Plug Depth	D Diameter of Plug at Small End	A Diameter at End of Socket	H Depth of Hole
0	.6246	.0520	2	.252	.356	$2\frac{1}{32}$
1	.5986	.0499	$2\frac{1}{8}$	.396	.475	$2\frac{3}{16}$
2	.5994	.0500	$2\frac{9}{16}$	.572	.700	$2\frac{5}{8}$
3	.6023	.0502	$3\frac{3}{16}$	.778	.938	$3\frac{1}{4}$
4	.6232	.0519	$4\frac{1}{16}$	1.020	1.231	$4\frac{1}{8}$
5	.6315	.0526	$5\frac{3}{16}$	1.475	1.748	$5\frac{1}{4}$
6	.6256	.0521	$7\frac{1}{4}$	2.116	2.494	$7\frac{3}{8}$
7	.6240	.0520	10	2.750	3.270	$10\frac{1}{8}$

When the compound slide is aligned with the axis of the cross slide and swiveled off the index in either direction, an angle is directly read off the cross slide centerline (Figure E12.8). Since the lathe centerline is 90 degrees from the cross slide centerline, the reading on the lathe centerline index is the complementary angle. So, if the compound is set off the axis of the cross slide $14\frac{1}{2}$ degrees, the lathe centerline index reading is $90 - 14\frac{1}{2} = 75\frac{1}{2}$ degrees, as seen in Figure E12.8.

Tapers of any angle may be cut by this method, but the length is limited to the stroke of the compound slide. Since tapers are often given in TPF, it is sometimes convenient to consult a TPF-to-angle conversion table, like Table E12.2. A more complete table may be found in the *Machinery's Handbook*.

If a more precise conversion is desired, the following formula may be used to find the included angle.

Divide the taper in inches per foot by 24; find the angle that corresponds to the quotient in a table of tangents and double this angle. If the angle with centerline is desired, do not double the angle.

FIGURE E12.7 An angle may be set off the axis of the lathe from this index.

FIGURE E12.8 The compound set $14\frac{1}{2}$ degrees off the axis of the cross slide.

EXAMPLE

What angle is equivalent to a taper of $3\frac{1}{2}$ in./ft?

$$\frac{3.5}{24} = .14583$$

The angle of this tangent is 8 degrees 18 minutes, and the included angle is twice this, or 16 degrees 36 minutes.

TABLE E12.2 Tapers and corresponding angles

Taper per Foot	Included Angle		Angle with Centerline		Taper per Inch
	Degrees	Minutes	Degrees	Minutes	
$\frac{1}{8}$	0	36	0	18	.0104
$\frac{3}{16}$	0	54	0	27	.0156
$\frac{1}{4}$	1	12	0	36	.0208
$\frac{5}{16}$	1	30	0	45	.0260
$\frac{3}{8}$	1	47	0	53	.0313
$\frac{7}{16}$	2	5	1	2	.0365
$\frac{1}{2}$	2	23	1	11	.0417
$\frac{9}{16}$	2	42	1	21	.0469
$\frac{5}{8}$	3	00	1	30	.0521
$\frac{11}{16}$	3	18	1	39	.0573
$\frac{3}{4}$	3	35	1	48	.0625
$\frac{13}{16}$	3	52	1	56	.0677
$\frac{7}{8}$	4	12	2	6	.0729
$\frac{15}{16}$	4	28	2	14	.0781
1	4	45	2	23	.0833
$1\frac{1}{4}$	5	58	2	59	.1042
$1\frac{1}{2}$	7	8	3	34	.1250
$1\frac{3}{4}$	8	20	4	10	.1458
2	9	32	4	46	.1667
$2\frac{1}{2}$	11	54	5	57	.2083
3	14	16	7	8	.2500
$3\frac{1}{2}$	16	36	8	18	.2917
4	18	56	9	28	.3333
$4\frac{1}{2}$	21	14	10	37	.3750
5	23	32	11	46	.4167
6	28	4	14	2	.5000

The Offset Tailstock Method

Long, slight tapers may be produced on shafts and external parts between centers. Internal tapers cannot be made by this method. Power feed is used, so good finishes are obtainable. The taper per foot or taper per inch must be known so that the amount of offset for the tailstock can be calculated. Since tapers are of different lengths, they would not be the same TPI or TPF for the same offset (Figure E12.9). When the taper per inch is known, the offset calculation is as follows: Where

$$TPI = \text{taper per inch}$$
$$L = \text{length of workpiece}$$

$$\text{Offset} = \frac{TPI \times L}{2}$$

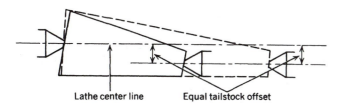

FIGURE E12.9 When tapers are of different lengths, the TPF is not the same with the same offset.

When the taper per foot is known, use the following formula:

$$\text{Offset} = \frac{\text{TPF} \times L}{24}$$

If the workpiece has a short taper in any part of its length (Figure E12.10) and the TPI or TPF is not given, use the following formula:

$$\text{Offset} = \frac{L \times (D - d)}{2 \times L_1}, \text{ where}$$

D = diameter at large end of taper
d = diameter at small end of taper
L = total length of workpiece
L_1 = length of taper

When you set up for turning a taper between centers, remember that the contact area between the center and the center hole is limited (Figure E12.11). Frequent lubrication of the centers may be necessary. You should also note the path of the

lathe dog bent tail in the drive slot (Figure E12.12). Check to see that there is adequate clearance.

To measure the offset on the tailstock, use either the centers and a rule (Figure E12.13) or the witness mark and a rule (Figure E12.14); both methods are adequate for some purposes. A more precise measurement is possible with a dial indicator, as shown in Figure E12.15. The indicator is set on the tailstock spindle while the centers are still aligned. A slight loading of the indicator is advised, since the first .010 or .020 in. movement of the indicator may be inaccurate or the mechanism loose due to

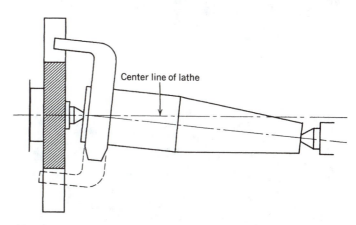

FIGURE E12.12 The bent tail of the lathe dog should have adequate clearance.

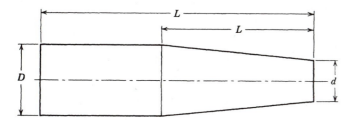

FIGURE E12.10 Long workpiece with a short taper.

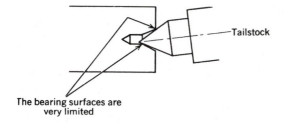

FIGURE E12.11 The contact area between the center hole and the center is small.

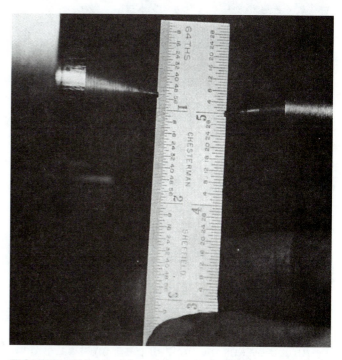

FIGURE E12.13 Measuring the offset on the tailstock by use of the centers and a rule.

FIGURE E12.14 Measuring the offset with the witness mark and a rule.

FIGURE E12.15 Using the dial indicator to measure the offset.

wear, causing fluctuating readings. The bezel is set at zero; the tailstock is loosened and moved toward the operator the calculated amount. Clamp the tailstock to the way. If the indicator reading changes, loosen the clamp and readjust. When cutting tapered threads such as pipe threads, the tool should be square with the centerline of the workpiece, not the taper (Figure E12.16).

When you have finished making tapers by the offset tailstock method, realign the centers to .001 in. or less in 12 in. When more than one part must be turned by this method, all parts must have identical lengths and center hole depths if the tapers are to be the same.

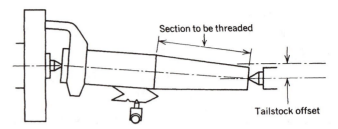

FIGURE E12.16 Adjusting the threading tool for cutting tapered threads. The tool is set square to the centerline of the work rather than the taper.

The Taper Attachment Method

The taper attachment features a slide independent to the ways that can be angled and will move the cross slide according to the angle set. Slight to fairly steep tapers ($3\frac{1}{2}$ in./ft) may be made, but length is limited to the stroke of the taper attachment. Centers may remain in line without distortion of the center holes. Work may be held in a chuck and both external and internal tapers may be made, often with the same setting for mating parts. Power feed is used. Taper attachments are graduated in taper per foot (TPF) or in degrees.

There are two types of taper attachments, the plain taper attachment (Figure E12.17) and the telescoping taper attachment (Figure E12.18). The crossfeed binding screw must be removed to free the nut when the plain attachment is set up. The depth of cut must then be made by using the compound feed screw handle. The cross feed may be used for depth of cut when using the telescoping taper attachment since the crossfeed binding screw is not disengaged with this attachment.

When a workpiece is to be duplicated or an internal taper is to be made for an existing external taper, it is often convenient to set up the taper attachment by using a dial indicator (Figure E12.19). The contact point of the dial indicator must be on the center of the workpiece. The workpiece is first set up in a chuck or between centers so that there is no runout when it is rotated. With the lathe spindle stopped, the indicator is moved from one end of the taper to the other. The taper attachment is adjusted until the indicator does not change reading when moved.

The angle, the taper per foot, or the taper per inch must be known to set up the taper attachment to cut specific tapers. If none of these are known, proceed as follows.

FIGURE E12.17 The plain taper attachment.

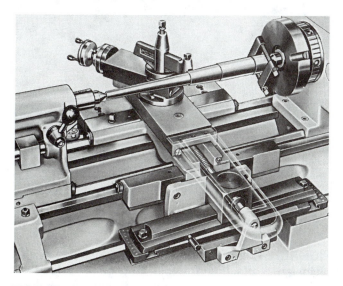

FIGURE E12.18 The telescopic taper attachment. (Clausing Machine Tools)

If the end diameters (*D* and *d*) and the length of taper (*L*) are given in inches:

$$\text{Taper per foot} = \frac{D - d}{L} \times 12$$

FIGURE E12.19 Adjusting the taper attachment to a given taper with a dial indicator.

If the taper per foot is given, but you want to know the amount of taper in inches for a given length, use the following formula:

$$\text{Amount of taper} = \frac{\text{TPF}}{12} \times \text{Given length of tapered part}$$

When the TPF is known, to find TPI divide the TPF by 12. When the TPI is known, to find TPF multiply the TPI by 12.

To set up the taper attachment (refer to Figure E12.20), proceed as follows:

Step 1. Clean and oil the slide bar (*a*).

Step 2. Set up the workpiece and the cutting tool on center. Bring the tool near the workpiece and to the center of the taper.

Step 3. Remove the crossfeed binding screw (*b*) that binds the crossfeed screw nut to the cross slide. *Do not remove* this screw if you are using a telescoping taper attachment. The screw is removed *only* on the plain type. Put a temporary plug in the hole to keep chips out.

Step 4. Loosen the lock screws (*c*) on both ends of the slide bar and adjust to the required degree of taper.

Step 5. Tighten the lock screws.

Step 6. Tighten the binding lever (*d*) on the slotted cross-slide extension at the sliding block, *plain type only.*

Step 7. Lock the clamp bracket (*e*) to the lathe bed.

Step 8. Move the carriage to the right so that the tool is from $\frac{1}{2}$ to $\frac{3}{4}$ in. past the start position. This should be done on every pass to remove any backlash in the taper attachment.

Step 9. Feed the tool in for the depth of the first cut with the cross slide unless you are using a plain

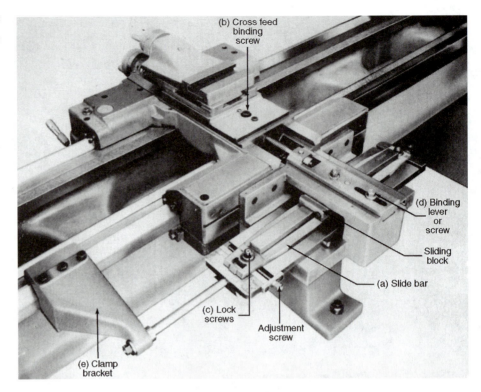

FIGURE E12.20 The parts of the taper attachment. (The Lodge & Shipley Company)

(b) Cross feed binding screw

(d) Binding lever or screw

Sliding block

(a) Slide bar

(c) Lock screws

Adjustment screw

(e) Clamp bracket

attachment. Use the compound slide for the plain type.

Step 10. Take a trial cut and check for diameter. Continue the roughing cut.

Step 11. Check the taper for fit and readjust the taper attachment, if necessary.

Step 12. Take a light cut, about .010 in., and check the taper again. If it is correct, complete the roughing and final finish cuts.

Internal tapers (Figure E12.21) are best made with the taper attachment. They are set up in the same manner as prescribed for external tapers.

Other Methods of Making Tapers

A tool may be set with a protractor to a given angle (Figures E12.22a and E12.22b) and a single plunge cut may be made to produce a taper. This method is often used for chamfering a workpiece to an angle, such as the chamfer used for hexagonal bolt heads and nuts. Tapered form tools sometimes are used to make V-shaped grooves. Only very short tapers can be made with form tools.

Tapered reamers are sometimes used to produce a specific taper such as a Morse taper. A roughing reamer is first used, followed by a finishing reamer. Finishing Morse taper reamers are often used to true up a badly nicked and scarred internal Morse taper.

Hole   Screw

FIGURE E12.21 Internal taper being made with a plain taper attachment. Note that the crossfeed nut locking screw has been removed and the hole has not been plugged. This hole *must* be plugged and the screw stored in a safe place.

METHODS OF MEASURING TAPERS

The most convenient and simple way of checking tapers is to use the taper plug gage (Figure E12.23) for internal tapers and the taper ring gage (Figure E12.24)

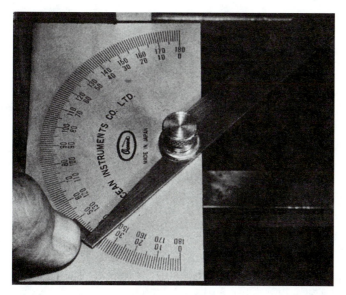

FIGURE E12.22a Tool is set up with protractor to make an accurate chamfer or taper.

FIGURE E12.23 Taper plug gage.

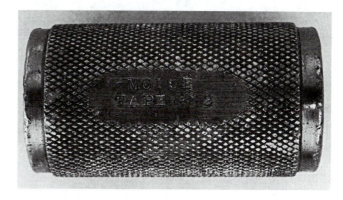

FIGURE E12.24 Taper ring gage.

FIGURE E12.22b Making the chamfer with a tool.

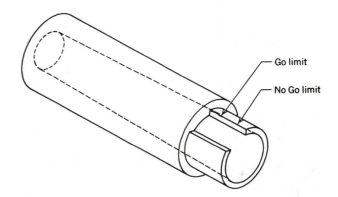

FIGURE E12.25 Go/no go taper ring gage.

for external tapers. Some taper gages have go and no go limit marks on them (Figure E12.25).

To check an internal taper, a chalk or Prussian blue mark is first made along the length of the taper plug gage (Figures E12.26a and E12.26b). The gage is then inserted into the internal taper and turned slightly. When the gage is taken out, the chalk mark will be partly rubbed off where contact was made. Adjustment of the taper should be made until the chalk mark is rubbed off along its full length of contact,

FIGURE E12.26a Chalk mark is made along a taper plug gage prior to checking an internal taper.

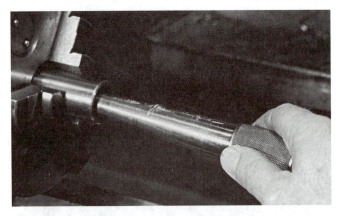

FIGURE E12.26b The taper has been tested and the chalk mark has been rubbed off evenly, indicating a good fit.

FIGURE E12.27b The ring gage is placed on the taper snugly and is rotated slightly.

FIGURE E12.27a The external taper is marked with chalk or Prussian blue before being checked with a taper ring gage.

FIGURE E12.27c The ring gage is removed and the chalk mark is rubbed off evenly for the entire length of the ring gage, which indicates a good fit.

indicating a good fit. An external taper is marked with chalk to be checked in the same way with a taper ring gage (Figures E12.27a to E12.27c).

The taper per inch may be checked with a micrometer by scribing two marks one inch apart on the taper and measuring the diameters (Figures E12.28a and E12.28b) at these marks. The difference is the taper per inch. A more precise way of making this measurement, in Figures E12.29a and E12.29b, is by using a surface plate with precision parallels and drill rods. However, the tapered workpiece has to be removed from the lathe.

Perhaps an even more precise method of measuring a taper is with the **sine bar** and gage blocks on the surface plate (Figure E12.30). When this is done, it is important to keep the centerline of the taper parallel to the sine bar and to read the indicator at the highest point.

FIGURE E12.28*a* Measuring the taper per inch (TPI) with a micrometer. The larger diameter is measured on the line with the edge of the spindle and the anvil of the micrometer contacting the line.

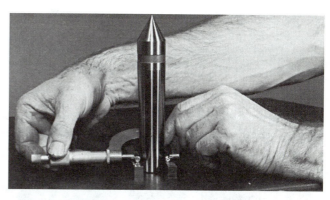

FIGURE E12.29*a* Checking the taper on a surface plate with precision parallels, drill rod, and micrometer. The first set of parallels is used so that the point of measurement is accessible to the micrometer. (Lane Community College)

FIGURE E12.28*b* The second measurement is taken on the smaller diameter at the edge of the line in the same manner. (Lane Community College)

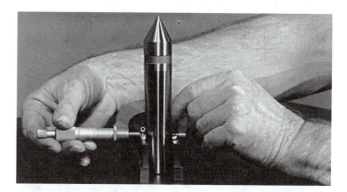

FIGURE E12.29*b* When one-inch-wide parallels are in place, a second measurement is taken. The difference is the taper per inch. (Lane Community College)

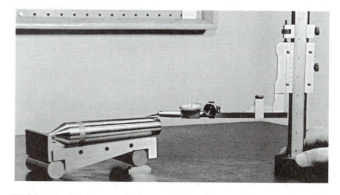

FIGURE E12.30 Using a sine bar and gage blocks with a dial indicator to measure a taper. (Lane Community College)

Tapers are also measured with taper micrometers. The outside taper micrometer (Figure E12.31) and inside taper micrometer (Figure E12.32) measure the taper directly and without removing the part from the lathe.

FIGURE E12.31 Outside taper micrometer. (Taper Micrometer Corporation)

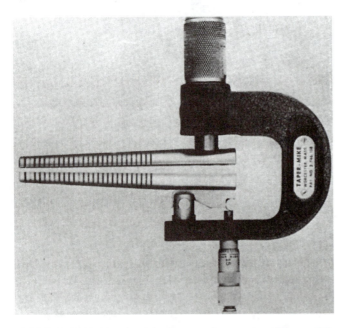

FIGURE E12.32 Inside taper micrometer. (Taper Micrometer Corporation)

SELF-TEST

1. State the difference in use between steep tapers and slight tapers.
2. In what three ways are tapers expressed (measured)?
3. Briefly describe the four methods of turning a taper in the lathe.
4. When a taper is produced by the compound slide method, is the reading in degrees on the compound swivel base the same as the angle of the finished workpiece? Explain.
5. If the swivel base is set to a 35-degree angle at the cross slide centerline index, what would the reading be at the lathe centerline index?
6. Calculate the offset for the taper shown in Figure E12.33. The formula is

$$\text{Offset} = \frac{L \times (D - d)}{2 \times L_1}$$

7. Name four methods of measuring the offset on the tailstock for making a taper.

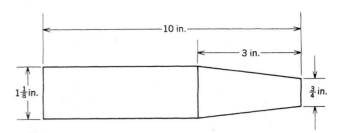

FIGURE E12.33

8. What are the two types of taper attachments and what are their advantages over other means of making a taper?
9. What is the most practical and convenient way to check internal and external tapers when they are in the lathe? Name four methods of measuring tapers.
10. Describe the kinds of tapers that may be made by using a form tool or the side of a tool.

Lathe Accessories

Many lathe operations would not be possible without the use of the steady and follower rests. These valuable attachments make internal and external machining operations on long workpieces possible on a lathe. Long, slender shafts cannot be turned or threaded even on CNC machines without steady or follower rests. This unit will familiarize you with these useful adjuncts to the engine lathe.

OBJECTIVES

After completing this unit, you should be able to:

1. Identify the parts and explain the uses of the steady rest.
2. Explain the correct uses of the follower rest.
3. Correctly set up a steady rest on a straight shaft.
4. Correctly set up a follower rest on a prepared shaft.

THE STEADY REST

On a lathe, long shafts tend to vibrate when cuts are made, leaving chatter marks. Even light finish cuts will often produce chatter when the shaft is long and slender. To help eliminate these problems, use a steady rest to support workpieces that extend from a chuck more than four or five diameters of the workpiece for turning, facing, drilling, and boring operations.

The steady rest (Figure E13.1) is made of a cast-iron or steel frame that is hinged so that it will open to accommodate workpieces. It has three or more adjustable jaws that are tipped with bronze, plastic, or ball bearing rollers. The base of the frame is machined to fit the ways of the lathe, and it is clamped to the bed by means of a bolt and crossbar.

A steady rest is also used to support long workpieces for various other machining operations such as threading, grooving, and knurling (Figure E13.2). Heavy cuts can be made by using one or more steady rests along a shaft.

Adjusting the Steady Rest

Workpieces should be mounted and centered in a chuck whether a tailstock center is used or not. If the shaft has centers and finished surfaces that

turn concentrically (have no runout) with the lathe centerline, setup of the steady rest is simple. The steady rest is slid to a convenient location on the shaft, which is supported in the dead center and chuck, and the base is clamped to the bed. The two lower jaws are brought up to the shaft finger tight only (Figure E13.3). A good high-pressure lubricant is applied to the shaft, and the top half of the steady rest is closed and clamped. The upper jaw is brought to the shaft finger tight, and then all three lock screws are tightened. Some clearance is necessary on the upper jaw to avoid scoring of the shaft. As the shaft warms or heats up from friction during machining, readjustment of the upper jaw is necessary.

A finished workpiece can be scored if any hardness or grit is present on the jaws. To protect finishes, brass or copper strips or abrasive cloth is often placed between the jaws and the workpiece. When using abrasive cloth, the abrasive side is placed outward against the jaws.

When there is no center in a finished shaft that is of uniform diameter, the setup procedure is as follows:

Step 1. Position the steady rest near the end of the shaft with the other end lightly chucked in a three- or four-jaw chuck.

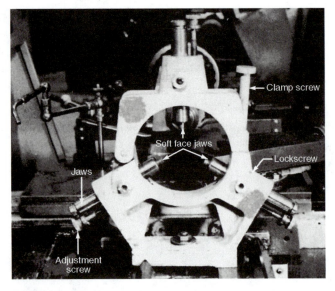

FIGURE E13.1   The parts of the steady rest.

FIGURE E13.3   Adjusting the steady rest jaws to a centered shaft.

FIGURE E13.2   A long, slender workpiece is supported by a steady rest near the center to limit vibration or chatter.

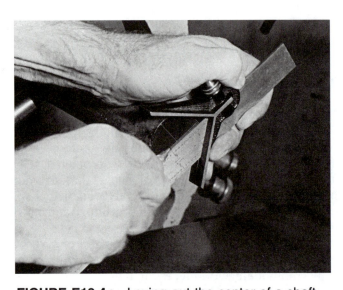

FIGURE E13.4a   Laying out the center of a shaft.

Step 2. Scribe two cross center lines with a center head on the end of the shaft and prick punch it (Figures E13.4a and E13.4b).

Step 3. Bring the dead center up near to the punch mark.

Step 4. Adjust the lower jaws of the steady rest so that the dead center is aligned to the center of the shaft.

Step 5. Tighten the chuck. If it is a four-jaw chuck, check for runout with a dial indicator.

Step 6. The steady rest may now be moved to any location along the shaft.

Stepped shafts may be set up by using a similar procedure, but the steady rest must remain on the diameter on which it is set up.

FIGURE E13.4b Aligning the shaft center with the dead center.

FIGURE E13.6 Using a cat head for supporting a square piece. Drilling and boring in the end of a heavy square bar requires the use of an external cat head.

Using the Steady Rest

A frequent misconception among students is that the steady rest may be set up properly by using a dial indicator near the steady rest on a rotating shaft. This procedure would never work since the indicator would show no offset or runout, no matter where the jaws were moved.

Steady rest jaws should never be used on rough surfaces. When a forging, casting, or hot-rolled bar must be placed in a steady rest, a concentric bearing with a good finish must be turned (Figure E13.5). Thick-walled tubing or other materials that tend to be out-of-round also should have bearing surfaces ma-

chined on them. The usual practice is to remove no more in diameter than necessary to clean up the bearing spot.

When the piece to be set up is very irregular, such as a square or hexagonal part, a cat head is used (Figure E13.6). The piece is placed in the cat head and the cat head is mounted in the steady rest while the other end of the workpiece is centered in the chuck. The workpiece is made to run true near the steady rest by adjusting screws on the cat head. In most cases the workpiece is given a center to provide more support for turning operations. A centered cat head (Figures E13.7a and E13.7b) is sometimes used when a permanent center is not required in the

FIGURE E13.5 Turning a concentric bearing surface on rough stock for the steady rest jaws.

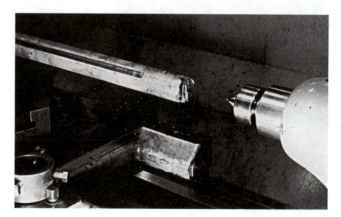

FIGURE E13.7a Using a centered cat head to provide a center when the end of the shaft or tube cannot be centered conveniently.

FIGURE E13.7b The cat head is adjusted over the irregular end of the shaft.

THE FOLLOWER REST

Long, slender shafts tend to spring away from the tool, vary in diameter, chatter, and often climb the cutting tool. To avoid these problems when machining a slender shaft along its entire length, a follower rest (Figure E13.9) is often used. Follower rests are bolted to the carriage and follow along with the tool. Most follower rests have two jaws placed so as to back up the work opposite to the tool thrust. Some types are made with different-size bushings to fit the work.

Using the Follower Rest

The workpiece should be 1 to 2 in. longer than the job requires to allow room for the follower rest jaws. The end is turned to be smaller than the finish size. The tool is adjusted ahead of the jaws about $1\frac{1}{2}$ in., and a trial cut of 2 or 3 in. is made with the jaws backed off. Then the lower jaw is adjusted finger tight (Figure E13.10) followed by the upper jaw. Both locking screws are tightened. A cutting oil should be used to lubricate the jaws.

workpiece. Internal cat heads (Figure E13.8) are used for truing to the inside diameter of tubing that has an irregular wall thickness so that a steady rest bearing spot can be machined on the outside diameter. These also have adjustment screws.

FIGURE E13.8 Tubing being set up with a cat head using a dial indicator to true the inside diameter.

FIGURE E13.9 A follower rest is used to turn this long shaft.

FIGURE E13.10 Adjusting the follower rest.

FIGURE E13.12 Both steady and follower rests being used.

The follower rest is often used when cutting threads on long, slender shafts, especially when cutting square or Acme threads (Figure E13.11). Burrs should be removed between passes to prevent them cutting into the jaws. Jaws with rolls are sometimes used for this purpose. On quite long shafts, sometimes both a steady rest and follower rest are used (Figure E13.12).

FIGURE E13.11 Long, slender Acme-threaded screw being machined with the aid of a follower rest.

SELF-TEST

1. When should a steady rest be used?
2. In what ways can a steady rest be useful?
3. How is the steady rest set up on a straight finished shaft when it has centers in the ends?
4. What precaution can be taken to prevent scoring of a finished shaft?
5. How can a steady rest be set up when there is no center hole in the shaft?
6. Is it possible to correctly set up a steady rest by using a dial indicator on the rotating shaft in order to watch for runout?
7. Should a steady rest be used on a rough surface? Explain.
8. How can a steady rest be used on an irregular surface such as square or hex stock?
9. When a long, slender shaft needs to be turned or threaded for its entire length, which lathe attachment could be used?
10. The jaws of the follower rest are usually one or two inches to the right of the tool on a setup. If the workpiece happens to be smaller than the dead center or tailstock spindle, how would it be possible to bring the tool to the end of the work to start a cut without interference by the follower rest jaws?

SECTION F
Milling Machines

Milling machines remove metal by means of rotating, multiple-point cutters, in contrast to the lathe, which removes metal from a rotating piece by means of a single-point cutter and which can only produce shapes that rotate about a single axis.

Both horizontal and vertical milling machines can produce a variety of shapes: grooves, squares, rectangular shapes, flat surfaces, and hollow pockets. However, generated forms that are very difficult and often impossible to be made on manual machines can be produced when these machines are computer controlled.

The vertical spindle milling machine has been one of the most versatile of machine tools and was readily adapted to CNC controls, some remaining in much the same configuration as the manual machine, while others were developed into vertical spindle machining centers.

The horizontal spindle milling machine also evolved into several kinds of machining centers. These CNC machines have, to a great extent, replaced many conventional manufacturing machines.

This section is designed to familiarize you with the principles and usage of horizontal and vertical milling machines.

Vertical Milling Machines

The vertical milling machine is a relatively recent development compared to the horizontal milling machine. The basic difference is that the spindle of the vertical milling machine is normally in a vertical position, and cutting tools are mounted at the end of the spindle. There is a similarity to the drill press in that the quill moves the spindle up and down, but there the similarity ends. The spindle and quill on the vertical milling machine are heavy duty and can usually be tilted or rotated in one or two axes in order to make angular cuts. The spindle has a power feed and, on some machines, the table is also equipped with power feed. These unique characteristics make the vertical milling machine one of the most versatile machines found in the shop. The knee and column principle has been adapted from the horizontal milling machine, which allows the milling table to be raised and lowered in relation to the spindle.

The transition from the vertical milling machine to the computer-operated vertical spindle machine was gradual. At first, the manually operated vertical mill was adapted for numerically controlled (NC) operation by using stepping motors and air cylinders controlled by punched tape. Then, similar vertical spindle machines were developed for computer numerically controlled (CNC) operations. (Figure F1.1).

OBJECTIVES

After completing this unit, you should be able to:

1. Identify safe vertical milling machine practices.
2. Identify the important parts of a vertical milling machine.
3. Identify and select from commonly used vertical milling machine cutting tools.

SAFE MILLING MACHINE PRACTICES

Aside from the safety instructions given earlier in this text—including proper dress, eye protection, jewelry removal, and alertness—there are a few safety instructions that pertain to the vertical milling machine. All machine guards should be in place prior to starting a machine. Observe other machines in operation around you to make sure they are guarded properly. Report any unsafe or missing guards to your supervisor.

Safety also involves keeping the machine and the area surrounding it clean. Any oil or coolant spills on the floor should be wiped up immediately to avoid slipping and falling. Chips should be swept up with a brush or broom and deposited in chip or trash containers. Do not handle chips with your bare hands or you will get cut. Dirty and oily rags should be kept in closed containers and should not accumulate in piles on the floor. Do not use an air hose to clean a machine. Flying chips can hurt you or those around you. When you lift heavy workpieces or machine attachments on or off the machine, ask someone to help you. When you are lifting anything, use proper lifting methods.

Be careful when handling tools and sharp-edged workpieces to avoid getting cut. Use a rag to protect your hand. Workpieces should be rigidly supported and tightly clamped to withstand the usually high cutting forces encountered in machining. When a workpiece comes loose while machining, it is usually ruined and often the cutter is ruined. The operator can also be hurt by flying particles from the cutter or workpiece.

The cutting tools should be securely fastened in the machine spindle to prevent any movement dur-

FIGURE F1.1 Vertical spindle CNC machine with tool changer. (Haas Automation Inc., Oxnard CA)

THE VERTICAL SPINDLE MILLING MACHINE

Identification

Knowing the names of milling machine parts and components is useful in locating trouble spots or in operating the machine controls. Figure F1.2 identifies many of the important parts of a light-duty vertical milling machine.

The column is the backbone of the machine; it rests on the base. The front or face of the column is accurately machined to provide a guide for the vertical travel of the knee. The top part of the column is machined to provide a swivel on which the ram can be rotated. The knee supports and guides the saddle. The saddle provides cross travel for the machine and is the support and guide for the table. The table provides lengthwise travel for the machine and supports the workpiece or workholding devices. The ram can be adjusted toward or away from the column to increase the working capacity of the milling machine. The toolhead on this machine

FIGURE F1.2 The important parts of a vertical milling machine.

ing the cutting operation. Cutting tools need to be operated at the correct revolutions per minute (RPM) and feedrate for any given material. Excessive speeds and feeds can break the cutting tools. On vertical milling machines, care has to be exercised when swiveling the workhead to make angular cuts. After loosening the clamping bolts that hold the workhead to the ram, retighten them lightly to create a slight drag. There should be enough friction between the workhead and ram that the head only swivels when pressure is applied to it. If the clamping bolts are completely loosened, the weight of the heavy spindle motor will flip the workhead upside down or until it hits the table, possibly injuring the operator's hand or a workpiece.

Measurements are frequently made during machining operations. Do not make any measurements until the spindle has come to a complete stop.

is attached to the end of the ram and can be swiveled on some milling machines in one or two planes. These six assemblies are the major components of the ram-style vertical milling machine.

Some of these components have controls or parts that are important for you to know. For instance, the toolhead (Figure F1.3) contains the motor, which powers the spindle. Speed changes are made with a variable-speed drive. On variable drives, the spindle has to be revolving while speed changes are made. When changing speeds to the high or low speed range, the spindle must be stopped. The same is true for V-belt or gear-driven speed changes on other milling machines. The spindle is contained in a quill. The quill can be extended from and retracted into the toolhead by a quill feed hand lever or handwheel. The quill feed hand lever is used to rapidly position the quill or for drilling holes. The quill feed handwheel gives a controlled slow manual feed as needed when boring holes.

Power feed to the quill is obtained by engaging the feed control lever. Different power feeds are available through the power feed change lever. The power feed is automatically disengaged when the quill dog contacts the adjustable micrometer depth stop (Figure F1.4). When feeding upward, the power feed disengages when the quill reaches its upper limit. The micrometer dial allows depth stop adjustments in .001 in. increments. The quill clamp is used to lock the quill in the head to get maximum rigidity when milling. The spindle brake or spindle lock is needed to keep the spindle from rotating when installing or removing tools from it. The toolhead is swiveled on the ram by loosening the clamping nuts on the toolhead and then turning the swivel adjustment until the desired angle is obtained.

The ram is adjusted toward or away from the column by the ram positioning pinion. The ram also can be swiveled on the column after the turret clamps are loosened.

The table can be moved manually with the table traverse handwheel. Table movement toward or away from the column is accomplished with the cross traverse handwheel. Raising and lowering the knee is done with the vertical traverse crank. Each of these three axes of travel can be adjusted in .001 in. increments with micrometer dials. The table, saddle, and knee can be locked securely in position with the

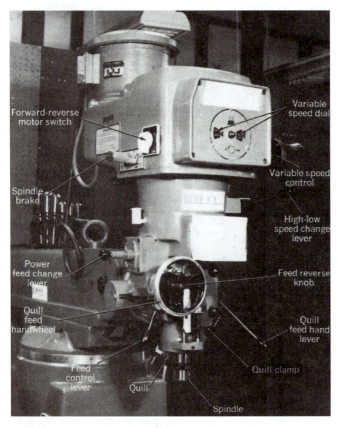

FIGURE F1.3   The toolhead.

FIGURE F1.4   Quill stop. (Lane Community College)

FIGURE F1.5 Clamping devices. (Lane Community College)

clamping levers shown in Figure F1.5. During machining, all table axes except the moving one should be locked. This will increase the rigidity of the setup.

Before operating any of the machine controls, the machine should be cleaned and lubricated. Lubrication is needed on all moving parts. Follow the machine manufacturer's recommendation as to the kind of lubricant required.

Cutting Tools for the Vertical Milling Machine

The most frequently used tool on a vertical milling machine is the end mill. End mills are made in either a right-hand or a left-hand cut. Identification is made by viewing the cutter from the cutting end. A right-hand cutter rotates counterclockwise. The helix of the flutes can also be left- or right-hand; a right-hand helix flute angles downward to the right when viewed from the side. Figure F1.6 is an illustration of the cutting end of a four-flute end mill, which is an example of a right-hand cut, right-hand helix end mill.

The end teeth of an end mill can vary, depending on the cutting to be performed (Figure F1.7). Two-flute end mills are center cutting, which means they can make their own starting hole. This is called *plunge cutting.* Four-flute end mills may have either center cutting teeth or a gashed or center-drilled end. End mills with center-drilled or gashed ends cannot be used to plunge cut their own starting holes. These

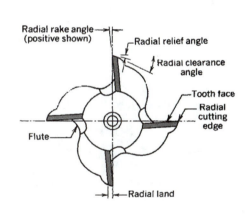

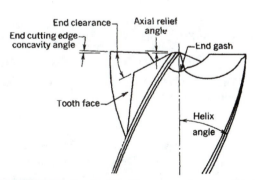

FIGURE F1.6 End mill nomenclature. (National Twist Drill & Tool Div., Lear Siegler, Inc.)

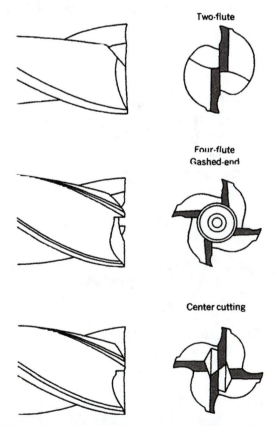

FIGURE F1.7 Types of end teeth on end mills. (National Twist Drill & Tool Div., Lear Siegler, Inc.)

end mills only cut with the teeth on their periphery. End mills can be single end (Figure F1.8) or double end (Figure F1.9). Double-end-type end mills are usually more economical because of the savings in tool material in their production.

End mills are manufactured with two, three, four, or more flutes and with straight flutes (Figure F1.10), and slow, regular, and fast helix angles. A slow helix is approximately 12 degrees, a regular helix is 30 degrees, and the fast helix is 40 degrees or more when measured from the cutter axis. Most general-purpose cutting is done with a regular helix angle cutter.

Aluminum is efficiently machined with a fast helix end mill and highly polished cutting faces to minimize chip adherence (Figure F1.11). The end mills illustrated so far are made from high-speed steel. High-speed steel end mills are available in a great variety of styles, shapes, and sizes as stock items. High-speed steel end mills are relatively low in cost when compared with carbide-tipped or solid-carbide end mills. But to machine highly abrasive or hard and tough materials, or in production milling, carbide tools should be considered.

Carbide tools may have carbide cutting tips brazed to a steel shank (Figure F1.12). This two-flute carbide-tipped end mill is designed to cut steel. It has a negative axial rake angle and a slow left-hand helix. Another kind of carbide-tipped end mill uses throwaway inserts (Figure F1.13). Each of the carbide inserts has three cutting edges; when all three cutting edges are dull, the insert is replaced with a new one. No sharpening is required.

When large amounts of material need to be removed, a roughing end mill (Figure F1.14) should be used. These end mills are also called *hogging end mills* and have a wavy thread form cut on their periphery. These waves form many individual cutting edges. The tip of each wave contacts the work and produces one short compact chip. Each succeeding

FIGURE F1.8 Single-end, helical-teeth end mill. (The Weldon Tool Company)

FIGURE F1.9 Two-flute, double-end, helical-teeth end mill. (The Weldon Tool Company)

FIGURE F1.10 Straight-tooth, single-end end mill. (The Weldon Tool Company)

FIGURE F1.11 Forty-five degree helix-angle, aluminum-cutting end mill. (The Weldon Tool Company)

FIGURE F1.12 Two-flute, carbide-tipped end mill. (Brown & Sharpe Mfg. Co.)

FIGURE F1.13 Inserted-blade end mill.

FIGURE F1.14 Roughing mill. (Illinois/Eclipse, Division Illinois Tool Works Inc.)

FIGURE F1.15 Three-flute tapered end mill. (The Weldon Tool Company)

wave tip is offset from the next one, which results in a relatively smooth surface finish. During the cutting operation, a number of teeth are in contact with the work. This reduces the possibility of vibration or chatter.

Tapered end mills (Figure F1.15) are used in mold making, die work, and pattern making, where precise tapered surfaces need to be made. Tapered end mills have included tapers ranging from 1 degree to over 10 degrees.

Ball-end end mills (Figure F1.16) have two or more flutes and form an inside radius or fillet between surfaces. Ball-end end mills are used in tracer milling and in die sinking operations. Round bottom grooves can also be machined with them. Precise convex radii can be machined on a milling machine with corner-rounding end mills (Figure F1.17). Dovetails are machined with single-angle milling cutters (Figure F1.18). The two commonly available angles are 45 degrees and 60 degrees. T-slots in machine tables and workholding devices are machined with T-slot cutters (Figure F1.19). T-slot cutters are made in sizes to fit standard T-nuts.

FIGURE F1.18 Single-angle milling cutter. (Illinois/Eclipse, Division Illinois Tool Works Inc.)

FIGURE F1.19 T-slot milling cutter. (The Weldon Tool Company)

FIGURE F1.16 Two-flute, single-end, ball end mill. (The Weldon Tool Company)

FIGURE F1.20 Woodruff keyslot milling cutter. (Illinois/Eclipse, Division Illinois Tool Works Inc.)

Woodruff key slots are cut into shafts to retain a woodruff key as a driving and connecting member between shafts and pulleys or gears. Woodruff keyslot cutters (Figure F1.20) come in many standard sizes. When larger flat surfaces need to be machined, a shell end mill (Figure F1.21) can be used. Shell end mills are more economical to produce because a smaller amount of the expensive tool material is needed to make a shell mill than for a solid shank end mill of the same size. Shell mills are also made with carbide in-

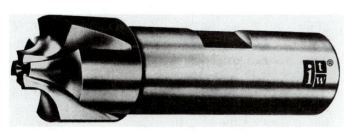

FIGURE F1.17 Corner-rounding milling cutter. (Illinois/Eclipse, Division Illinois Tool Works Inc.)

FIGURE F1.21 Shell end mill. (Illinois/Eclipse, Division Illinois Tool Works Inc.)

FIGURE F1.23 Shell mill arbor. (The Weldon Tool Company)

FIGURE F1.22 Shell mill with carbide inserts.

FIGURE F1.24 Fly cutter.

serts (Figure F1.22). The ease with which new sharp cutting edges can be installed makes this a very practical, efficient cutting tool. Shell end mills are mounted and driven by shell mill arbors (Figure F1.23). All the tools and toolholders will perform satisfactorily only if they are cleaned and inspected for nicks and burrs and corrections made before mounting them in the spindle.

The fly cutter (Figure F1.24) is an inexpensive face-milling tool. High-speed or carbide tool bits can be used if they are sharpened to have the correct clearance and rake angles.

All of these cutting tools must be mounted on and driven by the machine spindle. End mills or other straight shank tools can be held in collets. The most rigid type of collet is the solid collet (Figure F1.25). This collet has been precision ground with a hole that is concentric to the spindle and the exact size of the tool shank. Driving power to the tool is transmitted by one or two set screws engaging in flats on the tool shank.

Another type of frequently used collet is the split collet (Figure F1.26). The shank of the tool is held

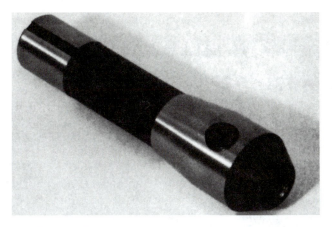

FIGURE F1.25 Solid collet.

FIGURE F1.26 Split collet.

FIGURE F1.27 Quick-change adapter and tool holders.

by the friction created when the tapered part of the collet is pulled into the taper of the spindle nose. When a heavy side thrust is created by a deep cut, a large feedrate, or a dull tool, helical-flute end mills have a tendency to be pulled out of the collet. With the solid collet, the set screws prevent any end movement of the cutting tool.

To speed up frequent tool changes, a quick-change toolholder (Figure F1.27) is used. Different tools can be mounted in their own holder, preset to a specific length, and interchanged with a partial turn of a clamping ring.

SELF-TEST

1. How are chips handled?
2. When is eye protection worn?
3. Why do machine guards need to be in place before operating a machine?
4. Why is a clean work area important?
5. What are danger points when handling tools?
6. How is a vertical milling machine head swiveled safely?
7. How are safe measurements made on a machine?
8. How is a right-hand-cut end mill identified?
9. What is characteristic of end mills that can be used for plunge cutting?
10. What is the main difference between a general-purpose end mill and one designed to cut aluminum?
11. When are carbide-tipped end mills chosen over high-speed steel end mills?
12. Name the six major components of a vertical milling machine.

Using the Vertical Milling Machine

Using the vertical milling machine involves the selection of cutting tools and workholding devices as well as cutting speed, feed, depth of cut, and coolant. Machine adjustments such as the alignment of the head and ram positions may also need to be made. The vertical milling machine is one of the most versatile machines found in a machine shop. This unit will illustrate some of the many possible vertical milling machine operations.

OBJECTIVES

After completing this unit, you should be able to:

1. Select cutting speeds and calculate RPM for end mills.
2. Select and calculate feedrates for end mills.
3. Use end mills to machine grooves and cavities.
4. Align vises and toolheads, locate the edges of workpieces, and find the centers of holes.
5. Identify and select vertical milling machine setups for a variety of different machining operations.

SPEED

One of the major requirements in efficient cutting with end mills and other cutting tools is the intelligent selection of the correct cutting speed. The cutting speed of a cutting tool is influenced by the tool material, condition of the machine, rigidity of the setup, and use of coolant. The most commonly used tool materials are high-speed steel and cemented carbide. After the cutting speed has been selected for a job, the cutting operation should be observed carefully so that speed adjustments can be made before a job is ruined. Table F2.1 gives starting values for some commonly used materials. As a rule, lower speeds are used for hard materials, tough materials, abrasive materials, heavy cuts, minimum tool wear, and maximum tool life. Higher speeds are used when machining softer materials, for better surface finishes, with smaller diameter cutters, for light cuts, for delicate workpieces, and nonrigid setups. When you calculate an RPM to use on a job, use the lower cutting speed value to start. The formula for this is

$$\text{RPM} = \frac{\text{CS} \times 4}{D}$$

D is the diameter of the cutting tool in inches.

FEED

Feed is another very important factor in efficient machining. The feed in milling is calculated by starting with a desired feed per tooth. The feed per tooth determines the chip thickness. The chip thickness affects the tool life of a cutter. Very thin chips dull the cutting edges very rapidly. Very thick chips produce high pressures on the cutting edges and cause a built-up edge (BUE) that produces rough finishes and leads to tool breakage. Commonly used feed-per-tooth values for end mills are given in Table F2.2. Usually the feed per tooth is the same for HSS and carbide end mills.

The values in Table F2.2 are only intended as starting points and may need to be adjusted up or down, depending on the machining conditions of the job at hand. The highest feed per tooth without BUE will usually give the longest tool life between sharpenings. Excessive feeds will cause tool breakage or the chipping of the cutting edges. When feed per tooth for a cutter is selected, the feedrate can be calculated. The feedrate on a milling machine is expressed in inches per minute (IPM); feed per tooth (f) in inches times the RPM times the number of teeth on the cutter (n). The formula for feedrate is

$$\text{Feedrate} = f \times \text{RPM} \times n$$

TABLE F2.1 Starting values for some commonly used materials (RPM)

Work Material	Tool Material	
	High-Speed Steel	Cemented Carbide
Aluminum	300–800	1000–2000
Brass	200–400	500–800
Bronze	65–130	200–400
Cast iron	50–80	250–350
Low-carbon steel	60–100	300–600
Medium-carbon steel	50–80	225–400
High-carbon steel	40–70	150–250
Medium-alloy steel	40–70	150–350
Stainless steel	30–80	100–300

EXAMPLE

Using the values given in Tables F2.1 and F2.2, calculate the RPM and feedrate for a $\frac{1}{2}$-in. diameter HSS two-flute end mill that is cutting aluminum.

$$\text{RPM} = \frac{\text{CS} \times 4}{D} = \frac{300 \times 4}{\frac{1}{2}} = \frac{1200}{0.5} = 2400$$

Feedrate = $f \times \text{RPM} \times n = .005 \times 2400 \times 2 = 24$ IPM

DEPTH OF CUT

The third factor to be considered in using end mills is the depth of cut. The depth of cut is limited by the amount of material that needs to be removed from the workpiece, by the power available at the machine spindle, and by the rigidity of the workpiece, tool, and setup. As a rule, the depth of cut in mild steel for an end mill should not exceed one-half of the diameter of the tool. But if deeper cuts need to be made, the feedrate needs to be reduced to prevent tool breakage. In softer metals such as aluminum, the depth of cut can be increased considerably, especially if the correct

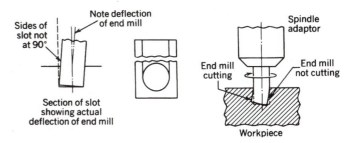

FIGURE F2.1 The causes of a leaning slot in end milling.

end mill is used for the material being cut. The end mill must be sharp and should run concentric in the end mill holder, and it should be mounted with no more tool overhang than necessary to do the job.

A problem that occasionally arises when using end mills to machine grooves or slots is a slot with non-perpendicular (leaning) sides. The causes for these out-of-square grooves are worn spindles, loose spindle bearings, dull end mills, and excessive feedrates. The leaning slot is produced by an end mill that is deflected by high cutting forces (Figure F2.1). To correct the problem, reduce the feedrate, use end mills with only a short projection from the spindle, and use end mills with straight or low-helix-angle flutes.

CUTTING FLUIDS

Cutting fluids should be used when high-speed steel cutters are used. The cutting fluid dissipates the heat generated while cutting by acting as a lubricant between the tool and chip. Higher cutting speeds can be used with cutting fluids. A stream of coolant also washes the chips away. Water-base coolants have very good cooling qualities and oil-base coolants produce very good surface finishes. Most milling with carbide cutters is done dry unless a large enough flow of coolant at the cutting edge can be maintained to

TABLE F2.2 Feeds for end mills (feed per tooth in inches)

Cutter Diameter	Aluminum	Brass	Bronze	Cast Iron	Low-Carbon Steel	High-Carbon Steel	Medium-Alloy Steel	Stainless Steel
$\frac{1}{8}$	.002	.001	.0005	.0005	.0005	.0005	.0005	.0005
$\frac{1}{4}$	.002	.002	.001	.001	.001	.001	.0005	.001
$\frac{3}{8}$	.003	.003	.002	.002	.002	.002	.001	.002
$\frac{1}{2}$	.005	.003	.003	.0025	.003	.002	.001	.002
$\frac{3}{4}$	.006	.004	.003	.003	.004	.003	.002	.003
1	.007	.005	.004	.0035	.005	.003	.003	.004
$1\frac{1}{2}$	.008	.005	.005	.004	.006	.004	.003	.004
2	.009	.006	.005	.005	.007	.004	.003	.005

keep the cutting edge from being intermittently heated and cooled, which usually results in **thermal cracking** and premature tool failure.

Some materials, such as cast iron, brass, and plastics, are commonly machined dry. A stream of compressed air can be used to cool tools and to keep the cutting area clear of chips, but precautions have to be taken to prevent flying chips from injuring anyone.

WORKHOLDING METHODS

The two most commonly used methods of holding workpieces for machining on the vertical milling machine are clamping the work to the machine table and using a machine vise. When a workpiece is fastened to the machine table, it must be aligned with the axis of the table. Milling machine tables are accurately machined and the table travels parallel to its outside surfaces and also parallel to its T-slots. Workpieces can be aligned by placing them against stops that fit snugly into the T-slots (Figure F2.2), or by measuring the distance from the edge of the table to the workpiece in a few places (Figure F2.3). More accurate alignments can be made when a dial indicator is used to indicate the edge of a workpiece (Figure F2.4). When a vise is used to hold the workpiece, the solid jaw of the vise should be indicated to insure its alignment with the axis of the table travel. See Unit 4 in this section, "Using the Horizontal Milling Machine," for explicit details on aligning machine vises.

FIGURE F2.2 Work aligned by locating against stops in T-slots.

FIGURE F2.3 Measuring the distance from the edge of the table to the workpiece.

FIGURE F2.4 Aligning a workpiece with the aid of a dial indicator.

ALIGNING THE TOOLHEAD

For precise machining operations, the toolhead needs to be aligned squarely to the top surface of the machine table. This is normally done when others have been using the machine before you, or you have a critical machining operation to perform. The following is the recommended procedure for aligning the toolhead:

Step 1. Fasten a dial indicator onto the machine spindle (Figure F2.5). The dial indicator should sweep a circle slightly smaller than the width of the table.

FIGURE F2.5 Aligning the toolhead square to the table with a dial indicator.

Step 2. Lower the quill until the indicator contact point is deflected .015 to .020 in. Lock the quill in this position.

Step 3. Tighten the knee clamping bolts. If this is neglected, the knee will sag on the front, causing an erroneous reading.

Step 4. Next, set the indicator bezel to read zero.

Step 5. Loosen the head clamping bolts one at a time and retighten each one to create a slight drag. This slight drag makes fine adjustments easier.

Step 6. Rotate the spindle by hand until the indicator is to the left of the spindle in the center of the table, and note the indicator reading.

Step 7. Rotate the spindle 180 degrees so that the indicator is to the right of the spindle, and note the indicator reading at that place. Be careful that the indicator contact does not catch and hang up when crossing the T-slots.

Step 8. Turn the head-tilting screw to correct the error between the right-hand reading and the left-hand reading. Tilt the head so that one-half the difference between these two readings is noted on the indicator.

Step 9. Check and compare the indicator reading at the left side of the table. If both readings are the same, tighten the head clamping bolts. If the readings differ, repeat Step 8.

Step 10. After the head clamping bolts are tight, make another comparison on both sides. Often, the tightening of the bolts changes the head location and additional adjustments need to be made.

Step 11. The next step is to align the head crosswise to the table (if the machine also tilts on that axis). The procedure is the same as for the lengthwise alignment. A final check should be made to make sure all clamping bolts are tight.

ALIGNING THE SPINDLE CENTERLINE TO THE WORK

When the workpiece edges are aligned parallel with the table travel and the toolhead is aligned square with the table top, it becomes necessary to locate the spindle centerline with the edges of the workpiece. A commonly used tool to locate edges on a milling machine is an offset edge finder (Figure F2.6). An offset edge finder consists of a shank and a tip that is held against the shank by an internal spring. The shank is usually $\frac{1}{2}$ in. in diameter, and the tip is either .200 or .500 in. in diameter. The edge finder is usually mounted in a collet when in use. With the spindle revolving at 600 to 800 RPM, the tip is moved off center so that it wobbles. An edge or locating point of the workpiece is then moved slowly toward the wobbling tip of the edge finder until it just touches (Figure F2.7). Continue to advance the work more slowly than before, gradually reducing the eccentric runout of the tip until it seems to turn without any wobble. Now, the spindle axis is exactly (that is, within one or two ten-thousandths of an inch) one-half of the tip diameter

FIGURE F2.6 Offset edge finder.

FIGURE F2.7 Work approaches the tip of the offset edge finder.

from the edge of the workpiece. If the tip diameter is .200 in., then the centerline of the spindle is .100 in. away from the workpiece edge. Set the micrometer dial of the now-adjusted machine axis .100 in. from zero, taking care that the backlash is removed in the direction you intend to move the work. You should practice the approach to the workpiece a few more times with the edge finder while observing the micrometer dial position until you feel secure in locating an edge with the edge finder. Also practice the edge-finding process for the other machine axis.

If an edge finder is not available in your shop, an edge can be located with the aid of a dial indicator in the following way. The dial indicator is mounted in the spindle. Rotate the spindle by hand and set the indicator contact point as close to the spindle centerline as possible. Lower the spindle so that the indicator contact point touches the workpiece edge and registers a .010- to .020-in. deflection (Figure F2.8). A slight rotating movement of the spindle forward and backward is used to find the lowest reading on the dial indicator. Set the dial indicator to register zero. Raise the spindle so that the indicator contact point is $\frac{1}{2}$ in. above the workpiece to clear it, and then turn the spindle 180 degrees from the original zeroed position. Hold a precision parallel against the edge of the workpiece so it extends above the workpiece. Lower the spindle until the indicator contact point is against the parallel (Figure

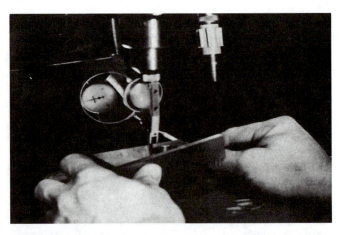

FIGURE F2.9 Indicator against parallel to locate the edge of a workpiece.

F2.9). Read the indicator value; it is easier if you use a mirror to read the indicator when it faces away from you. Now turn the table handwheel to move the table to where the indicator pointer is halfway between the present reading against the parallel and the zero on the indicator dial. Now set the dial on zero and check the position of the spindle, as in Figure F2.8. Continue to make corrections until both readings are the same.

To pick up the center of an existing hole, the indicator is mounted in the spindle and swiveled so that the contact point touches the side of the hole (Figure F2.10). The spindle should be rough centered, first in one table axis and then in the other. The spindle is then rotated by hand and the table adjustments are made until the same reading is obtained throughout a complete circle.

SETTING UP THE MACHINE

Step 1. Clamp the workpiece securely on the table, vise, or fixture.

Step 2. Mount the cutter in the spindle and set the correct RPM.

Step 3. Turn on the power to the spindle and check its rotation. If it is the wrong direction for the cutter, use the reverse rotation.

Step 4. Loosen the cross-slide lock and the knee locking lever.

Step 5. Bring the quill up into its housing and tighten the quill locking clamp. The quill should not be extended from its housing for rough milling work unless necessary, since it is less rigid when extended. Make sure the feed mechanism is disengaged.

Step 6. Position the table so that the cutter is directly over the workpiece.

FIGURE F2.8 Indicator used to locate the edge of a workpiece.

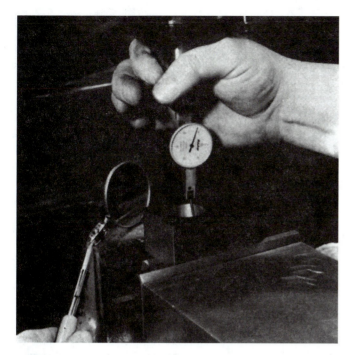

FIGURE F2.10 Dial indicator locating the center of a hole.

Step 7. Raise the knee slowly by turning the vertical hand feed crank until the cutter just touches the workpiece (in a place where metal will be removed so a finished surface will not be marred). If you cut below the surface, you have gone too far and should try again in another place.

Step 8. Set the micrometer dial on the knee feed screw to zero.

Step 9. Move the cutter clear of the work (unless you intend to plunge cut) and feed in the depth of cut on the micrometer dial. Remember, this should be no more than one-half the diameter of the end mill in steel.

Step 10. Tighten the knee and cross-slide locking levers. Always tighten **all** locks on a machine except the one that is being used in order to have a rigid setup and avoid chatter.

Step 11. Turn on the coolant.

Step 12. Set the table feed mechanism. If it is to be hand fed, observe the chip form, always producing a chip and backing off when no cutting is being done to avoid rubbing, which dulls the cutter. Avoid heavy feeds by hand since the erratic feedrate of hand feeding can cause momentary feedrates that are much too high, resulting in tool breakage.

Step 13. When the cut is completed, turn off the coolant and stop the spindle rotation before returning the table to the original position.

Step 14. Brush off the chips and wipe the workpiece clean.

Step 15. Make the necessary measurements before removing the workpiece from the machine. Additional cuts may be required or finish cuts may be needed.

VERTICAL MILLING MACHINE OPERATIONS

Many vertical milling machine operations such as the milling of steps are performed with end mills (Figure F2.11). Two surfaces can be machined in one setup, both square to each other. The ends of workpieces can be machined square and to a given length by using the peripheral teeth of an end mill (Figure F2.12). Center-cutting end mills make their own starting hole when used to mill a pocket or cavity (Figure F2.13).

Before making any milling cuts, the outline of the cavity should be accurately laid out on the workpiece as a guide or reference line. Only when finish cuts are made should these layout lines disappear. Good milling practice is to rough out the cavity to within .030 in. of finish size before making any finishing cuts.

When you are milling a cavity, the direction of the feed should be against the rotation of the cutter (Figure F2.14a). This assures positive control over the

FIGURE F2.11 Using an end mill to mill steps.

FIGURE F2.12 Using an end mill to square stock.

distance the cutter travels and prevents the workpiece from being pulled into the cutter because of backlash. When you reverse the direction of table travel, you will have to compensate for the backlash in the table feed mechanism. When doing conventional milling in a pocket, as in Figure F2.14*a*, an undercut in the corner is often made, especially when making heavy cuts. One way to overcome this problem is to take a light finishing cut by climb milling. However, the machine you are using must have an antibacklash device on the feed screws in order to use climb milling techniques. Climb milling without an antibacklash device will usually break cutters and damage the workpiece. Conventional and climb milling are illustrated in Figure F2.14*b*.

FIGURE F2.13 Using an end mill to machine a pocket.

During any milling operation, all table movements should be locked except the one that is being used in order to obtain the most rigid setup possible. Spiral (helical)-fluted end mills may work their way out of a split collet when deep, heavy cuts are made or when the end mill gets dull. As a precaution, to warn you that this is happening, you can make a mark with a felt-tip pen on the revolving end mill shank where it meets the collet face. Observing this mark during the cut will give you an early indication if the end mill is changing its position in the collet.

One often-performed operation with end mills is the cutting of **keyways** in shafts. It is very impor-

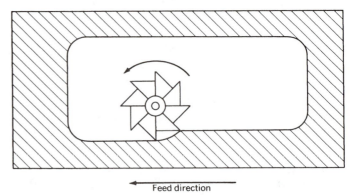
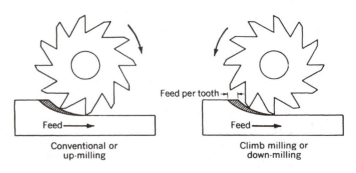

FIGURE F2.14*a* Feed direction is against cutter rotation when conventional milling procedures are used.

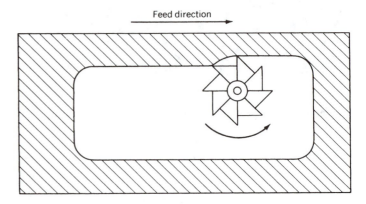

FIGURE F2.14*b* Conventional and climb milling.

tant that such a keyway be centrally located in the shaft. A very accurate method of doing this is by positioning the cutter with the help of the machine dials. After clamping the shaft in a vise, or possibly to the table, the quill is lowered so that the cutter is alongside the shaft but not touching it. Then, with the spindle motor off and the spindle rotated by hand, the table is moved until a paper feeler strip is pulled from your hand (Figure F2.15). At this point, the cutter is approximately .002 in. from the shaft. Zero the cross-slide micrometer dial to compensate for the .002 in. Raise the quill so that the cutter clears the workpiece. Move the cross slide a distance equal to half the shaft diameter plus half the cutter diameter. This will locate the cutter centrally over the shaft. Lock the cross-slide table at this position. Raise the quill to its top position and lock it there. Move the table lengthwise to position the cutter where the keyseat is to begin. Start the spindle motor and raise the knee until the cutter makes a circular mark equal to the cutter diameter (Figure F2.16a). Zero the vertical-travel micrometer dial. Turn on the coolant. Slowly move the knee upward a distance equal to half the cutter diameter plus an additional .005 in. Lock the knee. Cut the keyseat the required length (Figure F2.16b).

To machine a T-slot or a dovetail into a workpiece, two operations are performed. First, a slot is cut with a regular end mill, and then a T-slot cutter or a single-angle milling cutter is used to finish the contour (Figures F2.17 and F2.18). Angular cuts on workpieces can be made by tilting the workpiece in a vise with the aid of a protractor (Figure F2.19) and its built-in spirit level. The angle cut in Figure F2.19 can be made with an end mill, as shown in Figure F2.20. The workhead may also be tilted to

FIGURE F2.16*a*   Cutter centered on work and lowered to make a circular mark. The cutter was raised in order to show the spot.

FIGURE F2.16*b*   After plunging the cutter to the correct depth, table feed is engaged and the keyseat is cut.

FIGURE F2.15   Setting an end mill to the side of a shaft with the aid of a paper feeler.

produce an angle, as shown in Figure F2.21, but the length of cut is limited to the diameter of the cutter in this setup. To finish this angle, the tool must be raised, the table moved, and the cutter lowered to the line. However, in the tilted-head setup in Figure F2.22, an angle can be milled equal to the length of flutes on the cutter. A quite practical method of milling an angle can be seen in Figure F2.23, using a shell mill and making a series of cuts along the work to depth.

Accurate holes can be drilled at any angle the head can be swiveled to. These holes can be drilled by using the sensitive quill feed lever or the power feed mechanism (Figure F2.24) or, in the case of vertical holes, the knee can be raised.

Holes can be machine tapped by using the sensitive quill feed lever and the instant spindle reversal

FIGURE F2.17 Milling a slot and then the dovetail.

FIGURE F2.20 Machining an angle with an end mill.

FIGURE F2.18 First a slot is milled and then the T-slot cutter makes the T-slot.

FIGURE F2.21 Cutting an angle by tilting the work-head and using the end teeth of an end mill.

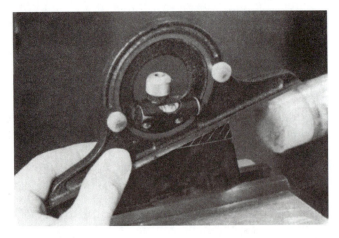

FIGURE F2.19 Setting up a workpiece for an angular cut with a protractor.

FIGURE F2.22 Cutting an angle by tilting the work-head and using the peripheral teeth of an end mill.

FIGURE F2.23 Using a shell mill to machine an angle.

FIGURE F2.25 Tapping in a vertical milling machine. (Cincinnati Milacron)

knob (Figure F2.25). When an offset boring head is mounted in the spindle, precisely located and accurately dimensioned holes can be bored (Figure F2.26). Circular slots can be milled when a rotary table is used (Figure F2.27). Precise indexing can be performed when a dividing head is mounted on the milling machine. Figure F2.28 shows a square being milled on the end of a shaft. On many vertical milling machines a shaping attachment is mounted on the rear of the ram. This shaping attachment can be brought over the machine table by swiveling the ram 180 degrees. Shaping attachments are used to machine irregular shapes on or in workpieces, such

FIGURE F2.24 Drilling of accurately located holes. (Cincinnati Milacron)

FIGURE F2.26 Boring with an offset boring head. (Cincinnati Milacron)

FIGURE F2.27 Using a rotary table to mill a circular slot. (Cincinnati Milacron)

FIGURE F2.29 A shaping head used to cut a square corner hole.

as the milled cavity being shaped to a square hole shown in Figure F2.29. When a right-angle milling attachment (Figure F2.30) is mounted on the spindle, it is possible to machine hard-to-get-at cavities that are often at difficult angles on workpieces.

FIGURE F2.28 A dividing head in use. (Cincinnati Milacron)

FIGURE F2.30 Using a right-angle milling attachment. (Cincinnati Milacron)

SELF-TEST

1. Name four factors that influence the cutting speed of milling cutters.
2. How is tool breakage caused by thick chips and high tool pressures?
3. The depth of cut can be varied according to the situation. As a rule of thumb, what should the depth be in mild steel for end milling?
4. With what tool material should coolant always be used (with a few exceptions)?
5. When setting up on a machine in which the previous operator used a vise and left it in place, should you assume the vise and toolhead are in perfect alignment?
6. When accurately setting up work in a vertical milling machine, the spindle centerline must usually be first aligned to the edge of the workpiece. What instrument is used for this purpose?
7. How can a cutter be aligned with the edge of a workpiece without any precision tools?
8. When milling a cavity or drilling precisely located holes where the direction of travel is reversed, what machine characteristic can cause considerable error and how can this problem be corrected?
9. How can angular cuts be taken on a vertical mill?
10. Name four attachments that can be used on vertical milling machines.

WORKSHEET: MAKING A KEYSEAT ON A SHAFT

Prior to starting each procedure for this project, study and complete the Post-Tests for:

1. Identifying cutters, collets, and machine parts: Section F, Unit 1.
2. Making the keyseat: Section F, Units 1 and 2, Vertical Milling Machines, and Using the Vertical Milling Machine.

Note: If you have already completed Post-Tests for any of these units, do not take the test again, but simply refer to the units to refresh your memory of the procedures. Also, your instructor may have an alternative keyseat for you to cut.

Objectives

1. Learn how to set up round stock on a milling machine table.
2. Be able to use an end mill to cut a keyseat.

Materials

5 inches of $1\frac{1}{2}$ in. dia. CR round stock and one $\frac{3}{8}$ in. two-flute, single end cutter, and two strap clamps.

Procedure for Making the Body

1. Place the piece of CR shafting near the center of the milling table on a middle groove. Mark the length of the keyseat to be cut with a scribe or grease pencil.
2. Using the two-strap clamps, one on each side of the shaft, clamp the shaft securely, leaving space clear where the keyseat will be cut.
3. Place the two-flute end mill in a collet (preferably a solid collet with a set screw; standard collets tend to loosen and allow the cutter to move out and gouge the work). Measure the $\frac{3}{8}$ in. cutter across the lands to make sure it is exactly .375 in.
4. Tighten the collet in place with the bolt on the top of the spindle. Tighten the set screw on the collet, making sure the set screw is seated on the flat on the cutter shank.
5. Set up the correct speed for a $\frac{3}{8}$ in. cutter.
6. If the machine is equipped with a feed mechanism, set the feed from Table F2.2 in Section F Unit 2. Remember, the table is for feed per tooth,

so you will have to double the amount in the table in this case. If there is no feed mechanism, then you will have to feed by hand, moving the table only enough to produce a chip, but not a thick one. Too much feed will break the cutter; too little feed will dull it.

7. Bring the cutter to the side of the workpiece. Using a strip of paper, turn the spindle by hand so the cutter flutes just graze the paper against the shaft. This locates the cutter within .003 in. of the edge.
8. Now zero the micrometer collar on the cross traverse handwheel and raise the quill so that the cutter can clear the top of the workpiece.
9. Move the cutter one-half the diameter of the shaft plus the radius of the cutter. This will place the cutter within two- or three-thousandths of an inch on the center of the shaft.
10. Turn on the machine and coolant (whether it be spray mist or from a coolant pump or with a pump can).
11. Just touch the cutter to the work near the outer end by bringing down the quill. Make sure to make a flat, just $\frac{3}{8}$ in. wide, but no deeper. This establishes the top of the keyseat, which will be one-half of the cutter width in depth. Lock the quill with the quill clamp.
12. Move the table so that the cutter clears the workpiece, and zero the micrometer dial on the vertical traverse. Then bring it up one-half the cutter diameter.
13. Lock the knee clamps.
14. With the coolant on, begin to make the cut to the prescribed length.
15. When the keyseat is finished, turn off the machine and coolant. Back out the cutter with the spindle stopped. Remove the chips with a paint brush and measure the width of the slot with the correct measuring instruments. It should be no more than two- or three-thousandths of an inch over the nominal size of .375 in.
16. Deburr the edges of the kcyseat and clean the workpiece.

Evaluation

Clean up the machine and show your work to your instructor for grading.

Horizontal Milling Machines

The horizontal milling machine serves the unique function of removing material from a workpiece with a rotating circular cutter. This metal-cutting system, unlike turning machines, makes possible the machining of flat surfaces, slots, grooves, gears, splines, and many other forms in metal and other materials. In addition, it provides a fairly high metal-removal rate as compared to other machine tools, while maintaining a relatively high level of precision. There are two basic types of horizontal milling machines found in the shop: plain and universal. The table of the plain horizontal machine (Figure F3.1) does not swivel as does the universal milling machine (Figure F3.2). The main difference between these machines is that the universal machine has an additional housing that swivels on the saddle and supports the table. This allows the table to be swiveled 45 degrees in either direction in a horizontal plane. The universal milling machine is specially designed to machine helical slots or grooves, as in twist drills and milling cutters. Other than these special applications, a universal mill and a plain milling machine can perform the same operations.

OBJECTIVES

After completing this unit, you should be able to:

1. Identify the important parts of the horizontal milling machine.
2. Identify various mounting devices used to drive milling cutters.
3. Identify by name and explain the application for a number of milling cutters.

Operators of machine tools should be familiar with the names and the location of machine tool components and parts. The major components of a horizontal milling machine are the column, knee, saddle, table, spindle, and overarm. Figure F3.2 shows a horizontal milling machine with the major parts identified.

The column is the main part of the milling machine. Its face is machined to provide an accurate guide for the up-and-down travel of the knee. The column also contains the main drive motor and the spindle. The spindle is hollow and holds and drives the various cutting tools, chucks, and arbors. The end facing the operator is called the *spindle nose;* it has a tapered hole with a standard milling machine taper, which is $3\frac{1}{2}$ in. per ft.

The overarm is mounted on top of the column and supports the arbor through an arbor support. The overarm slides in and out and can be clamped securely in any position. The knee can be moved vertically on the face of the column. The knee supports the saddle, which provides the sliding surface for the table. The saddle can move toward and away from the column to give crosswise movement of the machine table. The table provides the surface on which the workpieces are fastened. T-slots are machined along the length of the top surface of the table to align and hold fixtures and workpieces. Handwheels or hand cranks are used to manually position the table. Micrometer collars make possible positioning movements as small as .001 in.

Power feed levers control automatic feeds in three axes; the feedrate is adjusted by the feed change crank (Figure F3.3). On many milling machines, the power feed only operates when the spindle is turning. On others, however, the feed continues after the spindle is shut off if the motor switch is still on. This is an unsafe situation since the work can be forced into a cutter that is not turning, causing extensive damage. Two safety stops on each slide movement limit travel and thus prevent accidental damage to the feed mechanisms by providing automatic kickout of the power feed. In addition, two adjustable trip dogs for each axis allow the operator to preset specific power-feed-kickout travel distances. Rapid positioning of the table is accomplished with the rapid

FIGURE F3.1 Small, plain, automatic knee and column milling machine. Note that this machine has no provision for swiveling the table. (Cincinnati Milacron)

FIGURE F3.3 Feed change crank. (Cincinnati Milacron)

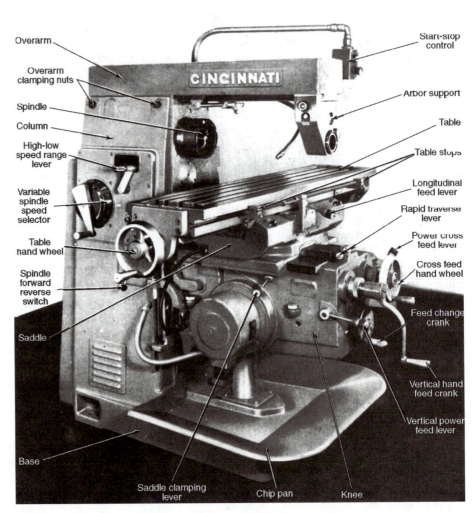

FIGURE F3.2 Universal horizontal milling machine. (Cincinnati Milacron)

traverse lever (Figure F3.4). The direction of the rapid advance is dependent on the position of the respective feed lever. When using rapid traverse, it is extremely important to see that the moving part is clear of obstructions to avoid damage to the machine and tooling.

Locking devices on the table, saddle, and knee are used to prevent unwanted movements in any or all of these axes. The locking devices should be released only in the axis in which power feed is used. The spindle can rotate clockwise or counterclockwise, depending on the position of the spindle forward-reverse switch. Spindle speeds are changed to a high or low range with the speed range lever (Figure F3.5). The variable spindle speed selector makes any spindle speed possible between the minimum and maximum RPM available in each speed range. The speed range lever has a neutral position between the high and low range. When the lever is in this neutral position, the spindle can easily be rotated by hand during machine setup.

The size of a horizontal milling machine is usually given as the range of movement possible and the power rating of the main drive motor of the machine. An example would be a milling machine with a 28-in. longitudinal travel, 10-in. cross travel, and 16-in. vertical travel with a 5 horsepower main drive motor. The larger the physical capacity of a machine, the

FIGURE F3.5 Speed change levers. (Cincinnati Milacron)

more power is available at the spindle through a larger motor.

Before any machine tool is operated, it should be lubricated. A good starting point is to wipe all sliding surfaces clean and apply a coat of a good way lubricant to them. Way lubricant is a specially formulated oil for sliding surfaces. Dirt, chips, and dust will act like a lapping compound between sliding members and cause excessive machine wear. Most machine tools have a lubrication chart that outlines the correct lubricants and lubrication procedures. When no lubrication chart is available, check all oil sight gages for the correct oil level and refill if necessary. Too much oil causes leakage. Lubrication should be performed progressively, starting at the top of the machine and working down. Machine points that are hand oiled should only receive a small amount of oil at any one time, but this should be repeated at regular intervals, at least daily. If the machine is equipped with a pump oiling system, a few pumps on the handle before starting the machine is sufficient in most cases.

Before operating a milling machine, you should be familiar with all control levers. Do not use force to engage or disengage controls or levers. Check that all operating levers are in the neutral position before the machine is turned on. Make sure that the locking levers are loosened on all moving slides. On a **variable-speed drive** milling machine, change spindle speeds only while the spindle motor is running. On

FIGURE F3.4 Rapid traverse lever. The table feed lever (top of picture) is in the left movement position. (Cincinnati Milacron)

FIGURE F3.7 CNC horizontal spindle machine (ghosted view). (Haas Automation Inc., Oxnard CA)

geared models, the spindle has to be stopped. Stop the spindle motor in either case when shifting from one range to another. All power feed levers should be in their neutral position before feed changes are made. Spindle rotation should be reversed only after the machine has come to a complete standstill.

FIGURE F3.6 CNC turning center (ghosted view). (Haas Automation Inc., Oxnard CA)

Horizontal spindle milling machines underwent a more drastic alteration than did vertical spindle machines when they became CNC machines (Figure F3.6). As extremely large machining centers, they have the capacity to remove great amounts of metal using face mills. Also, many have tool-changing devices that automatically select from many stored tools of differing configurations. Horizontal chucking machines are capable of producing uniform parts consistently from CNC programs (Figure F3.7).

SPINDLES AND ARBORS

Spindles

The spindle of the milling machine holds and drives milling cutters. Cutters can be mounted directly on the spindle nose, as with face milling cutters, or by means of arbors and adaptors that have tapered shanks that fit into the tapered hole or socket in the spindle nose.

Self-releasing tapers have a large included angle, generally over 15 degrees. This steep taper permits easy and quick removal of arbors from the spindle nose. Most manufacturers have adapted the standard national milling machine taper. This taper is $3\frac{1}{2}$ in. per ft (IPF), or about $16\frac{1}{2}$ degrees. National milling machine tapers are available in four standard sizes, numbered 30, 40, 50, and 60. Self-releasing taper-type shanks must be locked in the spindle socket with a draw-in bolt. Positive drive is obtained through two keys in the spindle nose that engage in keyways in the flange of arbors and adaptors. For most purposes these keys should **never** be removed.

Arbors

Two common arbor styles are shown in Figure F3.8. Style *A* arbor has a cylindrical pilot on the end opposite the shank. The pilot is used to support the free end of the arbor. Style *A* arbors are used mostly on small milling machines. They are also used on larger machines when a style *B* arbor support cannot be used because of a small-diameter cutter or interference between the arbor support and the workpiece.

Style *B* arbors are supported by one or more bearing collars and arbor supports. Style *B* arbors

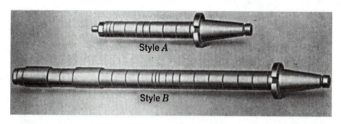

FIGURE F3.8 Arbors, styles *A* and *B*. (Cincinnati Milacron)

are used to obtain rigid setups in heavy-duty milling operations.

Style *C* arbors are also known as *shell end mill arbors* or as *stub arbors* (Figure F3.9). Shell end milling cutters are face milling cutters up to 6 in. in diameter. Because of their relatively small diameter, these cutters cannot be counterbored large enough to be mounted directly on the spindle nose, as face mills are, but are mounted on shell end mill arbors. Since face mills (Figure F3.10) are mounted directly on the spindle nose, they can be much larger than shell end mills and make heavier cuts, removing more metal.

The method by which arbors are mounted in the horizontal milling machine is illustrated in Figure F3.11. The draw-in bolt is screwed *by hand* into the arbor as far as it will go, then the arbor is pulled into the spindle nose by tightening the draw-in bar locknut. Note that the cutters are mounted close to the spindle and the first bearing support is close to the cutter. Cutters should always be mounted as near the bearing supports as possible to avoid chatter and to prevent them from springing the arbor.

Hardened collars are used to position cutters on a bar. Collars are manufactured to very close tolerances with their ends or faces being parallel and also square to the hole. It is very important that the collars and other parts fitting on the arbor are handled

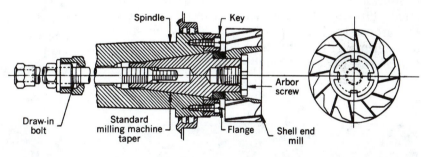

FIGURE F3.9 Style *C* arbor; shell end mill arbor. (Cincinnati Milacron)

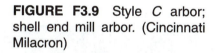

FIGURE F3.10 A face mill mounted on the spindle nose of a milling machine. (Cincinnati Milacron)

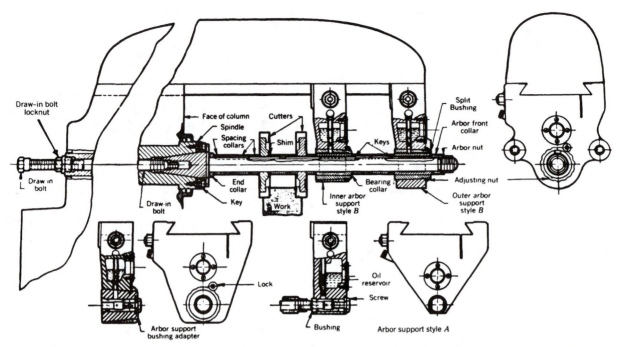

FIGURE F3.11 Section through arbor showing location of arbor collars, keys, bearing collars, and various arbor supports. (Cincinnati Milacron)

carefully to avoid damaging the collar faces. Any nicks, chips, or dirt between the collar faces will misalign the cutter or deflect the arbor and cause cutter runout. The arbor nut should be tightened or loosened only with the arbor support in place. Without the arbor support, the arbor can easily be sprung and permanently bent. Drive keys are used in a keyway in the arbor to drive the cutter.

Steps in Removing and Mounting an Arbor

Step 1. To remove an arbor, loosen the locknut on the draw-in bolt one full turn (Figure F3.12*a*).

Step 2. Tap the end of the draw-in bolt with a lead hammer. This releases the arbor shank from the spindle socket (Figure F3.12*b*).

FIGURE F3.12*a* Loosening the locknut on the draw-in bolt.

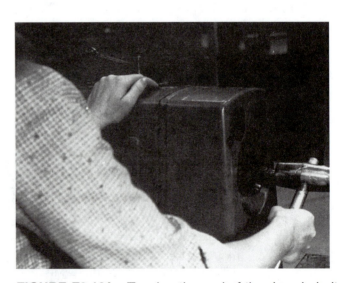

FIGURE F3.12*b* Tapping the end of the draw-in bolt with a lead hammer.

FIGURE F3.12*c* Have another person hold heavy arbors when you are loosening them.

FIGURE F3.12*e* Installing the arbor and tightening the draw-in bolt locknut.

FIGURE F3.12*d* Cleaning the external and internal taper prior to installation of the arbor.

FIGURE F3.12*f* Arbor storage.

Step 3. Some arbors are heavy; you may need someone to hold the arbor while you unscrew the draw-in bolt from the rear of the machine (Figure F3.12*c*).

Step 4. To install an arbor, the reverse process is used. Before inserting a tapered shank into the spindle socket, clean all mating parts and check for nicks and burrs, which should be removed with a honing stone (Figure F3.12*d*).

Step 5. When the arbor is in place in the spindle, screw the draw-in bolt in *by hand* as far as it will go, then tighten the locknut (Figure F3.12*e*).

Arbors should be stored in an upright position. Long arbors laying on their sides, if not properly supported, may bend (Figure F3.12*f*).

MILLING CUTTERS

Steps in Mounting a Cutter

Step 1. When changing cutters on an arbor, place the cutter and spacers on a smooth and clean area on the worktable to avoid damage to their accurate bearing or contact surfaces (Figure F3.13*a*).

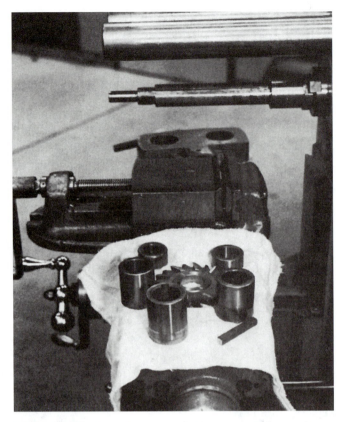

FIGURE F3.13a Parts should be placed on a clean area.

FIGURE F3.13b Stoning off nicks on the contact face of the spacers.

FIGURE F3.13c Stoning off nicks on the arbor.

Step 2. The cutter, spacing collars, and bearing collars should have a smooth, sliding fit on the arbor. Nicks should be stoned off. Clean all parts (Figures F3.13b and F3.13c).

Step 3. If possible, use an arbor length that does not give much arbor overhang beyond the outer arbor support. Arbor overhang may cause vibration and chatter. Mount cutters as close to the column as the work permits. Cutters are sharp; handle carefully with shop towels. Cutters may be assembled on the

FIGURE F3.13d Assembling the cutter on the arbor.

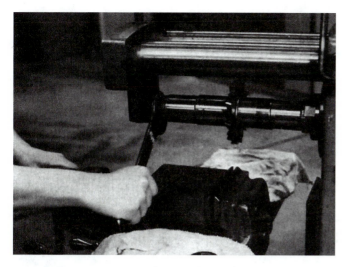

FIGURE F3.13*e* Tightening the arbor nut.

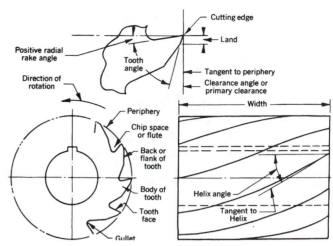

FIGURE F3.14 Nomenclature of plain milling cutter. (Cincinnati Milacron)

arbor when mounted in the machine or at the workbench (Figure F3.13*d*), but tighten the arbor nut only by hand.

Step 4. Tighten the arbor nut with a wrench that fits accurately *after* the arbor support is in place. Do not use a hammer to tighten the arbor nut. Overtightening will spring or bend the arbor. Never mount a dull cutter on an arbor. Dull cutters can produce poor results on the workpiece and may be ruined beyond repair if they are used in this condition (Figure F3.13*e*).

Arbor-Driven Milling Cutters

Milling cutters are the cutting tools of the milling machines. They are made in many different shapes and sizes. Each of these various cutters is designed for a specific application. A milling machine operator should be capable of selecting a cutting tool for the required application. One should be able to identify milling cutters by sight as well as know their capabilities and limitations. For the most part, milling cutters are made from solid high-speed steel, but large cutters are often made with inserted teeth of high-speed steel. Today, inserted cemented carbide cutters are more often used. Milling cutters can be divided into profile-sharpened cutters and form-relieved cutters. Profile-sharpened cutters are resharpened by grinding a narrow land (Figure F3.14) in back of the cutting edges. An example of this type is the plain milling cutter. Form-relieved cutters are resharpened by grinding the face of the tooth parallel to the axis of the cutter. This does not alter the form or shape of the cutter. An example of a form-relieved cutter is the concave or convex milling cutter.

Milling cutters are manufactured for either right-hand or left-hand helix. The hand of a milling cutter is determined by looking at the side of a cutter from the end of the table (Figure F3.15). If the flutes slope down and to the right, it is a right-hand helix; a left-hand helix slopes down and to the left. Cutters may be either right-hand or left-hand cut.

Plain Milling Cutters

Plain milling cutters are designed for milling plain surfaces where the width of the work is narrower than the cutter (Figure F3.16). Plain milling cutters less than $\frac{3}{4}$ in. wide have straight teeth. On straight-tooth cutters, the cutting edge will cut along its entire length at the same time. Cutting pressure increases until the chip is completed. At this time the sudden change in tooth load causes a shock that is transmitted through the drive and often leaves chatter marks or an unsatisfactory surface finish. Light-duty milling cutters have a large number of teeth,

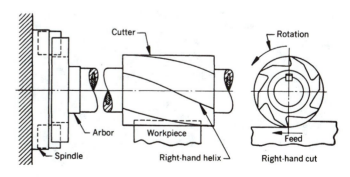

FIGURE F3.15 Plain milling cutter with right-hand helix and right-hand cut. (Cincinnati Milacron)

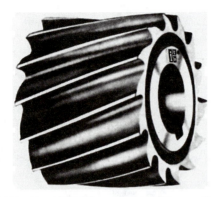

FIGURE F3.16 Light-duty plain milling cutter. (Illinois Eclipse, Division Illinois Tool Works Inc.)

FIGURE F3.18 Helical plain milling cutter. (Illinois/Eclipse, Division Illinois Tool Works Inc.)

which limits their use to light or finishing cut because of insufficient chip space for heavy cutting.

Heavy-duty plain mills (Figure F3.17) have fewer and much coarser teeth than light-duty mills, giving them stronger teeth with ample chip clearance. These are therefore used when heavy cutting with a high removal rate is desired. The helix angle of heavy-duty mills is about 45 degrees. The helical form enables each tooth to take a cut gradually, which reduces shock and lowers the tendency to chatter. Plain milling cutters are also called *slab mills*. Plain milling cutters with a helix angle over 45 degrees are known as *helical mills* (Figure F3.18). These milling cutters produce a smooth finish when used for light cuts or on intermittent surfaces. Plain milling cutters do not have side cutting teeth and should not be used to mill shoulders or steps on workpieces.

Side Milling Cutters

Side milling cutters are used to machine steps or grooves. These cutters are made from $\frac{1}{4}$ to 1 in. in width. Figure F3.19 shows a straight-tooth side

milling cutter. To cut deep slots or grooves, a staggered-tooth side milling cutter (Figure F3.20) is preferred because the alternate right-hand and left-hand helical teeth reduce chatter and give more chip space for higher speeds and feeds than is possible with straight-tooth side milling cutters. To cut slots over 1 in. wide, two or more side milling cutters may be mounted on the arbor simultaneously. Shims between the hubs of the side mills can be used to get any precise width cutter combination or to bring the cutter again to the original width after sharpening.

FIGURE F3.19 Side milling cutter. (Lane Community College)

FIGURE F3.17 Heavy-duty plain milling cutter. (Illinois/Eclipse, Division Illinois Tool Works Inc.)

FIGURE F3.20 Staggered-tooth milling cutter. (Illinois/Eclipse, Division Illinois Tool Works Inc.)

FIGURE F3.22 Plain metal slitting saw (Illinois Tool Works Inc.)

Half side milling cutters are designed for heavy-duty milling where only one side of the cutter is used (Figure F3.21). For straddle milling, a right-hand and a left-hand cutter combination is used.

Plain metal slitting saws are designed for slotting and cutoff operations (Figure F3.22). Their sides are slightly relieved or "dished" to prevent binding in a slot. Their use is limited to a relatively shallow depth of cut. These saws are made in widths from $\frac{1}{32}$ to $\frac{5}{16}$ in.

To cut deep slots or when many teeth are in contact with the work, a side-tooth metal slitting saw (Figure F3.23) will perform better than a plain metal slitting saw. These saws are made from $\frac{1}{16}$ to $\frac{3}{16}$ in. wide.

Extra-deep cuts can be made with a staggered-tooth metal slitting saw (Figure F3.24). Staggered-tooth saws have greater chip carrying capacity than other saw types. All metal slitting saws have a slight clearance ground on the sides toward the hole to prevent binding in the slot and scoring of the walls of the slot. Staggered-tooth saws are made from $\frac{3}{16}$ to $\frac{5}{16}$ in. wide.

Angular milling cutters are used for angular milling, such as cutting of dovetails, V-notches, and

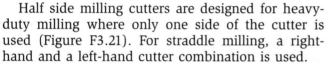

FIGURE F3.21 Half side milling cutter (Illinois Tool Works Inc.).

FIGURE F3.23 Side-tooth metal slitting saw (Illinois Tool Works Inc.)

FIGURE F3.24 Staggered-tooth metal slitting saw (Illinois Tool Works Inc.).

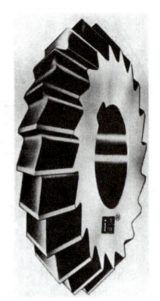

FIGURE F3.26 Double-angle milling cutter (Illinois Tool Works Inc.).

serrations. Single-angle cutters (Figure F3.25) form an included angle of 45 or 60 degrees, with one side of the angle at 90 degrees to the axis of the cutter.

Double-angle milling cutters (Figure F3.26) usually have an included angle of 45, 60, or 90 degrees. Angles other than those mentioned are special milling cutters.

Convex milling cutters (Figure F3.27) produce concave bottom grooves, or they can be used to make a radius in an inside corner. Concave milling cutters (Figure F3.28) make convex surfaces. Corner-rounding milling cutters (Figure F3.29) make rounded corners. The cutters illustrated in Figures F3.27 to F3.30 are form-relieved cutters.

Involute gear cutters (Figure F3.30) are commonly available in a set of eight cutters for a given pitch, depending on the number of teeth for which the cutter is to be used. The ranges for the individual cutters are given in Table F3.1. These eight cutters are designed so that their forms are correct for

FIGURE F3.25 Single-angle milling cutter (Illinois Tool Works Inc.).

FIGURE F3.27 Convex milling cutter (Illinois Tool Works Inc.).

FIGURE F3.28 Concave milling cutter (Illinois Tool Works Inc.).

FIGURE F3.29 Corner-rounding milling cutter (Illinois Tool Works Inc.)

FIGURE F3.30 Involute gear cutter. (Lane Community College)

TABLE F3.1 Gear cutters

Number of Cutter	Range of Teeth
1	135 to rack
2	55 to 134
3	35 to 54
4	26 to 34
5	21 to 25
6	17 to 20
7	14 to 16
8	12 and 13

FIGURE F3.31 Inserted-tooth carbide cutter. (Lovejoy Tool Company Inc. "Milling Cutters with Indexable Blades" Lovejoy 1993 SPC34315)

the lowest number of teeth in each range. If an accurate tooth form near the upper end of a range is required, a special cutter is needed.

The high-speed milling cutters shown here are not much used in CNC machines. Many different styles of carbide, inserted-tooth cutters are used. Figure F3.31 is an example of an inserted-tooth carbide cutter that is used for manually operated horizontal milling machines.

SELF-TEST

1. What are the two basic kinds of milling cutters with reference to their tooth shape?
2. What is the difference between a light-duty and a heavy-duty plain milling cutter?
3. Why are plain milling cutters not used to mill steps or grooves?
4. What kind of cutter is used to mill grooves?

5. How does the cutting action of a straight-tooth side milling cutter differ from a staggered-tooth side mill?
6. When are metal slitting saws used?
7. Give two examples of form-relieved milling cutters.
8. When are angular milling cutters used?
9. What kind of cutters are mounted directly on the spindle nose?
10. Milling machine spindle sockets have two classes of taper. What are they?
11. What is the amount of taper on a national milling machine taper?
12. Why is it important to carefully clean the socket, shank, and arbor spacers prior to mounting them on a milling machine?
13. When is a style A arbor used?
14. Where should the arbor supports be in relation to the cutter?
15. What is a style C arbor?
16. Why should the arbor support be in place before the arbor nut is tightened or loosened?

Using the Horizontal Milling Machine

When a student or apprentice begins to use the horizontal milling machine, there are several things to consider before turning on the power and starting a cut. The first consideration is that of personal safety, and the second is care of the machine so it will not be damaged. It also involves the selection and alignment of a workholding device, milling cutter, speed, feed, and depth of cut.

OBJECTIVES

After completing this unit, you should be able to:

1. Observe safety precautions on a horizontal milling machine.
2. Align workholding devices.
3. Mill flat surfaces, grooves, and slots.
4. Mill surfaces square to each other.

HORIZONTAL MILLING MACHINE SAFETY

As mentioned earlier in Section A, loose-fitting clothing may catch in rotating machinery and pull the operator into it. Rings, bracelets, and watches can be very dangerous because they may catch in a cutter and tear off a finger or pull your hand or arm into the cutter, with obvious consequences. Long or loose hair can be caught in a cutter or even be wrapped around a rotating smooth shaft, resulting in a quick, painful scalping. Persons with long hair should wear a cap or hairnet in the machine shop.

A milling machine should not be operated while wearing gloves because of the danger of getting caught in the machine. When gloves are needed to handle sharp-edged materials, the machine should be stopped. Eye protection should be worn at all times in machine shops. Eye injuries can be caused by flying chips, tool breakage, or cutting fluid sprays. Keep your fingers away from the moving parts, such as the cutter, gears, spindle, or arbor. Do not reach over a rotating spindle or cutter.

Safe operation of a machine tool requires that you think before you do something. Before starting up a machine, know the location and operation of its controls. Operate all controls on the machine yourself; do not have another person start or stop the machine for you. While operating a milling machine, observe the cutting action at all times so that you can stop the machine immediately when you see or hear something unfamiliar. *Always stay within reach of the controls while the machine is running.* An unexpected emergency may require quick action on your part. *Never* leave a running machine unattended.

Before operating the rapid traverse control on a milling machine, loosen the locking devices on the machine axis to be moved. Check that the handwheels or hand cranks are disengaged, or they will spin and injure anyone near them when the rapid traverse is engaged. The rapid traverse control will move any machine axis that has its feed lever engaged. Do not try to position a workpiece too close to the cutter with rapid traverse, but approach the final two inches by using the handwheels or hand cranks.

A person concentrating on a machining operation should not be approached quietly from behind, since it may annoy and alarm the person, thus causing a ruined workpiece or injury. Do not lean on a running machine; moving parts can hurt you. Signs posted on a machine indicating a dangerous condition or a repair in progress should only be removed by the person making the repair or by a supervisor.

Measurements should only be taken on a milling machine after the cutter has stopped rotating and after the chips have been cleared away. Milling machine chips are dangerously sharp and often hot and contaminated with cutting fluids. They should not be handled with bare hands. Chips should be removed with a brush. Compressed air should not be used to clean off chips from a machine because it will make small missiles out of chips that can injure a person at quite a distance away. A blast of air will also force small chips into the ways and sliding surfaces of the milling machine where they will cause scoring and rapid premature wear. Cleaning chips and cutting fluids from the machine or workpiece should only be done after the cutter has stopped turning. Keep the area around the machine clean of chips, oil spills, cutting fluids, and other obstructions to prevent the operator from slipping or stumbling. Wash your hands and arms thoroughly after being splashed with cutting fluid to prevent getting dermatitis, a skin disease.

Injuries can be caused by improper setups or the use of wrong tools. Use the correct size wrench when loosening or tightening nuts or bolts, preferably a box wrench or a socket wrench. An oversized wrench will round off the corners on bolts and nuts and prevent sufficient tightening or loosening; a slipping wrench can cause mashed fingers or other injuries to the hands or arms. Milling machine cutters have very sharp cutting edges; handling cutters carefully and with a cloth will avoid cuts on the hands.

SETTING SPEEDS AND FEEDS ON THE HORIZONTAL MILLING MACHINE

Speeds

To get maximum use from a milling cutter and to avoid damaging it, it is important that it is operated at the correct cutting speed. Cutting speed is expressed in surface feet per minute, which varies for different work materials and cutting tool materials. The cutting speed of a milling cutter is the distance the cutting edge of a cutter tooth travels in 1 minute. Table F4.1 is a table of cutting speeds for a number of commonly used materials.

The cutting speeds given in Table F4.1 are only intended to be starting points. These speeds represent experience in instructional settings where cutter and machine conditions are often less than ideal. Different hardnesses within each materials group account for the wide range of cutting speeds. Generally,

TABLE F4.1 Cutting speeds for milling

Material	Cutting Speed SFM	
	High-Speed Steel Cutter	Carbide Cutter
Free-machining steel	100–150	400–600
Low-carbon steel	60–90	300–550
Medium-carbon steel	50–80	225–400
High-carbon steel	40–70	150–250
Medium-alloy steel	40–70	150–350
Stainless steel	30–80	100–300
Gray cast iron	50–80	250–350
Bronze	65–130	200–400
Aluminum	300–800	1000–2000

speeds should be lower for hard materials, abrasive materials, and deep cuts. Speeds are higher for soft materials, better finishes, light cuts, delicate workpieces, and light setups. Cutting edges will dull rapidly if the cutting speed is too high for the material being machined. Excess speed can destroy an expensive cutter in a few seconds by burning the cutting edges beyond saving by resharpening. If the cutting speed is too slow, cutting action will be inefficient, causing a relatively slow removal rate; also, a built-up edge will form on the cutting edge, which increases tool pressure and often results in tool breakage.

It is a good practice to use the lower cutting speed value as it is given in Table F4.1 to begin with, and then, if the setup allows, increase it toward the higher speed until chatter develops or the coolant begins to vaporize excessively, indicating heat buildup. To use these cutting speed values on a milling machine, they have to be expressed in RPM. The formula used to convert cutting speed into RPM is

$$RPM = \frac{CS \times 4}{D}$$

CS = cutting speed found in Table F4.1
4 = constant
D = diameter of cutter in inches

EXAMPLE

Calculate the RPM for a 3-in. diameter high-speed steel cutter to be used on cast iron.

$$RPM = \frac{CS \times 4}{D}$$

$$RPM = \frac{50 \times 4}{3} = \frac{200}{3} = 67$$

Tool materials other than high-speed steels, such as cast alloys or cemented carbides, can be used at higher cutting speeds because they retain a sharp cutting edge at elevated temperatures. As a rule, if we consider high-speed steel cutting speeds as 100 percent, then the cutting speed for cast alloys can be 150 percent and cemented carbides can have a cutting speed rated between 200 to 600 percent. Determining the speeds and feeds accurately for milling is complicated by the fact that the cutting edge does not remain in the work continuously and that the chip being made varies in thickness during the cutting period. The type of milling being done (slab, face, or end milling) also makes a difference, as does the way that the heat is transferred from the cutting edge.

Feeds

Only secondary to speed in efficient milling machine operation is the feed. It is expressed as a feedrate and given in inches per minute (IPM). Most milling machines have two separate drive motors: one to power the spindle and one to power the feed mechanism. Independent changes in feedrate and spindle speed are made possible by having two motors.

The feedrate is the product of the feed per tooth times the number of teeth on the cutter times the RPM of the spindle. You have already determined how to calculate the RPM of a cutter; by counting the number of teeth in a cutter and knowing the feed per tooth, you can determine the feedrate.

Table F4.2 is a chart of commonly used feeds per tooth (**FPT**). As you can see in Table F4.2, there is only a slight difference in the feed per tooth allowance between high-speed steel cutters and carbide cutters.

EXAMPLE

Calculate the feedrate for a 3-in. diameter six-tooth helical mill that is cutting free-machining steel, first for a high-speed steel tool and then for a carbide tool. The formula is:

$$\text{FPT} \times N \times \text{RPM} = \text{feedrate in inches per minute (IPM)}$$
$$\text{FPT} = \text{feed per tooth}$$
$$N = \text{number of teeth on cutter}$$
$$\text{RPM} = \text{revolutions per minute of cutter}$$

For a starting point, calculate the RPM (refer to Table F4.1):

$$\text{RPM} = \frac{\text{CS} \times 4}{D} = \frac{100 \times 4}{3} = \frac{400}{3} = 134 \text{ RPM}$$

Table F4.2 gives an FPT of .002 in. The cutter has six teeth. The feedrate is

$$.002 \times 6 \times 134 = 1.608 \text{ IPM}$$

for the high-speed steel cutter.

TABLE F4.2   Feed in inches per tooth (instructional setting)

Type of Cutter	Aluminum		Bronze		Cast Iron		Free-Machining Steel		Alloy Steel	
	HSS	Carbide	HSS	Carbide	HSS	Carbide	HSS	Carbide	HSS	Carbide
Face mills	.007 to .022	.007 to .020	.005 to .014	.004 to .012	.004 to .016	.006 to .020	.003 to .012	.004 to .016	.002 to .008	.003 to .014
Helical mills	.006 to .018	.006 to .016	.003 to .011	.003 to .010	.004 to .013	.004 to .016	.002 to .010	.003 to .013	.002 to .007	.003 to .001
Side cutting mills	.004 to .013	.004 to .012	.003 to .008	.003 to .007	.002 to .009	.003 to .012	.002 to .007	.003 to .009	.001 to .005	.002 to .008
End mills	.003 to .011	.003 to .010	.003 to .007	.002 to .006	.002 to .008	.003 to .010	.001 to .006	.002 to .008	.001 to .004	.002 to .007
Form-relieved cutters	.002 to .007	.002 to .006	.001 to .004	.001 to .004	.001 to .005	.002 to .006	.001 to .004	.002 to .005	.001 to .003	.001 to .004
Circular saws	.002 to .005	.002 to .005	.001 to .003	.001 to .003	.001 to .004	.002 to .006	.001 to .003	.001 to .004	.005 to .002	.001 to .004

Note: The feedrate should be adjusted according to the capability of the setup. Lower feedrates should be used at first, then adjusting to higher feedrates as the situation permits.

EXAMPLE

For a carbide cutter the RPM is

$$\text{RPM} = \frac{\text{CS} \times 4}{D} = \frac{400 \times 4}{3} = \frac{1600}{3} = 534 \text{ RPM}$$

The FPT is .003 in. The cutter has six teeth. The feedrate is

$$.003 \times 6 \times 534 = 9.612 \text{ IPM}$$

for the carbide cutter.

To calculate the starting feedrate, use the lower figure from the feed-per-tooth table and, if conditions permit, increase the feedrate from there. The most economical cutting takes place when the greatest amount (expressed in cubic inches of metal per minute) is removed and a long tool life is obtained. The tool life is longest when a relatively low speed and high feedrate is used. Try to avoid feedrates of less than .001 in. per tooth, because this will cause rapid dulling of the cutter. An exception to this limit is the use of small-diameter end mills on harder materials. The depth and width of cut also affect the feedrate. Wide and deep cuts require a smaller feedrate than do shallow, narrow cuts. Roughing cuts are made to remove material rapidly. The depth of cut may be $\frac{1}{8}$ in. or more, depending on the rigidity of the machine, the setup, and the horsepower available. Finishing cuts are made to produce precise dimensions and acceptable surface finishes. The depth of cut for a finishing cut should be between .015 and .030 in. A depth of cut of .005 in. or less will cause the cutter to rub instead of cut, which also results in excessive cutting edge wear.

Cutting fluids should be used when machining most metals with high-speed steel cutters on a milling machine. A cutting fluid cools the tool and the workpiece. The lubricating action of the coolant reduces friction between the tool face and chip. Cutting fluids prevent rust and corrosion and, if applied in sufficient quantity, will flush away chips. Because of these characteristics and the use of higher cutting speeds, increased production is made possible through the use of good cutting fluids.

Most milling with carbide cutters is done dry unless a large constant flow of cutting fluid can be directed at the cutting edge. An interrupted coolant flow on a carbide tool causes thermal cracking and results in subsequent chipping of the tool.

WORKHOLDING DEVICES ON THE MILLING MACHINE

The Machine Vise

Probably the most common method of workholding on a milling machine is a vise. Vises are simple to use and can quickly be adjusted to the size of the workpiece. A vise should be used to hold work with parallel sides if it is within the size limits of the vise, because it is the quickest and most economical workholding method. The plain vise (Figure F4.1) is bolted to the machine table. Alignment with the table is provided by two slots at right angles to each other on the underside of the vise. These slots are fitted with removable keys that align the vise with the table T-slots either lengthwise or crosswise. A plain vise can be converted to a swivel vise (Figure F4.2) by mounting it on a swivel plate. The swivel plate is

FIGURE F4.1 Plain vise. (Cincinnati Milacron)

FIGURE F4.2 Swivel vise. (Cincinnati Milacron)

graduated in degrees. This allows the upper section to be swiveled to any angle in the horizontal plane. When swivel bases are added to a plain vise, the versatility increases, but rigidity is lessened.

For work involving compound angles, a universal vise (Figure F4.3) is used. This vise can be swiveled 90 degrees in the vertical plane and 360 degrees in the horizontal plane. The most rigid setup is the one where the workpiece is clamped close to the table surface.

Air or hydraulically operated vises are often used in production work; but in general toolroom work, vises are opened and closed by cranks or levers. To hold workpieces securely without slipping under high cutting forces, a vise must be tightened by striking the crank with a lead hammer (Figure F4.4).

Whenever possible, position the vise so that the cutting pressure will be against the solid jaw (Figure F4.5). Often references are made to the "solid jaw" of a vise (for example, "alignment is made on the solid jaw"). The solid jaw will not move or change when the vise is tightened, although the movable jaw will align itself to some degree with the work. The vise should never be aligned with a dial indicator on the movable jaw since it can move on its guide a few thousandths of an inch.

A good machine vise is an accurate and dependable workholding device. When milling only the top of a workpiece, it is not necessary that the vise be square to the column or parallel to the table travel. However, when the job requires that the outside surface be parallel to a step or groove in the workpiece, the vise must be precisely aligned and positioned on the table.

The base of the vise should be located with keys (Figure F4.6) that fit snugly into the T-slots on the

FIGURE F4.5　Cutting pressure against solid jaw.

FIGURE F4.3　Universal vise. (Cincinnati Milacron)

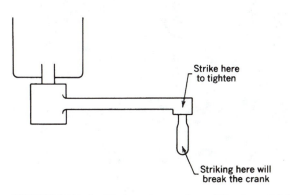

Strike here to tighten

Striking here will break the crank

FIGURE F4.4　Tightening a vise. (Cincinnati Milacron)

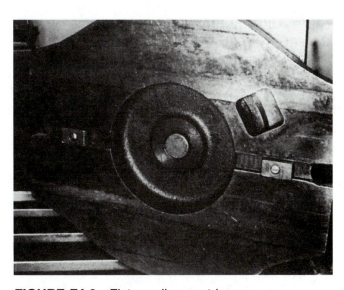

FIGURE F4.6　Fixture alignment keys.

milling machine table. This normally positions the solid jaw of the vise parallel with or square to the face of the column. Before mounting a vise or other fixture on a machine table, inspect the base carefully for small chips and nicks and remove any that you find. When the base is clean, fasten the vise to the table.

MACHINE SETUP

Preparing a machine tool prior to machining is called *setting up* the machine. Before a setup can be made, the machine should be cleaned, especially all sliding surfaces such as the ways and the machine table. After wiping the table clean, use your hand to feel for nicks or burrs. If you find any, use a honing stone to remove them. Workpieces must be fastened securely for the machining operation. They can be held in a vise, clamped to an angle plate, or clamped directly to the table. Odd-shaped workpieces may be held in a fixture designed for that purpose. On a universal milling machine it is good practice to check the alignment of the table before mounting a vise or fixture on it.

Setting Up the Arbor and Cutter

Prior to assembly of the spacing collars and cutter on the arbor, all pieces should be cleaned. Keys should always be used to drive the cutter. Do not depend on the friction between the spacers and cutter. The drive keys should extend into the spacing collars on both sides of the cutter. Tighten and loosen the arbor locknut only when the arbor support is in place, and do not use a hammer on the wrench. Select the proper cutter for the job.

Table Alignment on a Universal Milling Machine

Step 1. Clean the face of the column and the machine table.

Step 2. Fasten a dial indicator to the table with a magnetic base or other mounting device (Figure F4.7).

Step 3. Preload the indicator to approximately one-half revolution of its dial and set the bezel to zero.

Step 4. Move the table longitudinally with the handwheel to indicate across the column.

Step 5. If the indicator hand moves, loosen the locking bolts on the swivel table and adjust the table one-half of the indicated difference.

FIGURE F4.7 Aligning the universal milling machine.

Step 6. Tighten the locking bolts and indicate across the column again; make another adjustment if needed.

Never indicate the table with the indicator mounted on the column, as this would always show alignment even though the table was off.

There are a number of different methods of aligning a vise on a table.

Aligning a Vise Parallel with the Table Travel

1. Fasten a dial indicator with a magnetic base to the arbor (Figure F4.8), set the dial indicator to

FIGURE F4.8 Aligning the vise parallel to the table.

the solid vise jaw, and preload the indicator contact point to one-half revolution of the dial. Set the bezel to zero.

2. Move the table so that the indicator slides along the solid jaw. Record any indicator movement.
3. Loosen the holddown bolts and lightly retighten. Lightly tap the vise with a lead or soft-faced hammer to move the vise one-half the distance of the indicator movement.
4. Indicate the solid jaw again to check the alignment. **Always** take another indicator reading after securely tightening the clamping bolts. Often the final tightening will move a vise, fixture, or workpiece.

Aligning a Vise at a Right Angle to the Table Travel

1. Fasten a dial indicator with a magnetic base to the arbor (Figure F4.9) and preload the indicator.
2. Move the table with the cross feed handwheel and indicate the solid jaw.
3. Loosen the vise holddown bolts and make any necessary correction.
4. Indicate the solid jaw again to check the alignment. **Always** take another indicator reading after securely tightening the clamping bolts. Often the final tightening will move a vise, fixture, or workpiece.

If no indicator is available to align a vise on a table, a combination square may be used, as shown in Figure F4.10. The beam of the square is slid along the machined surface of the column until contact is

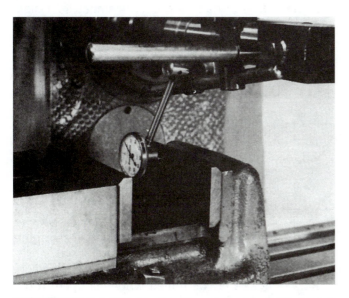

FIGURE F4.9 Aligning the vise square to table travel.

FIGURE F4.10 Using a square to align a vise on the table.

made with the solid jaw of the vise. Two strips of paper used as feeler gages help in locating the contact point. A soft-headed hammer or lead hammer is used to tap the vise into position.

Aligning a Vise at an Angle Other Than 90 Degrees to the Table Travel

Occasionally a vise needs to be mounted on the table at an angle other than square to the table travel. This can be done with a protractor (Figure F4.11). Paper strips are used as feeler gages, the angular setting being correct when both strips contact the protractor blade and the vise jaw at the same time. This is not a precise method of setting an angle because of the limitations in setting an angle accurately with a protractor, maintaining the level of the protractor blade, and accurately testing the "drag" on the paper strips.

Workholding on the Machine Table

Before machining a workpiece to size on a milling machine, several important decisions need to be made. One consideration is how to hold the workpiece while it is being machined. Large workpieces can be clamped directly to the table (Figure F4.12). Workpieces tend to move on the table from the cutting pressure against them. This movement can be prevented by clamping a stop block on the table and placing the workpiece against it.

FIGURE F4.11   Using a protractor to align a vise on the table.

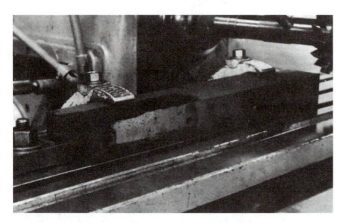

FIGURE F4.12   Workpiece clamped to the table.

FIGURE F4.13   Work clamped to the table with T-slot bolts and clamps. (Cincinnati Milacron)

FIGURE F4.14   Shafts are often clamped in T-slots when keyseats are being cut. (Lane Community College)

T-slots, which run lengthwise along the top of the table, are accurately machined and parallel to the sides of the table. These T-slots are used to retain the clamping bolts. Workpieces can also be aligned against snug-fitting parallels that are set into the T-slot; the workpiece is pushed against these parallels while the work is being clamped. Figure F4.13 shows the workpiece clamped to the table with T-slot bolts and clamps. The bolts are placed close to the workpiece; the block supporting the outer end of the clamp is the same height as the shoulder being clamped. When the bolt is closer to the work than to the clamp support block, maximum leverage is obtained. The support block should never be lower than the work being clamped.

When workpieces with finished or soft surfaces are clamped, care must be taken to protect those surfaces from damage by clamping. A shim should be placed between the work surface and the clamp. Before placing rough castings or weldments on a machine table, protect the table surface with a shim. This shim can be paper, sheet metal, or even plywood, depending on the accuracy of the machining to be performed.

A workpiece should have a support directly underneath the location where a clamp exerts pressure. Clamping an unsupported workpiece may cause it to bend or spring, and it will bend back after clamping pressure is released. If the workpiece material is brittle, clamping pressure may break it.

Round stock may be milled while clamped to the table in the T-slot groove for alignment (Figure F4.14). Keyways are often milled on long shafting in

FIGURE F4.15 Use of angle plate to mill ends of work-piece.

FIGURE F4.16 Setup of a workpiece to machine an end square.

this manner, the overhang being supported on a stand or with a hoist. Keyways on short shafts are sometimes milled while they are clamped in a vise. Flats and square ends may be milled on shafts by rotating them 90 degrees after each cut.

Angle plates are sometimes used for a rigid setup when the end of a piece is machined (Figure F4.15). The vise can also be used to set up relatively short pieces for milling the ends square (Figure F4.16). When work must be clamped off-center in a vise, a spacer must be used (Figure F4.17).

A dividing head is often used for machining parts that require accurate spacing on a circle. Figure F4.18 shows a gear being cut from a blank that is mounted on a mandrel supported by centers on the head and the footstock and driven by a dog. Dividing heads are also used for milling splines, squares, and hexagons on ends of shafts.

Workholding with a Vise

Many workpieces can be held securely in a machine vise (Figure F4.19). If the workpiece is high enough, seat it on the bottom of the vise. If it is not, use parallels to raise it. It is the friction between the

FIGURE F4.17 Work clamped off-center needs a spacer.

FIGURE F4.18 Dividing head and foot stock. (Cincinnati Milacron)

FIGURE F4.19 Workpiece held in a vise for slab milling.

vise jaws and the workpiece that holds the workpiece and the more contact area there is, the better it is gripped.

MILLING OPERATIONS

After the cutter for the job has been selected, the speed and feed can be calculated and set on the machine. The depth of cut depends largely on the amount of material that is to be removed and the rigidity of the setup.

Conventional and Climb Milling

Figure F4.20 shows one cutter operated in a conventional milling mode and the other cutter in a climb milling operation. In conventional milling modes (sometimes called *up-milling*), the workpiece is forced against the cutter and the teeth of the cutter try to lift the workpiece up, especially at the beginning of a cut. In climbing (or down-milling), the cutter tends to hold the workpiece down.

Climb milling should only be performed on machines equipped with an antibacklash device. Backlash, which is slack between the table drive screw and nut assembly, would let the cutter pull the workpiece under it, break the cutter, and ruin the workpiece. **Always** use the conventional or up-milling mode unless the machine you are using is equipped for climb milling.

Milling a Slot

The diameter of the cutter to be used on a job depends on the depth of the slot or step. As a rule, the smallest-diameter cutter that will do the job should be used, as long as sufficient clearance remains between arbor and work and between arbor support and vise (Figure F4.21). Use only a sharp cutter to minimize cutting pressures and to get a good surface finish. Resharpen a cutter when it becomes slightly dull. A slightly dull cutter can be resharpened easily and quickly.

Milling cutters with side cutting teeth are used when grooves and steps have to be machined on a workpiece. The size and kind of cutter to be used depends largely on the operation to be performed. Full side mills with cutting teeth on both sides are used when slots or grooves are cut (Figure F4.22).

Cutters usually make a slot that is slightly wider than the nominal width of the cutter. Cutters that wobble because of dirt or chips between the arbor

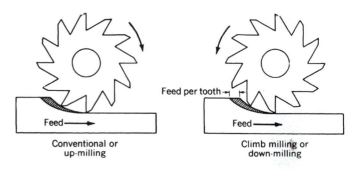

FIGURE F4.20 Conventional and climb milling.

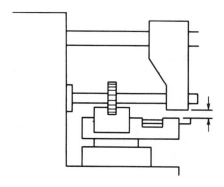

FIGURE F4.21 Check for clearance.

spacers will cut slots that are considerably oversize. A slot will become wider when more than one cut is made through it. If a slot needs to be .375 in. wide, a $\frac{3}{8}$-in. cutter probably cannot be used because it may cut a slot .3755 to .376 in. wide. A trial cut in a piece of scrap metal will tell the exact slot width. It may be necessary to use a $\frac{5}{16}$-in.-wide cutter and make two or more cuts.

The width of a slot from a given cutter is also affected by the amount of feed used. A very slow feed will tend to allow the cutter to cut more clearance for itself and make a wider groove. A fast feed crowds

the cutter, which means a narrower slot. A cutter tends to cut a wider slot in soft material than it will in harder material. The width of slots is measured with venier calipers or with adjustable parallels. If it is a keyway, the key itself can be used as a gage to test the slot width.

Squaring a Block

Slab milling (Figure F4.23) cutters are used in a setup where a uniform, flat surface is needed and where square or rectangular workpieces are milled with sides parallel and square. Shims are often needed in a vise to achieve true squareness since most milling machine vises become slightly out of square from heavy use.

A good machinist will mark the workpiece with layout lines before fastening it in a vise. The layout should be an exact outline of the part to be machined. The reason for making the layout prior to machining is that reference surfaces are often removed by machining. After the layout has been made, make diagonal lines on the portion to be machined away. This helps in identifying on which side of the layout line the cut is to be made (Figure F4.24). The cutter can be accurately positioned on the workpiece with the hand feed cranks and the micrometer dials. When the finished outside surface of a workpiece must not be marred or scratched by a revolving cutter, a paper strip is held between the workpiece and the cutter (Figure F4.25). The power is turned off and the spindle is rotated by hand. Make sure the paper strip is long enough so that your hands are not near the cutter. Carefully move the table toward the revolving cutter. When the cutter pulls the paper strip from your

FIGURE F4.22 Full side (staggered-tooth) milling cutter machining a groove.

FIGURE F4.23 Location of shim to square up work.

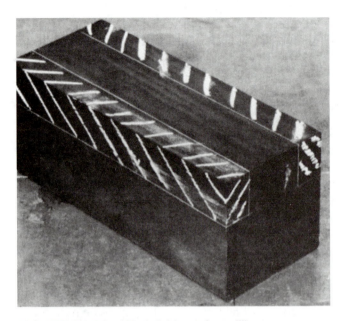

FIGURE F4.24 Work laid out for milling.

FIGURE F4.26 Using a paper strip to set the depth of cut.

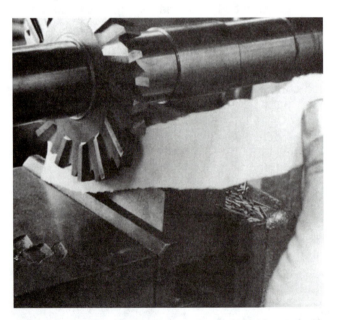

FIGURE F4.25 Setting up a cutter by using a paper strip.

FIGURE F4.27 Positioning a cutter using a steel rule.

fingers, the cutter is about .002 in. from the workpiece. At this time set the cross feed dial to zero. Lower the knee until the cutter clears the top of the workpiece. Then, by using the cross feed handwheel, position the work where the cut is to be made. The same method will work in positioning a cutter above a workpiece and establishing the zero position for the depth of cut without actually touching the workpiece with the cutter (Figure F4.26).

A quicker, but not as accurate, method is illustrated in Figure F4.27. A steel rule is used to position a side mill a given distance from the outside edge of a workpiece. The end of the rule is held firmly against the side cutting edge of a tooth. The distance is indicated by the edge of the workpiece. The micrometer dials of the cross feed and knee controls should be zeroed when the cutter contacts the side and top of the workpiece. When these zero positions are established, additional cuts can easily be made by positioning from these points with the micrometer dials.

After the first cut is made, the distance of the side of the step or groove to the outside of the workpiece should be measured. A measurement made on both ends shows if the cut is parallel to the sides of the workpiece. If the step is not parallel, the vise needs to be aligned or, on a universal milling machine, the table may need aligning.

Making a Step

Good milling practice is to take a roughing cut and then a finish cut. Better surface finishes and higher dimensional accuracy are achieved when roughing and finishing cuts are made. The depth of the roughing cut is often limited by the horsepower of the machine or the rigidity of the setup. A good starting point for roughing is .100 to .200 in. deep. The finishing cut should be .015 to .030 in. deep. Depth of cut less than .015 in. deep should be avoided, because a milling cutter, especially in conventional or up-milling, has a strong rubbing action before the cutter actually starts cutting. This rubbing action causes a cutter to dull rapidly. Assuming that a cut .100 in. deep has to be taken, the following steps outline the procedure to be used:

Step 1. Loosen the knee locking clamp and the cross slide lock.

Step 2. Turn on the spindle and check its rotation.

Step 3. Position the table so that the workpiece is under the cutter.

Step 4. Raise the knee slowly by turning the vertical hand feed crank until the cutter just touches the workpiece. (If the cutter cuts a groove, you have gone too far and should try again on a different place on the workpiece.)

Step 5. Set the micrometer dial on the knee feedscrew to zero.

Step 6. Lower the knee by approximately one-half revolution of the hand feed crank. (If the knee is not lowered, the cutter will leave tool marks on the workpiece in the following operation.)

Step 7. Move the table lengthwise until the cutter is clear of the workpiece. Move the cutter to that side of the workpiece, which will result in a conventional milling cut.

Step 8. Raise the knee past the zero mark to the 100 mark on the micrometer dial.

Step 9. Tighten the knee lock and the cross slide lock. Always tighten all locking clamps prior to start-

ing a machining operation, except the one that would restrict table movement while cutting. This aids in making a rigid, chatterfree setup.

Step 10. The machine is now ready for the cut. Turn on the coolant. Move the table slowly into the revolving cutter until the full depth of cut is obtained before engaging the power feed.

Step 11. When the cut is completed, disengage the power feed, stop the spindle rotation, and turn off the coolant before returning the table to its starting position. If the revolving cutter is returned over the newly machined surface, it will leave cutter marks and mar the finish.

Step 12. After brushing off the chips and wiping the workpiece clean, the workpiece should be measured while it is still fastened in the machine. If the workpiece is parallel at this time, additional cuts can be made if more material needs to be removed.

Cutting a Keyseat on a Shaft

Keyseats are commonly milled on horizontal milling machines with plain or staggered-tooth cutters. Standard key sizes for shafts are given in Table F4.3. Keyseats are cut on the end of a shaft or at locations along the shaft. When a cut is started on the shaft, away from the end, the table lock must be tightened

TABLE F4.3 Shaft diameter and key size

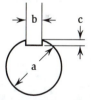

Shaft Diameter 1 in. a	Nominal Size Square Key-in. b	Keyseat Depth in. c
$\frac{1}{2}$	$\frac{1}{8}$	$\frac{1}{16}$
$\frac{5}{8}$	$\frac{3}{16}$	$\frac{3}{32}$
$\frac{3}{4}$	$\frac{3}{16}$	$\frac{3}{32}$
$\frac{7}{8}$	$\frac{3}{16}$	$\frac{3}{32}$
1	$\frac{1}{4}$	$\frac{1}{16}$
$1\frac{1}{4}$	$\frac{5}{16}$	$\frac{5}{32}$
$1\frac{1}{2}$	$\frac{3}{8}$	$\frac{3}{16}$
$1\frac{3}{4}$	$\frac{1}{2}$	$\frac{1}{4}$
2	$\frac{1}{2}$	$\frac{1}{4}$
$2\frac{1}{2}$	$\frac{5}{8}$	$\frac{5}{16}$
3	$\frac{3}{4}$	$\frac{3}{8}$

while making the plunge cut to the depth of the key-seat. The table lock is then loosened and the table feed is engaged. The shaft is clamped in a vise if it is short. If it is a long shaft, it is often clamped in a T-slot groove. The cutter is centered over the shaft by touching its side with the cutter using a paper strip for a feeler gage (Figures F4.28a and F4.28b). The micrometer collar is set to zero and the cutter is moved a distance equal to the radius of the shaft plus half the width of the cutter. The work is brought to the rotating cutter and a spot is cut equal to the

width of the cutter so that the correct depth can be made from the side of the slot. The collar on the knee feedscrew is zeroed and the cutter relocated away from the end of the shaft. The correct depth is set and the cut is started (Figure F4.29).

Coolants and Cutting Fluids

Most milling machines are equipped with a coolant tank and pump (Figure F4.30). Usually some kind of soluble oil-water mix is used as a cutting fluid. Unless the soluble oil contains a bactericide, the coolant often becomes rancid and can cause skin rashes. There are many synthetic cutting fluids available today that eliminate this problem. Smaller machines are often equipped with a coolant spray

FIGURE F4.28a Cutter positioned to shaft using paper strip for feeler gage.

FIGURE F4.29 Starting the cut.

FIGURE F4.28b Cutter positioned over shaft. A spot is made equal to the width of the cutter and the index zeroed for vertical travel. The shaft has been lowered for viewing.

FIGURE F4.30 Coolant cools the cutter and flushes away chips.

device that is quite effective. The operator should try to avoid breathing the spray by standing on the opposite side of the nozzle. Sulfurized (black) cutting oil from a pump can should not be used at all on a milling machine since the soluble oil coolant would be contaminated by it.

SELF-TEST

1. Name at least three hazards to a person around or near a revolving cutter on a milling machine.
2. When cleaning chips off the machine and when taking measurements, should the cutter be rotating? Should compressed air be used for cleaning?
3. A cut is to be made on low-carbon steel with a 4-in. diameter high-speed cutter with 16 teeth. A cutting speed of 90 FPM is selected from Table F4.1. What is the correct RPM?

4. A feed per tooth of .003 in. is chosen from Table F4.2 for selecting the table feed of the cutter in Question 3. What is the correct table feed?
5. What is the most common workholding device used on the milling machine?
6. The milling machine vise should be aligned to what part?
7. Name two ways that round stock may be set up for milling a keyseat.
8. On older milling machines or on those not equipped with an antibacklash device, should a conventional or climb milling mode be used?
9. How can backlash in a machine cause the cutter to be broken and the workpiece ruined?
10. Coolant is always used to flood the cutter and workpiece with high-speed cutters and sometimes with carbide cutters. When is it best not to use coolant with carbide cutters?

WORKSHEET: WHEEL PULLER

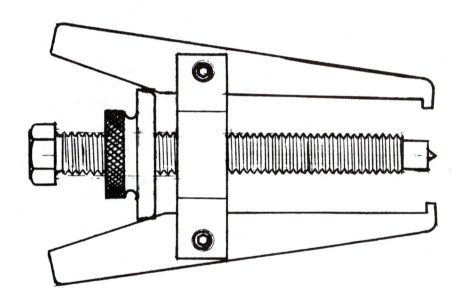

Objectives

1. Be able to turn a disc shape on a lathe and tap an internal thread.
2. Learn to mill slots on a horizontal milling machine.
3. Use a steady rest on a lathe.

Outline for Study

Prior to starting each procedure for this project, study and complete Post-Tests for:

1. Making the body: Section F, Units 3 and 4; Refer to Section E, Unit 9.

2. Making the nut: Refer to Section E, Unit 11.
3. Making the screw: Refer to Section E, Units 11 and 13.
4. Making the legs: Refer to Section A, Unit 3.

Procedures

Begin the procedures for this project by completing the following worksheets.

Worksheet 1. Making the Body
Worksheet 2. Making the Nut
Worksheet 3. Making the Screw
Worksheet 4. Making the Legs

WORKSHEET: MAKING THE BODY

Materials

Three $\frac{1}{4}$-20 Soc. HD capscrews; $4\frac{7}{8}$ in. of $3\frac{1}{2}$ in. HRMS round bar.

Procedure

1. Saw the material for the body from $3\frac{1}{2}$ in. MS round stock. Cut off a piece about $\frac{7}{8}$ in. long.
2. Set up a three-jaw chuck with the jaws set to grip the outside of this piece. Place the material in the chuck with the jaws gripping about $\frac{1}{8}$ in. True up the piece in the setup by measuring from the chuck face with a rule in several locations.

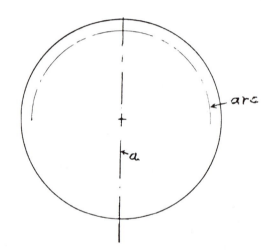

FIGURE F4.31

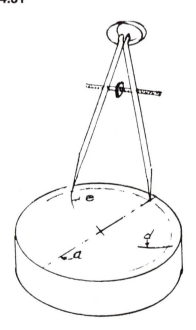

FIGURE F4.32

3. Take a facing cut to clean up one side.
4. Turn the OD to the size given in the drawing.
5. Remove the part from the chuck and apply layout dye to the finished surface.
6. To lay out the part, find the center and scribe a centerline a across the surface (Figure F4.31). Prick punch the center.
7. Set the dividers to approximately $1\frac{1}{2}$ in. Put one leg in the prick punch mark and scribe an arc of about 180 degrees.
8. With the dividers at the same setting, place one leg where the centerline and the arc coincide. This will mark points 60 degrees on either side of the centerline (d and e, Figure F4.32).
9. Scribe a centerline (b and c, Figure F4.33) through the center to the two short arcs (d and e).
10. On the four lines that now extend to the outside edge of the circle, measure in $\frac{7}{8}$ in. and scribe a short line perpendicular to the centerline to outline the bottom of the slots.
11. Measure $\frac{1}{8}$ in. on either side of the centerlines and scribe parallel lines to meet the short cross line (Figure F4.33).
12. Scribe parallel lines $\frac{1}{4}$ in. from either side of the slots. Mark their length from the outer edge $\frac{11}{16}$ in.
13. Connect the ends of these lines to form the remainder of the outline of this part. The layout is now complete.
14. Place the part in the lathe chuck with the finished face in. Grip it this time as far back into the jaws as it will go. Be sure the seating surfaces of the jaws are clean and free from chips.
15. Face to the thickness given in the drawing. This should remove the extra $\frac{1}{8}$ in. width where the jaws were holding the part in the last operation. Chamfer the OD.

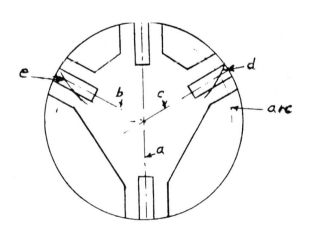

FIGURE F4.33

16. Centerdrill and then drill through with a pilot drill or a drill slightly smaller than the tap drill.
17. Drill through with a tap drill for a $\frac{3}{4}$-16 NF thread.
18. Chamfer the hole.
19. Set up the tap with the dead center supporting it and the tap handle resting on the carriage or on the compound, not on the ways. Release the spindle so it is free wheeling. Turn the chuck by hand to make the thread while following with the dead center by gently turning the tailstock handwheel. *Do not turn on the machine for this operation.* Use sulfurized cutting oil. Remove the work when finished and clean the machine.
20. To continue with this project, you should be familiar with the use of a small horizontal milling machine. You will need to be able to set up the cutter, vise, and work. You must also be able to set your speeds and feeds. You will need to study Section F, Units 3 and 4, to be ready for this operation.
21. Cut a slot first, then make the two cuts on each side of the slot (Figure F4.34) Use cutting fluid. Some machines are capable of doing this in one cut, however, some light duty machines may only be capable of .200 to .400 in. depth per pass.
22. Turn the work and set up the next centerline with a square head. Repeat the cutting procedure in step 21.
23. When all four slots and side cuts have been made, rotate the work so that the line on the spaces in between is level or parallel with the solid jaw of the vise.
24. Remove all excess material by taking cuts with the same milling cutter (Figure F4.35).
25. When the milling is finished, remove the part, deburr and chamfer all edges.
26. Lay out for the four holes in the ears for the capscrews and center punch them.
27. Set up in the drill press with one ear level with the table.

28. Drill through both halves of the ear with a $\frac{1}{4}$-20 tap drill.
29. Drill through the *top ear only* with a $\frac{1}{4}$-in. drill.
30. Rotate the piece to the next ear and repeat Steps 27 through 29.
31. After all four ears are drilled, deburr and tap through the ears with a $\frac{1}{4}$-20 tap. Clean and apply a light film of oil.

WORKSHEET: MAKING THE NUT

Material

$1\frac{1}{4}$ inches of $2\frac{1}{2}$ HRMS round stock.

Procedure

1. Cut off a piece of mild steel round stock slightly larger than $2\frac{1}{4}$ in. diameter. The next size larger may be $2\frac{1}{2}$ in. HRMS round bar. The length should be $1\frac{1}{4}$ in.
2. Since there are some very heavy tool pressures in this operation such as knurling, grooving, and drilling, it is best to set up the work in a four-jaw chuck. Leave $\frac{7}{8}$ in. extending out of the chuck and roughly true up the work with no more than 0.010 in. runout.
3. The operation with the greatest work force should be done first so that if the work moves it can be more easily corrected. Therefore, the knurling should be done first, but the part must first be machined to the knurl diameter.
4. When you have turned the nut to the knurl diameter in as far as the shoulder ($\frac{9}{16}$ in.), then set up a medium knurl and knurl the part.
5. Grind a tool bit with a $\frac{1}{8}$ in. radius.
6. Machine the groove with the lowest speed on the lathe. Use cutting oil. This plunge cut, like parting off, may produce some chatter. If the feed is too light, it will almost always chatter, so feed in

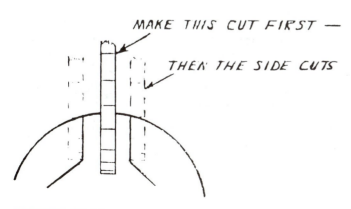

FIGURE F4.34

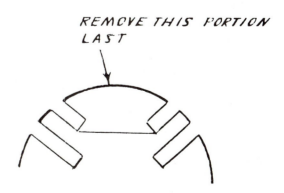

FIGURE F4.35

by hand at a rate that will just produce a chip. Too much feed will also cause chatter or jamming of the tool in the work, causing the work to come out of the chuck or the machine to stop.

7. Face the piece and chamfer both sides of the knurled part. Finish turn both faces and the shoulder again to $\frac{9}{16}$ in. from the end to correct any runout.
8. Centerdrill, pilot drill, and tap drill with the same procedure as with the body.
9. Tap $\frac{3}{4}$-16.
10. Remove the nut from the chuck and change to a three-jaw chuck.
11. Chuck a threaded stub mandrel having a $\frac{3}{4}$-16 NF thread. Your instructor may have this mandrel already made up; if not, you will have to make one.
12. Screw on the nut with the knurled part next to the chuck.
13. Turn the OD to $2\frac{1}{4}$ in. Face to the $\frac{13}{16}$ in. length. Mark a 2 in. diameter line on the face.
14. Free hand machine a slight radius from the OD to the 2 in. line as shown on the drawing.
15. Finish the radius with a file and deburr the edge nearest the chuck. Chamfer the threaded hole. Clean and apply a light film of oil.

WORKSHEET: MAKING THE SCREW

Material

$7\frac{1}{2}$ inches of 1 in. hexagonal stock

Procedure

1. Cut off a piece of 1 in. hexagonal stock $7\frac{1}{4}$ in. long. The extra length is for turning off the center at a later time.
2. Chuck the piece in a three-jaw chuck and face both ends. Chamfer one end to a point slightly below the flats on the hexagon stock. Centerdrill the other end. Chuck $\frac{1}{2}$ in. of the chamfered end and support the other end in the tailstock center.
3. Turn the diameter to size. Check for tapering and correct if needed.
4. Set up for threading and cut the $\frac{3}{4}$-16 thread. Check for fit with the nut you have just made. Turn the $\frac{1}{2}$ in. diameter. This should now be about $\frac{3}{8}$ in. longer than it will be when finished.
5. With one end still in the chuck, set up diameter. Move the tailstock away. Be sure the carriage is to the right of the steady rest so you can turn the end of the piece.
6. Machine off the center hole and face to length being careful to leave material in the center for the point.
7. Remove the part from the lathe. At this point the part could be case hardened for additional strength if your instructor wishes to do this for you or have it done.
8. Clean up the part and apply light oil.

WORKSHEET: MAKING THE LEGS

Material

$18\frac{1}{2}$ inches of $\frac{1}{4} \times 1$ in. HRMS flat bar

Procedure

1. Cut off three pieces of $\frac{1}{4} \times 1$ in. MS flat bar $6\frac{1}{16}$ in. long.
2. Lay out as shown on the drawings.
3. Drill the $\frac{1}{4}$ in. holes. Chamfer.
4. Saw along the layout lines with the vertical band saw, leaving a minimum of material to file. Use a push stick for safety when sawing. Ask your instructor to help you with this operation since information on using the vertical band saw is not in this book.
5. Finish file the legs and chamfer all edges.
6. Case harden all three legs by the roll method with the help of your instructor. Clean and apply a light film of oil.

Procedure for Assembly

1. Obtain three $\frac{1}{4}$-20 socket head capscrews $\frac{3}{4}$ in. long. Assemble the legs on the body.
2. Install the nut on the screw and turn the screw into the body.
3. This tool may also be used in the alternate position with two legs opposed.

Show your completed wheel puller to your instructor for evaluation.

The drawings are shown at approximately 42% of full size.

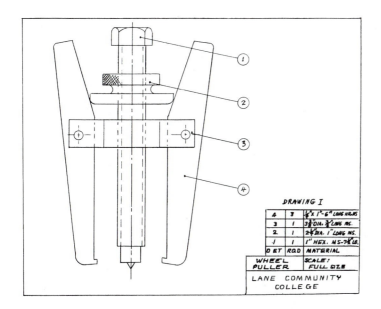

DRAWING I

DET	RQD	MATERIAL
4	3	¼"x 1"-6" LONG HR.MS
3	1	3½ DIA. ⅞ LONG MS.
2	1	2¼ DIA. 1" LONG MS.
1	1	1" HEX. MS-7½ LG.

WHEEL PULLER	SCALE: FULL SIZE

LANE COMMUNITY COLLEGE

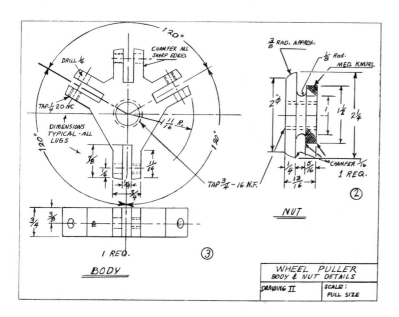

120°

CHAMFER ALL SHARP EDGES

DRILL ¼

TAP ¼ 20 NC

DIMENSIONS TYPICAL - ALL LUGS

120°

120°

11/16 R.

TAP ¾ - 16 N.F.

⅞

¼

11/16

¼

¾

NUT

⅜ RAD. APPROX.

⅛ Rad.
MED. KNURL

2 Ø

1

1½

2¼

¼

5/16

13/16

CHAMFER 1/16
1 REQ.

②

¾

⅜

¾

1 REQ.

BODY

③

WHEEL PULLER BODY & NUT DETAILS	
DRAWING II	SCALE: FULL SIZE

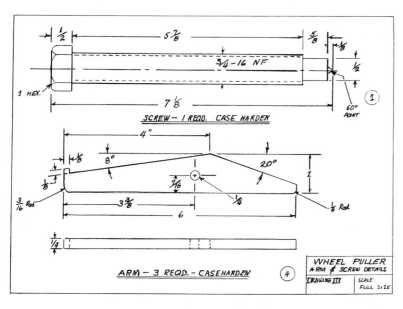

½

5⅞

⅝

⅛

¾ - 16 NF

½

1 HEX.

7⅛

60° POINT

①

SCREW - 1 REQD. CASE HARDEN

4"

8°

20°

⅛

⅛

3/16

¼

1

3/16 Rad

3⅜

6

⅛ Rad

¼

ARM - 3 REQD. - CASEHARDEN

④

WHEEL PULLER ARM & SCREW DETAILS	
DRAWING III	SCALE: FULL SIZE

SECTION G
Grinding Machines

Both cylindrical and surface grinding machines are extremely high-precision tools that are often used for finishing operations on parts that have been produced on other machine tools such as lathes and milling machines. After a part has been hardened by heat treatments, grinding is the preferred method of finishing to size. Grinding machines can hold tolerances within tenths of thousands of an inch.

However, some machines are capable of removing large amounts of metal for forming gears, splines, and other shapes. Creep-feed grinding is a process that removes metal at a fairly high rate.

As with other machine tools, grinding machines have been adapted to CNC operation (Figures GI.1 and GI.2). Specialty grinding machines are made for one particular operation, such as surfacing engine blocks and heads (Figure GI.1).

Tool and cutter grinders are special machine tools that are extremely versatile and are used for sharpening milling machine cutters and reamers (Figures GI.2 and GI.3).

This section should prepare you to use manually operated grinding machines so that as a CNC machine operator, you will understand their principles of operation, functions, and limitations.

FIGURE GI.1 Special grinding machine surfacing an engine head.

FIGURE GI.3 Sharpening a helical milling cutter.

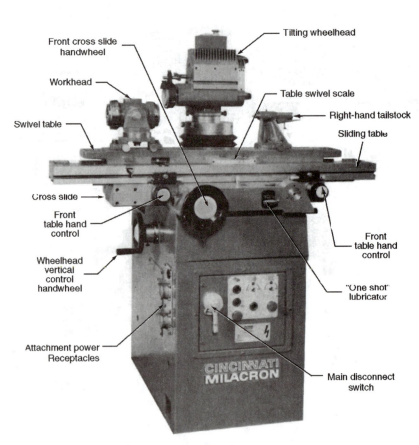

FIGURE GI.2 Components of the cutter and tool grinder. (Cincinnati Milacron)

Front cross slide handwheel

Tilting wheelhead

Workhead

Table swivel scale

Swivel table

Right-hand tailstock

Sliding table

Cross slide

Front table hand control

Front table hand control

Wheelhead vertical control handwheel

"One shot" lubricator

Attachment power Receptacles

Main disconnect switch

CINCINNATI MILACRON

Selection and Identification of Grinding Wheels

A grinding wheel is both a many-toothed cutting tool and a toolholder. Selecting a grinding wheel is somewhat more complicated than selecting a lathe tool or a milling cutter because there are more factors to be considered. These factors, including size, shape, and composition of the wheel, are expressed in a system of symbols (numbers and letters) that you must be able to interpret and apply. Whether you are selecting a wheel for a surface grinder or a pedestal grinder, the selection process is the same if you expect to get the right wheel for the job.

OBJECTIVES

After completing this unit, you should be able to:

1. List the five principal abrasives with their general areas of best use.
2. List the four principal bonds with the types of applications where they are most used.
3. Identify by type number and name, from unmarked sketches or from actual wheels, the four most commonly used shapes of grinding wheels.
4. Interpret wheel shape and size markings together with the five basic symbols of a wheel specification into a description of the grinding wheel.
5. Given several standard, common grinding jobs, recommend the kind of abrasive, approximate grit size and grade, and bond.

IDENTIFYING FACTORS

Six or seven factors must be considered in selecting a grinding wheel for a particular job. All these factors—wheel shape, wheel size, kind of abrasive, abrasive grain size, hardness, grain spacing, and bond—are expressed in symbols consisting of numbers and letters, most of which are easy to understand and interpret. For example, the size and shape of the wheel are usually determined by the type and size of the grinder. The abrasive, of which there are five major kinds, is determined primarily by the material being ground. The abrasives are not equally efficient on all materials. The size of the abrasive particles selected (the largest are perhaps $\frac{1}{8}$- to $\frac{1}{4}$-in. long and the smallest less than .001-in. long) depends on the amount of stock to be removed and the finish desired.

Selecting the grade or hardness of the wheel, which is a measure of the force required to pull out the abrasive grains, is typically a choice between one of two or, at most, three grades. Grain spacing or structure is most often standard, according to the grain or grit size and the grade of the wheel. Finally, there is the bond that holds the wheel together. Although there are four common bonds, the choice is usually made clear from the job.

Color can also be useful for identification, and it is used to some extent. One of the best abrasives for grinding tools is white when it is manufactured, and a more commonly used abrasive is gray or brown in color. Some wheels are green, pink, or black. Among a given number of grinding wheels, there are wheels of different diameters, thicknesses, or hole sizes. Different wheels are made of different sizes of abrasive grain or with different proportions of grain and the bond that holds the grain together in the wheel.

A code of numbers and letters has been developed by the grinding wheelmakers that provides the needed information about a given wheel in just a few letters and numbers. Within a given group of symbols, the order of listing is important.

Grinding wheels are designed for grinding either on the periphery (outside diameter), which is a curved surface, or on the flat side, but rarely on both. It is not a safe practice to grind on the side of a wheel designed for peripheral grinding. However, there are some exceptions to this rule. The shape of the wheel determines the type of grinding performed.

WHEEL SHAPES

The shapes of grinding wheels are designated according to a system published in an American National Standard, *Specifications for Shapes and Sizes of Grinding Wheels,* whose number is ANSI B74.2-1974. The various shapes have been given numbers ranging from 1 to 28, but only five are important for you now. These are described below.

Type 1 (Figure G1.1) is a peripheral grinding wheel, a straight wheel with three dimensions: diameter, thickness, and hole, in that order. A typical wheel for cylindrical grinding is 20 in. (diameter) × 3 in. (thickness) × 5 in. (hole). Probably most wheels are of this type.

A *Type 2* or cylinder wheel (Figure G1.2) is a side grinding wheel, to be mounted for grinding on the side instead of on the periphery. This also has three dimensions; for example, 14 in. (diameter) × 5 in. (thickness) × $1\frac{1}{2}$ in. (wall). This, of course, might also be called a 14 × 5 × 11 (14 in. *D* minus 2 times $1\frac{1}{2}$—the two wall thicknesses), but the wall thickness is more important than the hole size. Hence the change.

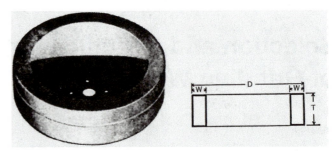

FIGURE G1.2 Cylinder or Type 2 wheel, whose grinding face is the rim or wall end of the wheel. Has three dimensions: diameter, thickness, and wall thickness. (Bay State Abrasives, Dresser Industries, Inc.)

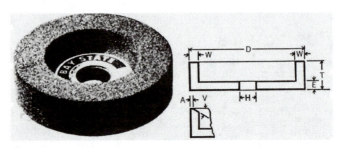

FIGURE G1.3 Straight cup or Type 6 wheel, whose grinding face is the flat rim or wall end of the cup. (Bay State Abrasives, Dresser Industries, Inc.)

A *Type 6* or straight cup wheel (Figure G1.3) is a side grinding wheel with one side flat and the opposite side deeply recessed. It has four essential dimensions: the diameter, thickness, hole size (for mounting), and wall.

A *Type 11* or flaring cup wheel (Figure G1.4) is a side grinding wheel that resembles a Type 6, except that the walls flare out from the back to the diameter and are thinner at the grinding face than at the back. This introduces a couple of new dimensions: the diameter at the back, called the "J" dimension, and the recess diameter at the back, the "K" dimension. This is mentioned only to emphasize that the "D" dimension, the diameter, is always the largest diameter of any wheel.

The *Type 12* dish wheel (Figure G1.5a) is essentially a very shallow Type 11 wheel, mostly for side grinding. The big difference is that the dish wheel has a secondary grinding face on the periphery, the "U" dimension, so that it is an exception to the rule of grinding *only* on the side or the periphery.

This factor of grinding on the side or the periphery of a wheel is important because it affects the grade of the wheel to be chosen. The larger the area, the softer

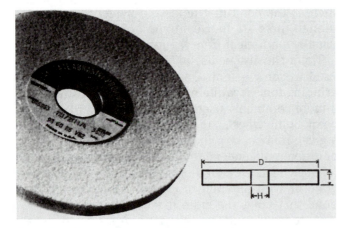

FIGURE G1.1 Straight or Type 1 wheel, whose grinding face is the periphery. This wheel usually comes with the grinding face at right angles to the sides, in what is sometimes called an "A" face. (Bay State Abrasives, Dresser Industries, Inc.)

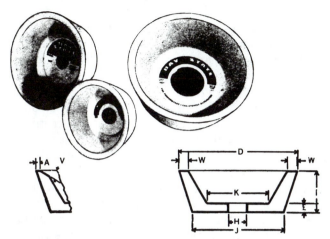

FIGURE G1.4 Flaring cup or Type 11 wheel, whose grinding face is also the flat rim or wall of the cup. Note that the wall of the cup is tapered. (Bay State Abrasives, Dresser Industries, Inc.)

FIGURE G1.5*b* An assortment of mounted wheels most often used for deburring and other odd jobs. (Bay State Abrasives, Dresser Industries, Inc.)

the wheel should be. In peripheral grinding, the contact is always between the arc of the wheel and either a flat (in surface grinding) or another arc (in cylindrical grinding). This makes for small areas of contact and somewhat harder wheels. On the other hand, if the flat side of the wheel is grinding the flat surface of a workpiece, then the contact area is larger and the wheel can be still softer.

It is important to understand that any grinding machine grinds either a flat surface or a round or cylindrical surface. The first group of grinders is collectively called *surface grinders;* the second group is called *cylindrical grinders,* whether the workpiece is held between centers or not and whether the grinding is external or internal. Various forms can be cut into the grinding faces of peripheral grinding wheels, and these can then be ground into either flat or cylindrical surfaces.

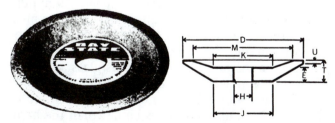

FIGURE G1.5*a* Dish or Type 12 wheel, similar to Type 11, but with a narrow, straight peripheral grinding face in addition to the wall grinding face. This is the only wheel of those shown that is considered safe for both peripheral and wall or rim grinding. (Bay State Abrasives, Dresser Industries, Inc.)

Mounted wheels like the ones in Figure G1.5*b* can be used in a variety of ways around a shop. Often they are used in portable grinders for jobs like deburring or breaking the edges of workpieces where the tolerances are not too critical. They are also used in internal grinding.

STANDARD MARKING SYSTEM

The description of a grinding wheel's composition is contained in a group of symbols known as the *standard marking system.* That is, the basic symbols for the various elements are standard, but they are usually amplified by individual manufacturer's symbols, so that it does not follow that two wheels with the same basic markings, but made by two different suppliers, would act the same. However, it is a useful tool for anyone concerned with grinding wheels.

There are five basic symbols. The first is a letter indicating the kind of abrasive in the wheel, called the **abrasive type.** The second is a number to indicate the approximate size of the abrasive; this is commonly called **grit size.** In the third position, a letter symbol indicates the *grade* or relative hardness of the wheel. The fourth, *structure,* is a number describing the spacing between abrasive grains. The fifth is a letter indicating the *bond,* the material that holds the grains together as a wheel. Thus, a basic toolroom wheel specification (Figure G1.6) might be

A60-J8V

But, since most wheelmakers have, for example, a number of different aluminum oxide or other types of

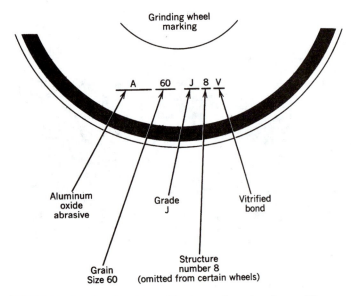

FIGURE G1.6 Sketch to illustrate the wheel specification that will be used as an illustration in following pages.

abrasives and a number of different vitrified or other bonds, the symbol sometimes appears cluttered:

9A80-K7V22

This means that the wheelmaker is using a particular kind of aluminum oxide (A) abrasive, which is indicated by the "9," and a particular vitrified (V) bond, which is indicated by the "22."

The wheel markings for diamond or cubic boron nitride wheels are a little different and are not standard enough for a simple explanation.

First Symbol—Type of Abrasive (A60-J8V)

The symbol in the first position, as suggested above, denotes the type of abrasive. Basically, there are five:

A Aluminum oxide
C Silicon carbide
D Natural diamond
MD or SD Manufactured diamond (sometimes called *synthetic diamond)*
B Cubic boron nitride

The first two are "cents-per-pound" abrasives, cheap enough so that it is practical to make whole wheels of the abrasive. Both are made in electric furnaces that hold literally tons of material, and both are crushed and graded by size for grinding wheels and other uses.

Diamond, both natural and manufactured, and cubic boron nitride are expensive enough that most wheels are made of a layer of abrasive around a core of other material. However, they have made a place for themselves because they will grind materials that no other abrasive will touch and because they stay sharp and last so long that they are actually less expensive per piece of parts ground. Natural or mined diamond that is definitely diamond but is of less than gem quality is crushed and sized. Both manufactured diamond and cubic boron nitride are made by a combination of high heat (in the range of 3000°F) and tremendous pressure (1 million or more PSI). This temperature, incidentally, is somewhat less than that needed for other manufactured abrasives, which require something in the range of 4000°F for fusing or crystallization.

Each of these abrasives generally has an area in which it excels, but there is no one abrasive that is first choice for all applications. Aluminum oxide is best for grinding most steels, but on the very hard tool steel alloys it is outclassed by cubic boron nitride. However, aluminum oxide (Al_2O_3) may have close to 75 percent of the market, because it is used in foundries for grinding castings and steel mills for billet grinding and in other high volume applications. Silicon carbide, which does poorly on most steels, is excellent for grinding nonferrous metals and nonmetallic materials. Diamond abrasive excels on cemented carbides, although green silicon carbide is occasionally used. Green silicon carbide was the recommended abrasive for cemented carbides until the introduction of diamonds. Finally, cubic boron nitride (CBN) is superior for grinding high-speed steels. CBN is a hard, sharp, cool-cutting, long-wearing abrasive.

There is a lot still unknown about why certain abrasives grind well on some materials and not on others, and this makes for some interesting speculation. All abrasives are harder than the materials ground. But their relative hardness is apparently only one of the factors in effectiveness. Aluminum oxide is three to perhaps five times harder than most steels, but it grinds them easily. Silicon carbide is harder than aluminum oxide but not at all effective on steel; on the other hand, it grinds glass and other nonmetallics that are as hard or harder than many steels. Diamond, which is much harder than cubic boron nitride and many times harder than the hardest steel, does not grind steels well, either. The best theory is that there are chemical reactions between certain abrasives and certain materials that make some abrasives ineffective on some metals and other materials.

The basic symbol in the first position, A, C, D, or B, is often preceded and sometimes followed by a manufacturer's symbol that indicates which abrasive within the group is meant, for example, 9A, 38A, AA.

Second Symbol—Grit Size (A60-J8V)

The symbol in the second position of the standard marking represents the size of the abrasive grain, usually called *grit size*. This is a number ranging from 4 to 8 on the coarse side to 500 or higher on the fine side. The number is derived from the approximate number of openings per linear inch in the final screen used to size the grain; the larger the number, the smaller the abrasive grain. Any standard grit size contains grain of smaller and larger sizes, whose amounts are strictly regulated by the federal government, because it would be very expensive to reduce the mix to just the size indicated.

Although there is no real agreement as to what is coarse or fine, for general purposes anything from 46 to 100 might be considered medium, with everything 36 and lower considered as coarse and anything 120 and higher considered as fine. Selection of grit in any shop depends on the kind of work it is doing. Thus, where the job is to remove as much metal as fast as possible, 46 or 60 grit size would be considered very fine. On the other hand, in a shop specializing in fine finishes and close tolerances, 240 might be considered coarse.

A final point is in order about grit size. Most standard symbols end in a "0," particularly the three-digit sizes 100 and finer. However, every abrasive grain manufacturer also makes for special uses, some combinations that are not standard and these usually end in a "1" or other low digit. Thus, a 240 grit is the finest that is sized by screening. Finer grits are sized by other means. On the other hand, 241 grit is a coarse 24 grit in a "1" combination.

Coarse grain is used for fast stock removal and for soft, ductile materials. Fine grain is used to obtain good finishes and for hard or brittle materials. Some materials are hard enough that fine grain removes as much stock as coarse, and neither removes very much. In a general machine shop or toolroom, most of the wheels used will be between 46 and 100 grit.

Third Position—Grade or "Hardness" (A60-J8V)

In the third position is a letter of the alphabet called *grade*. The later the letter, the "harder" the grade. Thus, a wheel graded F or G would be considered

"soft," and one graded R to Z would be very "hard." Actually, these descriptions are put into quotation marks because the words are as close as we can come but are still not quite accurate. What is being measured is the hold that the bond has on the abrasive grain (Figure G1.7); the greater the proportion of bond to grain, the stronger the hold and the harder the wheel. Precision grinding wheels tend to be on the soft side, because it is necessary to have the grains pull out as they become dull; otherwise, the wheel glazes and its grinding face becomes shiny, but the abrasive is dull. On high-speed, high-pressure applications like foundry **snagging,** the pressures to pull out the grains are much greater, and a harder wheel is needed to hold in the grains until they have lost their sharpness. Ideally, a wheel should

Weak holding power

Medium holding power

Strong holding power

FIGURE G1.7 Three sketches illustrating (from top down) a soft, medium, and hard wheel. This is the "grade" of the wheel. The white areas are voids with nothing but air, the black lines are the bond, and the others are the abrasive grain. The harder the wheel, the greater the proportion of bond and, usually, the smaller the voids. (Bay State Abrasives, Dresser Industries, Inc.)

be self-sharpening. The bond should hold each grain only long enough for it (the grain) to become dull. In practice, this is difficult to achieve.

One thing, however, should be mentioned. Grade is a much less measurable thing than type or size of abrasive or, as you will see later, than bond. Grade depends on the formula for the mix used in the wheels, but it must be checked after the wheel is finished.

Fourth Position—Structure (A60-J**8**V)

Following the grade letter, in the fourth position of the symbol is a number from 1 (dense) to 15 (open); this number describes the spacing of the abrasive grain in the wheel (Figure G1.8). Structure is used to provide chip clearance, so that the chips of ground material have some place to go and will be flung out of the wheel by centrifugal force or washed out by the coolant. If the chips remain in the wheel, then the wheel becomes loaded (Figure G1.9), stops cutting, and starts rubbing; it then has to be resharpened, or *dressed*, which is the trade term.

Structure is also a result of grain size and proportion of bond, similar to grade. Quite often, large-grain wheels tend to have open structure, while smaller-sized abrasive grain is often associated with dense structure. On the open side, 11 to 12 and up, the openness is aided by the inclusion of something in the mix like ground-up walnut shells, which will burn out as the wheel is fired and leave definite open spacing.

However, for many grit size and grade combinations, a "best" or standard structure has been worked out through experience and research, and so the structure number may be omitted.

Fifth Position—Bond (A60-J8**V**)

The fifth position of the wheel marking is a letter indicating the bond used in the wheel. This is always a letter, as follows: vitrified—V; resinoid—B (originally the bakelite bond); rubber—R (rubber was used well before resinoid); and shellac—E (originally the "elastic" bond, and also preceded by the now obsolete silicate bond). These are really general bond groups; each wheelmaker uses extra symbols to indicate, for instance, which vitrified bond he has used in a particular wheel, and there is no standardization in these extra symbols.

The bond used has important influence on both the manufacturing process and on the final use of the wheel. Vitrified-bonded wheels are fired at temperatures between 2000 and 2500°F; for that reason, no steel inserts can be used. If such inserts are needed, they must be cemented in afterward. The

FIGURE G1.8 Three similar sketches showing structure. From the top down, dense, medium, and open structure or grain spacing. The proportions of bond, grain, and voids in all three sketches are about the same. (Bay State Abrasives, Dresser Industries, Inc.)

others, also grouped together as organic bonds, are all baked at around 400°F, and inserts may be molded in without problems.

Vitrified, resinoid, and shellac wheels are all pressed in molds after mixing. Rubber-bonded wheels, on the other hand, are mixed in a process similar to that of making dough for cookies; they are then reduced to thickness by passing the mass or grain-impregnated rubber between precisely spaced rolls. For that reason, it is possible to make much thinner wheels with rubber bond than can be made with any other bond. Thin grinding wheels with either resinoid or rubber bonds can be used in cutoff operations such as abrasive saws, but the rubber wheels can be made thinner than the resinoid-bonded wheels.

In general, vitrified wheels are used for precision grinding; resinoid wheels are used for rough grinding with high wheel speeds and heavy stock removal; and rubber or shellac wheels are used for

FIGURE G1.9 The wheel at the left, with small bits of metal imbedded in its grinding face, is called "loaded." It is probably too dense in structure or perhaps has too fine an abrasive grain. At the right, the same wheel has been dressed to remove all the loading. (Desmond-Stephan Mfg. Co.)

more specialized applications. The first two bonds monopolize over 90 percent of the market.

Bond also determines the maximum safe wheel speed. Vitrified bonded wheels, with a few minor exceptions, are limited to about 6500 SFPM, or a little over a mile a minute. The others run much faster, with some resinoid wheels getting up to 16,000 SFPM or more. Of course, all grinding wheels must be properly guarded to protect the operator in the unlikely case of wheel breakage. Grinding machines in general, however, are safe machines. Many will not operate unless the guards are properly in place and closed.

FACTORS IN GRINDING WHEEL SELECTION

Generally, there are seven or eight factors to consider in the choice of a grinding wheel. Some people group together the factors of the amount of stock removal and the finish required. Separating them, however, seems to be more logical.

Of these two groups of factors, three are concerned with the workpiece, which thus change frequently, and five are concerned with the grinder, which are more constant.

Variable Factors

In the first group are things such as composition, hardness of the material being ground, amount of stock removal, and finish.

The composition of the material generally determines the abrasive to be used. For steel and most steel alloys, use aluminum oxide. For the very hard high-speed steels, use cubic boron nitride. For cemented carbides, use diamond. Whether this is natural or manufactured is too specialized a question to be discussed here, but the trend appears to be toward the use of manufactured diamond. For cast iron, nonferrous metals, and nonmetallics, use silicon carbide. On some steels, aluminum oxide may be used for roughing and silicon carbide for finishing. Between aluminum oxide and cubic boron nitride, because of the greatly increased cost, the latter is generally used only when it is superior by a wide margin or if a large percentage of work done on the machine is in the very hard steel alloys like T15.

Material hardness is of concern in grit size and grade. Generally, for soft, ductile materials, the grit is coarser and the grade is harder; for hard materials, finer grit and softer grades are the rule. Of course, it is understood that for most machine shop grinding these "coarse" grits are mostly in the range of 36 to 60, and the finer grits are perhaps in the range of 80 to 120. Likewise, the "soft" wheels are probably something in the range of F, G, H, and perhaps I, and the "hard" wheels are in the range of J, K, or L, and maybe M or N. Too coarse a grit might leave scratches that would be difficult to remove later. And sometimes on very hard materials, coarse grit removes no more stock than fine grit, so you use fine grit. Too soft a wheel will wear too fast to be practical and economical. Too hard a wheel will glaze and not cut.

If *stock removal* is the only objective, then you can use a very coarse (30 and coarser) resinoid-bonded wheel. However, in machine shop grinding, you are probably in the 36 to 60 grit sizes mentioned above and definitely in vitrified bonds.

For *finishing* on production jobs, the wheel may be rubber, shellac, or resinoid bonded. But resinoid bonds are often softened by coolants and therefore rarely used. And for finishing, fine grit sizes are usually preferred; however, as you will learn later when you study wheel dressing (sharpening or renewing the grinding face of a wheel), it is possible to dress a wheel so that a comparatively coarse grit like 54 or 64 will finish a surface as smoothly as 100 or 120 grit.

Fixed Factors

For any given machine the following five factors are likely to remain constant.

Horsepower of the machine, of course, is a fixed consideration. Grinding wheel manufacturers and grinding machine builders are constantly pressing for higher horsepower because that gives the machine and the wheels the capacity to do more work, but that is a factor, usually, only in the original purchase of the machine. In wheel selection, it affects only grade. The general rule is the higher the horsepower, the harder the wheel that can be used.

The *severity of the grinding* also remains pretty constant on any given machine. This affects the choice of a particular kind of abrasive within a general group. Thus, you would probably use a regular or intermediate aluminum oxide on most jobs. But,

in the toolroom, where pressures are low, you would probably want an easily fractured abrasive, white aluminum oxide. On the other end of the scale, for very severe operations like foundry snagging, you need the toughest abrasive you can get, probably an alloy of aluminum oxide and zirconium oxide. For most machine shop grinding, you will probably look first at white wheels.

The *area of grind contact* is also important, but, again, it remains constant for a given machine. The rule is finer grit sizes and harder wheels for small areas of contact; and coarser grit sizes and softer wheels for larger areas of contact. All of this, of course, is within the grit size range of about 36 to 120, or 150 or 180, and within a grade range from E or F to L or N.

It is easy to understand that on a side grinding wheel, where a flat abrasive surface is grinding a flat surface, the contact area is large and the wheels are fairly coarse grained and soft (Figure G1.10). However, on peripheral grinding wheels, it is a different story. The smallest area is in ball grinding, where the contact area is a point, the point where the arc of the grinding wheel meets the sphere or ball. Thus ball grinding wheels are very hard and very fine grained; for instance, 400 grit, Z grade. In cylindrical grinding the contact area is the line across the thickness of the wheel, usually where the arc of the wheel meets the arc of the workpiece (Figure G1.11). Here grit sizes of 54, 60, or 80, with grades of K, L, or M, are common. A still larger area is in surface grinding with peripheral grinding where the line of contact is slightly wider because the wheel is cutting into a flat surface (Figure G1.12). And a combination like 46 I or 46 J is

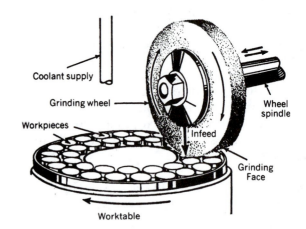

FIGURE G1.10 With the flat wall or rim of the wheel grinding a flat surface, as shown here, the wheel must be soft in grade and can have somewhat coarser abrasive grain. The area of contact between wheel and work is large. (Bay State Abrasives, Dresser Industries, Inc.)

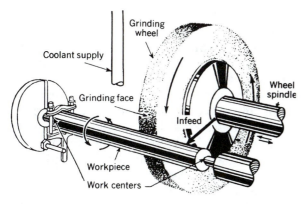

FIGURE G1.11   In center-type cylindrical grinding, as shown here, the arc of the grinding face meets the arc of the cylindrical workpiece, making the area of contact a line. This requires a "harder" wheel than in Figure G1.10. (Bay State Abrasives, Dresser Industries, Inc.)

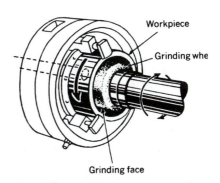

FIGURE G1.13   The contact area of the *OD* grinding face of the wheel and the *ID* surface of the workpiece creates a still larger area of contact and requires a somewhat "softer" grinding wheel than the two previous examples. (Bay State Abrasives, Dresser Industries, Inc.)

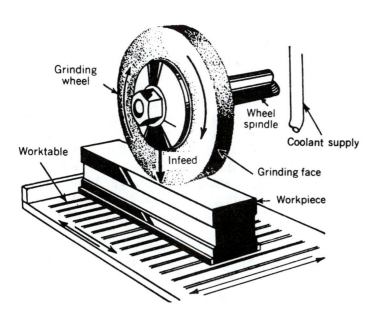

FIGURE G1.12   The contact area between the arc of the grinding face and the flat surface of the workpiece in surface grinding makes a somewhat wider line of contact than in cylindrical grinding. (Bay State Abrasives, Dresser Industries, Inc.)

FIGURE G1.14   The blotter on the wheel, besides serving as a buffer between the flange and the rough abrasive wheel, provides information as to the dimensions and the composition of the wheel, plus its safe speed in RPMs. This wheel is 7 in. in diameter × 1/4 in. thick × 1 1/4 in. hole. It is a white aluminum oxide wheel, 100 grit, I grade, 8 structure, vitrified 52 bond. It can be run safely at up to 3600 RPMs. (Bay State Abrasives, Dresser Industries, Inc.)

not unusual. An internal grinding wheel where the *OD* of the wheel grinds the *ID* of the workpiece may have just a shade more area of grinding contact (Figure G1.13). And then when you get to side grinding wheels (cylinder Type 2), cup wheels, and segmental wheels, which are flats grinding flats, you get grit sizes and grades like 30 J or 46 J (see Figure G1.10). Of course, you have to realize that there may be other factors important enough to override that of contact area. For example, for grinding copper, you might use

a grit size and grade like 14 J; the softness of the metal is probably the key factor.

Wheel speed is a factor that can be dealt with quickly. You must always stay within the safe speeds, which are shown on the blotter or label on every wheel of any size (Figure G1.14). Vitrified wheels generally have a maximum safe speed of 6500 SFPM or a little more; organic wheels (resinoid, rubber, or shellac) go up to 16,000 SFPM or sometimes higher, but these speeds are generally set by the machine designer, and they are safe speeds for the recommended wheels.

Wet or dry grinding is a factor only in that using a coolant will usually permit the use of about one grade harder wheel than would be used for dry grinding, without as much concern about burning the workpieces. Burning is a discoloration of the workpiece surface caused by overheating. The most common cause is usually the use of a wheel that is too hard.

In any shop, however, unless you are really starting from scratch, there will be some information on what wheels have been used and how they have worked. If a factor seems to need changing, it will probably be grit size or grade. You must remember that the shop probably handles a range of work and that it does not pay to switch wheels all the time. Change only one element, either grade or grit size, at a time.

SELF-TEST

1. In the course of a week's grinding you might come up with some of each of the following to grind: bronze valve bodies, steel fittings, tungsten carbide tool inserts, and high-speed steel tools. If you could pick the ideal abrasive for each metal, what would you use? List four abrasives. If you were limited to three, which one of the four could be eliminated most easily?

2. Straight (Type 1) and cylinder (Type 2) wheels both have three dimensions: diameter, thickness, and a third. What is the third dimension for each and why is it stated that way?

3. Five shapes of grinding wheels are described in this unit. Four are for side grinding and two are for peripheral grinding. List the wheels in the two groups either by name or shape number.

4. Tungsten used in the points of automobile engines is very expensive, which makes it necessary to use the thinnest abrasive cutting wheels possible, 6 × .008 × 1 in. What bond would be used and why?

5. Area of contact between wheel and workpiece is probably the most important factor in picking a wheel grade. Five different sets of grinding conditions are discussed, ranging from flat surfacing with a cylinder wheel to ball grinding. List the five in order by wheel grade, starting with the hardest.

6. Here are two wheel specifications, both for straight (Type 1) wheels: (a) A14-Z3 B, and (b) C14-J6V. Describe the composition of each wheel in a sentence or two and suggest the material to be ground by each.

7. Here are two more specifications: (a) C36-K8V and (b) C24-H9V, one for peripheral grinding and one for side grinding. From these specifications, tell which is which.

8. Here is an actual wheel specification: 32A46-H8VBE. Describe the wheel's composition, stating at least the abrasive used, the size of the abrasive, the grade, structure, and bond.

9. A wheel specification for cylindrical grinding of a hard steel fitting with a straight wheel is: A54-L5 V. If you were grinding a flat piece of the same steel with a straight wheel, what elements of the specification might change? Which way? For flat grinding of the same material with a segmental or a cup wheel, what further changes might be made?

10. Write one or two sentences about each of the following to show what elements of a wheel specification it affects.
    a. Material to be ground.
    b. Hardness of the material.
    c. Amount of stock to be removed.
    d. Kind of finish required.

Grinding Wheel Safety

An old saying among grinding wheel manufacturers is: A grinding wheel doesn't break, but it can be broken. Grinding machine builders go to great lengths to design guards to hold wheel fragments in case of breakage without cutting down on the ease of operation or the productivity of the grinder. They have succeeded to the extent that the modern grinder, as a machine tool, is a pretty safe piece of equipment. A modern grinder, treated with due care and respect, is a safe machine, but it does require a knowledge and the practice of safe methods of operation.

OBJECTIVES

After completing this unit, you should be able to:

1. List the steps in and, if possible, demonstrate checking a grinding wheel for soundness.
2. List and, if possible, demonstrate the preliminary steps in mounting a wheel and starting a machine safely.
3. List at least six things you should do or not do when grinding, aside from those covered above.
4. Demonstrate safe practices in handling and mounting grinding wheels.
5. Demonstrate safe personal practices around grinding machines.
6. Given a wheel diameter, calculate in revolutions per minute its maximum safe speed (6500 SFPM or less). Convert RPMs to SFPMs for given wheel diameters.

THE GRINDING MACHINE AND SAFETY

Here is the situation on practically all machine shop grinders. In the machine, spinning at 5000 FPM or more, is a vitrified abrasive wheel made of the same material as dishes. The wheel is susceptible to shocks or bumps. It can easily be cracked or broken. If that happens, even though the machine has been designed with a safety guard that will contain most of the pieces (Figures G2.1 and G2.2), there is a possibility of broken pieces from the wheel flying around the shop. That could be, at the least, unpleasant and even dangerous. Fortunately, there is not much chance that it will happen; but it could.

Other grinding wheels with organic bonds can be operated safely at speeds over 15,000 SFPM, but these are mostly for rough grinding. The wheels are built for it, but the principles are the same. *Every grinding wheel, wherever used, has a safe maximum speed, and this should never be exceeded.*

Any shop grinding machine, properly handled, is a safe machine. It has been designed that way. It should be maintained to be safe. It should be respected for its possibility of causing injury, even though that possibility is low.

Much of the image of grinding wheels and their breakage comes from portable grinders, which are often not well maintained and are sometimes operated by unskilled and careless people.

DETERMINING WHEEL SPEED

Grinding wheels are always marked with a maximum safe speed, but, because of the importance of wheel speed in grinding wheel safety, it is important to know how it is calculated. This quantity is expressed in surface feet per minute (SFPM), which is the distance a given spot on a wheel travels in a minute. It is calculated by multiplying the diameter (in inches) by 3.1416, dividing the result by 12 to convert to feet, and multiplying that result by the number of revolutions per minute (RPM) of the wheel. Thus, a 10-in.-diameter wheel traveling at 2400 RPM would be rated at approximately 6283 SFPM, under the safe speed of

FIGURE G2.1 Safety guard on a surface grinder. Note that the guard is somewhat squared off and covers well over half the wheel. (The DoAll Company)

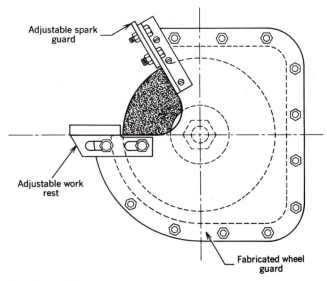

FIGURE G2.2 Safety guard for high-speed wheel. The work is handheld against the exposed peripheral grinding face of the wheel on top of the work rest. The squared corners tend to retain fragments in case of wheel breakage.

most vitrified wheels of 6500 SFPM. To find the safe speed in RPMs of a 10-in.-diameter wheel, the formula becomes

$$\frac{SFPM}{D \times 3.1416 \div 12}$$

or,

$$\frac{6500}{10 \text{ in.} \times 3.1416 \div 12} = 2483 \text{ RPM}$$

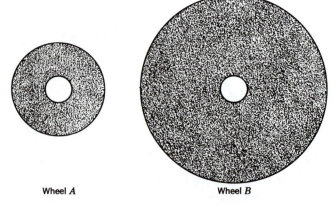

FIGURE G2.3 Wheel speed in SFPM, at a given RPM, increases as the wheel diameter increases. If the diameter of wheel *B* is twice that of wheel *A*, then the SFPM of wheel *B* is twice that of wheel *A*, at the same RPM.

Most machine shop-type flat surface or cylindrical grinders are preset to operate at a safe speed for the largest grinding wheel that the machine is designed to hold. As long as the machine is not tampered with and no one tries to mount a larger wheel on the machine than it is designed for, there should be no problem.

It should be clear that with a given spindle speed (RPM), the speed in SFPM increases as the wheel diameter is increased (Figure G2.3), and it decreases with a decrease in wheel diameter. Maximum safe speed may be expressed in either way, but on the wheel blotter it is usually expressed in RPM.

THE RING TEST

The primary method of determining whether or not a wheel is cracked is to give it the **ring test.** A crack may or may not be visible; of course, if there is a visible crack you need go no further and should discard the wheel immediately.

The test is simple. All that is required is to hold the wheel on your finger if it is small enough (Figure G2.4) or rest it on a clean, hard floor if it is too big to be held (Figure G2.5) and strike it about 45 degrees either side of the vertical centerline preferably with a wooden mallet or a similar tool. If it is sound, it should give forth a clear ringing sound. If it is cracked, it will sound dead. The sound of a vitrified wheel is clearer than the sound of any other, but there is always a different sound between a solid wheel and a cracked one.

There are a few permissible variations. Some operators prefer to hold the wheel on a stick or a

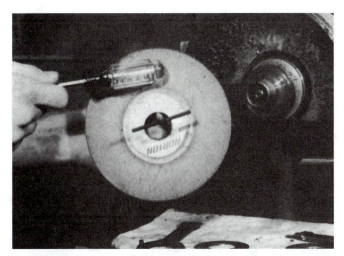

FIGURE G2.4 Making a ring test on a small wheel.

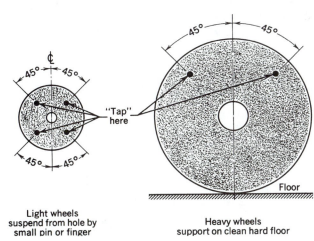

FIGURE G2.5 Sketch shows where to tap wheels for ring test, 45 degrees off centerline and 1 to 2 in. in from the periphery. After the first tapping, rotate the wheel about 45 degrees and repeat the test. (Bay State Abrasives, Dresser Industries, Inc.)

metal pin instead of a finger. Some shops prefer to suspend large wheels by a sling (chain or cable covered by a rubber hose to protect the wheel) instead of resting them on the floor (Figure G2.6).

The important point is that the test be done when each lot of wheels is received from the supplier, just before a wheel is mounted on a grinder, and again each time before it is remounted. If a wheel sounds cracked, or even questionable, then discard it or set it aside to be checked by the supplier and get another wheel, which, of course, must also be ring tested.

FIGURE G2.6 Wheel being lifted by a sling for ring testing. (The DoAll Company)

WHEEL STORAGE AT THE MACHINE

It should be clear that the wheels ought to be stacked carefully and separated from each other by corrugated cardboard or another buffer (Figure G2.7). Tools and other materials, particularly metals, should never be stacked on top of the wheels in transit or at the machine. In short, *handle with care.*

General wheel storage is a responsibility of shop management, but at the machine this is usually the operator's concern. Operator carelessness can ruin an otherwise good wheel-storage plan. On the other hand, care by the operator can do much to help out a substandard shop plan.

If the shop provides you with a proper storage area, all you have to do is to use it and keep it clean. If the shop does not provide such an area, you may need some ingenuity. Here are some suggestions.

1. Keep on hand only the wheels you really need. Do not allow extra wheels to accumulate around your machine. They can become cracked or broken and

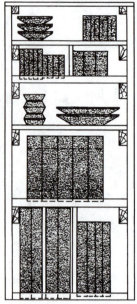
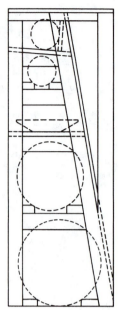

FIGURE G2.7 A recommended design for the safe storage of grinding wheels. Corrugated paper cushions should be placed between grinding wheels that are stored flat. The storage racks should be designed to handle the type and sizes of wheels used.

FIGURE G2.8 Storing extra wheels at the machine on pegs is often convenient and practical. The main requirement is to keep the wheels separate or protected and off the floor. (The DoAll Company)

may get in your way. If you do not need a wheel, take it back to the general storage.

2. Store wheels above floor level, either on a table or under it, on pegs in the wall (Figure G2.8), or in a cabinet. Be especially careful to protect wheels from each other and from metal tools or parts. Use corrugated cardboard, cloth, or even newspaper to keep wheels apart. *Keep wheels off the floor.*

MACHINE SAFETY REQUIREMENTS

Mounting flanges must be clean and flat and equal in size, at least one-third the wheel diameter (Figure G2.9). Between each flange and the wheel there should be a blotter, a circular piece of compressible paper to cushion the flange from the rough wheel and distribute the flange pressure. The side of the flange facing the wheel has a flat rim, but the rest is hollowed out so that there is no pressure at the hole, which is the weakest part of the wheel.

The wheel guard must be in place. Depending on the machine design, the safety guard covers half or more of the wheel (Figure G2.10). On older-model grinders the guard tends to be circular in shape and fits closely around the wheel. However, the latest models have square-shaped safety guards, sometimes called *cavernous,* on the theory that pieces of

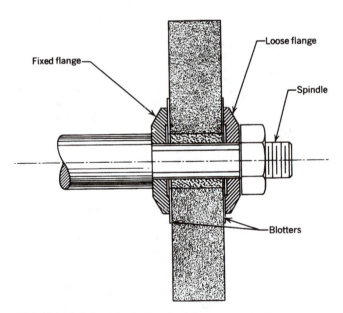

FIGURE G2.9 Typical set of flanges with flat rims and hollow centers with blotters separating the wheel and the flanges. Tightening the nut too much could spring the flanges and perhaps even crack the wheel.

FIGURE G2.10 Well-designed guard for a bench grinder. Note that the side of the guard, like the one on the surface grinder, is easily removed for access to the wheel.

any wheel that breaks will be retained in the corners instead of being slung out against the machine or possibly the operator.

MOUNTING THE WHEEL SAFELY; STARTING THE GRINDER

Any time you mount a wheel on the grinder or start the machine, good safety practice requires a set routine like the following.

1. Ring test the wheel as shown in Figure G2.4, then check the safe wheel speed printed on the blotter with the spindle speed of the machine. The spindle speed must never exceed the safe wheel speed, which is established by the wheel manufacturer after considerable research.
2. The wheel should fit snugly on the spindle or mounting flange. Never try to force a wheel onto the spindle or enlarge the hole in the wheel. If it is too tight or too loose, get another wheel.
3. Make sure the blotters on the wheel are larger in diameter than the flanges, and make sure the flanges are flat, clean, and smooth. Smooth up any nicks or burrs on the flanges with a small abrasive stone. Do not overtighten the mounting nut. It just needs to be snug.
4. Always stand to one side when starting the grinder; that is, stand out of line with the wheel.
5. Before starting to grind, let the wheel run at operating speed for about a minute with the guard in place.

Steps 4 and 5 are essentially a double check on the ring test. If anything is going to happen to a wheel, it usually happens very quickly after it starts to spin. Jog the switch when starting a new wheel.

OPERATOR RESPONSIBILITIES

In grinding, there are a number of operator duties that have long been a part of company safety policy and state safety regulations and that, with the passage of the OSHA regulations, also became a matter of federal regulations. These are mostly designed for your protection and not, as some operators have thought, to make the job more awkward or less productive.

1. Wear approved safety glasses or other face protection when grinding (Figure G2.11). This is the first and most important rule.
2. Do not wear rings, a wristwatch, gloves, long sleeves, or anything that might catch in a moving machine.
3. Grind on the side of the wheel *only* if the wheel is designed for this purpose, such as Type 2 cylinder wheels, Types 6 and 11 cup wheels, and Type 12 dish wheels (Figure G2.12).
4. If you are grinding with coolant, turn off the coolant a minute or so before you stop the wheel. This prevents coolant from collecting in

FIGURE G2.11 Approved safety glasses are required for all grinding. Today, as a matter of fact, they are usually required for anyone, even visitors, in the shop area. (Sellstrom Manufacturing Co., Palatine, Ill.)

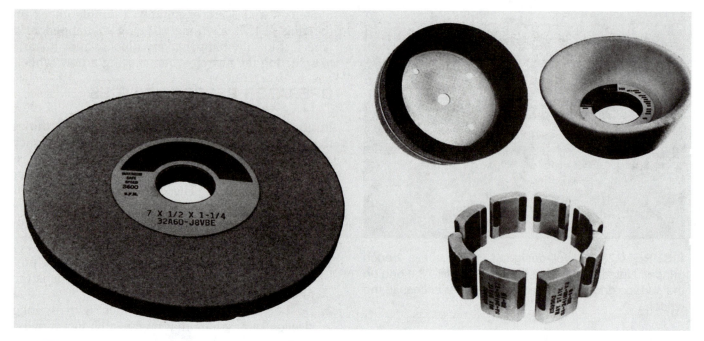

FIGURE G2.12 Straight wheels are designed for grinding on the periphery. Never grind on the side of one of these, because it is not considered safe. On the other hand, cylinder wheels, cup wheels (both straight and flaring), and segments or segmental wheels (shown here without the holder) are all designed and safe for side grinding. (Bay State Abrasives, Dresser Industries, Inc.)

the bottom half of the wheel while it is stopped and throwing it out of balance.

5. Never jam work into the grinding wheel. This applies particularly in off-hand grinding on a bench grinder.

SELF-TEST

1. A 12-in. vitrified wheel is available for mounting on a 2000 RPM grinder. Can this wheel be used safely on the grinder? Show the calculations to prove your answer.

2. List at least three important requirements for a ring test.

3. There are at least three times when a wheel should be ring tested. What are they, and why is the test needed each time?

4. Handling wheels carefully is most important. List at least four specific precautions to be taken in handling grinding wheels.

5. Wheel flanges are not always considered as important as they really are. Put down at least four points to be considered in their selection and care.

6. There are five or six essential steps to be followed in starting up a grinder. List them in order.

7. Providing a safe place to work is an employer's responsibility, but an operator also has safety responsibilities. List at least four things that an operator should or should not do to operate a grinder safely.

8. Explain the relationship of wheel diameter and spindle speed (RPM) to wheel speed in SFPM.

9. At the factory, grinding wheels are tested running free at 150 percent of safe operating speed. Why, then, isn't it all right to operate the wheels at the same higher speed?

10. Safe grinding practice requires that you not wear certain things (clothing and similar items) around a grinding machine. Name at least three.

Using the Surface Grinder

The small horizontal-spindle reciprocating-table surface grinder is considered to be basic to most general purpose machine shops. With accessories, it is extremely versatile and can do a remarkable variety of precision work. It is important that you learn the nomenclature and functions of this type of grinder and that you are aware of the many accessories that can be obtained to make this machine very useful in a wide variety of situations.

OBJECTIVES

After completing this unit, you should be able to:

1. Name the components of the horizontal-spindle surface grinder.
2. Define the functions of the various component parts of this grinder.
3. Name and describe the functions of at least four accessory devices used to increase the versatility of the surface grinder.

The small horizontal-spindle reciprocating-table surface grinder (Figure G3.1) must be considered the basic grinder. There is at least one in nearly every machine shop, toolroom, or school shop. Operating one on straight flat work can be relatively simple. In fact, it has been said that three-fourths or more of all grinding work could be done on a small 6-by-12-in.-capacity surface grinder.

In a sense, of course, all grinding is done on a surface, but a surface grinder is one for producing flat surfaces (or, with these horizontal spindle machines, formed surfaces) as long as the hills and valleys of the form run parallel with the path of the wheel as the magnetic chuck moves the work under it. Figure G3.2 shows a simple form. Vertical-spindle machines, however, are usually limited to flat surfaces because the grinding is done on the flat of the wheel with the abrasive "scratches" making an overlapping, circular path.

It is also possible on the horizontal-spindle reciprocating-table grinder, with proper fixturing, to grind surfaces that are not parallel but are either flat or formed. Finally, with accessories, it is possible to do almost any kind of grinding: center-type, cylindrical, centerless, and internal. This is usually incidental work and small workpieces. But these show that the small hand-operated surface grinder is a very versatile machine.

The best place to start learning the part names of a hand-operated horizontal-spindle reciprocating-table grinder is the wheel-work interface, the point where the action is. In this unit, you should understand that the term *grinder* means this type of machine.

The cutting element is the periphery, or OD, of the grinding wheel. The wheel moves up and down under the control of handwheel A, as shown in Figure G3.1. The doubleheaded arrow A shows the direction of this movement. The wheel, together with its spindle, motor drive, and other necessary attachments, is often referred to as the *wheelhead*.

MAGNETIC CHUCKS

The work usually is held on a flat magnetic chuck clamped to the table, which is supported on saddle ways. The magnetic chuck, the most common accessory for this grinder, holds iron and steel firmly enough to be ground. Using blocking of iron or steel, it will also hold nonmagnetic metals like aluminum, brass, or bronze. Only very rarely is it necessary to fasten a workpiece directly to the machine table.

In industrial shops, the electromagnetic chuck is the basic workholder for the surface grinder. Only large machines have the electromagnetic type; the permanent magnetic chuck is often used on small machines. There is no basic difference in principle

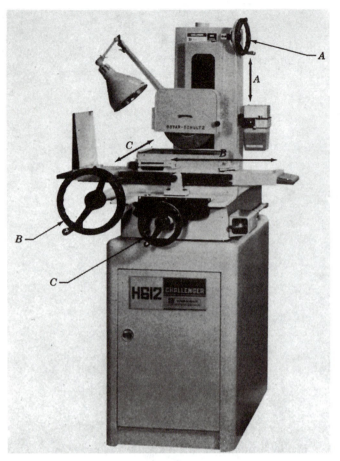

FIGURE G3.1 Surface grinder with direction and control of movements indicated by arrows. Wheel *A* controls down feed *A*. Large wheel *B* controls table traverse *B*. Wheel *C* controls cross feed *C*. (Boyar–Schultz Corporation, An Esterline Company, Broadview, Ill.)

FIGURE G3.2 Grinding a simple rounded form. Note that the wheel is dressed to the reverse of the form in the finished workpiece.

FIGURE G3.3 Common type of magnetic chuck for reciprocating surface grinder. The guards at the back and left side are usually adjustable and help keep work from sliding off the chuck. (The DoAll Company)

(Figure G3.3). Permanent magnetic chucks are made up of a series of alternating plates composed of powerful alnico magnets and a nonmagnetic material. Some improved types use ceramic magnets alternating with stainless steel plates. They exert a more concentrated holding force and can be used for milling as well as grinding. The electromagnetic chuck does have one advantage over the permanent magnet type, however, in that the electric control for the electromagnet usually has a built-in de-mag (demagnetization) mode that will de-mag the workpiece after grinding. One disadvantage of the electromagnetic chuck is the possibility of a power failure or a shorted cord. The result would be a damaged workpiece and wheel breakage.

So, for many workpieces, all that is needed is to clean off the chuck surface and the workpiece, place the part, turn on the magnet, check for firm workholding, and start grinding (assuming that the wheel on the grinder is suited to the workpiece). On all such pieces, it is imperative that the surface to be ground is parallel with the surface resting on the magnetic chuck. The surface of the

FIGURE G3.4 Magnetic sine chuck needed for grinding nonparallel surfaces. It is, of course, adjustable. (Hitachi Magna-Lock Corp.)

chuck and the "grinding line" of the wheel must always be parallel.

However, surfaces that are not parallel can be ground. As a simple illustration, consider a beveled edge to be ground around all four sides of the top of a rectangular block. You have two choices. One is to use a fixture that can be set to the angle you want, which is a magnetic sine chuck, as shown in Figure G3.4. The other is to **dress** the angle you want on the wheel with an attachment called a sine dresser and then grind in the usual fashion. The choice is up to the operator.

PROCEDURES

The chuck moves from left to right and back again (traverse), as shown in Figure G3.1, arrow *B*. On many grinders, traverse and crossfeed increments are controlled hydraulically with the table moving back and forth between preset stops. When the hydraulic control is actuated, the traverse wheel is automatically disengaged. This movement is controlled by large handwheel *B*. The chuck also moves toward and away from the operator (cross feed), as shown by arrow *C*, with the motion controlled by handwheel *C*. The grinding wheel, incidentally, always runs clockwise.

On a completely hand-operated machine, the operator stands in front of the machine with the left hand on the traverse wheel and the right hand on the crossfeed wheel (if right-handed), swinging the table and the magnetic chuck back and forth with the left

FIGURE G3.5 Operator in position at grinder. On many small surface grinders the cross feed and traverse are hydraulic.

hand and cross feeding with the right hand at the end of each pass across the workpiece (Figure G3.5). The wheel clears the workpiece at both ends of the pass, and the cross feed is always less than the width of the wheel so that there is an overlap. At the end of each complete pass across the entire surface to be ground, the operator feeds the wheel down with the downfeed wheel on the column.

The traverse handwheel is not marked, because all that is necessary is to clear both ends of the surface. However, both of the other wheels have very accurately engraved markings so that down feed and cross feed can be accurately measured. The downfeed handwheel (Figure G3.6) has 250 marks around it, and turning the wheel from one mark to the next lowers or raises the wheel .0001 in. This is one ten-thousandth of an inch, known in the shop usually as a "tenth." The crossfeed handwheel has 100 marks, and moving the wheel from one mark to the next moves the workpiece toward or away from the operator .001 in.

With the 250 marks on the downfeed handwheel and .0001 in. movement per mark, a complete revolution of the handwheel moves the grinding wheel .025 in. (250 × .0001 in.). A complete revolution of the crossfeed wheel, which has 100 marks, moves the table and the chuck .100 in. (100 × .001 in.).

FIGURE G3.6 Closeup of downfeed handwheel. Moving from one mark to the next lowers or raises the grinding wheel .0001 in. (The DoAll Company)

FIGURE G3.7 Same control wheel with slip ring set to zero. Now it is simpler for the operator to down feed the grinding wheel as it grinds. (The DoAll Company)

The other feature needing mention is the zeroing slip ring on the downfeed handwheel or the crossfeed handwheel. You cannot predict in advance just where on the scale either wheel will be when the grinding wheel first contacts the workpiece and where the control wheels ought to be when the surface is ground. These can be figured out mathematically, but with considerable chance of error. However, with the zeroing slip ring, the starting point on the scale on each wheel is simply set at zero, locked in place (Figure G3.7), and then ground until the required amount of stock has been removed. Allowance must be made for wheel wear, especially where considerable stock is removed by grinding. On most small surface grinders, downfeed for roughing should be no more than .002 in. per pass and no more than .0005 in. per pass for finishing operations.

On a hand-operated grinder a skilled operator develops a rhythm as the workpiece traverses back and forth under the grinding wheel, cross feeding at the end of each pass and down feeding when the whole surface has been covered. This is a knack you develop with experience. Using machines with hydraulic traverse and cross feed eliminates the need for physical coordination on the part of the operator, but skill is necessary to choose good traverse speeds and the amount of cross feed for each pass for first-class grinding results. Generally, combinations of large crossfeed movements on the order of one-half the wheel width and relatively small amounts of down feed are preferred because wheel wear is distributed better this way.

Actually, with the wheelhead, the magnetic chuck, and the three control wheels, you have the basic parts of a surface grinder. You also need a dressing device, which may be just a holder with the diamond mounted at the proper angle that can be mounted on the magnetic chuck or a built-in dresser (Figure G3.8). But everything else on the machine is either for support (for instance, the table, saddle ways, base, and column that holds the wheelhead) or an accessory that makes it possible to do something you could not do otherwise or that makes the job easier to do.

TRUEING THE WHEEL

Chances are that you will most often be **trueing** the wheel so that the peripheral surface is concentric with the center of the machine spindle and parallel to the centerline of the spindle and with a single-point diamond dresser. The dresser may be mounted above

FIGURE G3.8 Built-in wheel dresser. The lever traverses the dresser across the wheel. (The DoAll Company)

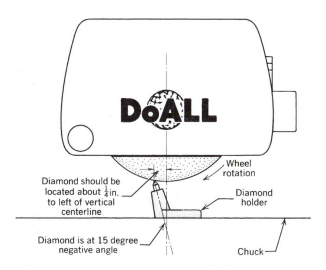

FIGURE G3.9 This is one of several ways of mounting a dresser on a surface grinder. The dresser with its diamond is simply spotted on the clean magnetic chuck. Note, however, that the diamond is slanted at a 15-degree angle and slightly past the vertical centerline of the wheel, as the wheel turns. (The DoAll Company)

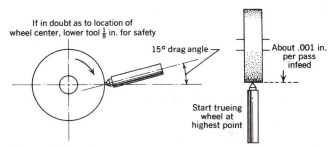

FIGURE G3.10 This illustrates the idea of wheel dressing instead of any specific setup. Dressing is rarely done freehand. Note that the diamond is always a little past the centerline and on an angle. (The Desmond–Stephan Manufacturing Co., Urbana, Ohio)

the wheel (Figure G3.8), on the magnetic chuck that is the workholder of the machine (Figure G3.9), or to the side (Figure G3.10). In any of these positions, it is possible to move the dresser back and forth across the face of the wheel. This is called **traversing** the dresser.

The diamond is mounted at an angle of 15 degrees so that it contacts the wheel just after the low or high point of the wheel or just below the centerline, depending on the location of the dresser (Figures G3.8, G3.9, and G3.10). Thus, the wheel is cutting toward the point of the diamond. Remember that dressing with a diamond is always a two-way operation; that is, the diamond is kept sharp while it sharpens the wheel grinding face. Often there is an arrow or other indicator on the dresser to indicate its position. The diamond always points in the direction of the wheel's rotation.

Any new wheel, or a wheel that has just been reflanged, must first be trued. In trueing, the wheel and the dresser are brought together so that the dresser is touching the high point of the wheel. Oth-

erwise, the traversing of the diamond might cause it to dig too deeply into the wheel, which could ruin the diamond. On the other hand, if you start at the high point, which is the point furthest from the center, the cross cut or traverse is short at first and gradually becomes longer until finally you are dressing the entire width of the wheel.

Infeed of the diamond into the wheel should be light, about .001 in. per pass; if dressing is being done dry, there should be frequent pauses, after every three or four passes, to allow the diamond to cool off. A hot diamond can be shattered if a drop of water or other liquid hits it. Turn the diamond frequently. This helps to keep it sharp.

Trueing the wheel is accomplished, then, by moving the dresser back and forth across the grinding face of the wheel while the wheel is rotating at operating speed. It is preferably done wet. If trueing is done wet, continuous coolant must be assured.

The speed of traverse is probably the remaining point of concern. Generally, the faster the dresser traverses across the wheel's grinding face, the sharper the dress will be and the better suited the wheel will be for rough grinding. Slower traverse means that the diamond does more cutting on the abrasive, dulling it a little bit, so that the wheel is better suited to finishing than to roughing work. If the diamond is traversed back over a newly dressed wheel without infeed, the wheel will be dulled, causing it to burn or glaze the workpiece. Sometimes, however, when a wheel is intended for finishing, it is good to take a few passes across the wheel without any infeed. The point is that with a little experience you can tell the degree of sharpness that is needed in the wheel face and dress the wheel accordingly, depending on what you want to do with it afterward. As grain becomes dulled, it tends to polish more and cut less, and size becomes less of a factor in the action of the grain.

Generally, a coarse-grain wheel is recommended for cutting and material removal and fine-grain is recommended for finishing. This is true, provided both are in the same degree of sharpness. It can now be said that by proper dressing, it is possible to make a finish with comparatively coarse grain like that of a much finer grain. For example, a 46 grit wheel, dressed to dull the grain, could give the same finish as a 120 grit wheel. The reverse is not true, however.

GRINDING FLUIDS

Grinding produces very high temperatures. Temperatures at the interface (the small area where the abrasive grains are actually cutting metal) are reliably estimated to be over 2000°F (1093°C), and that is enough to warp even fairly thick workpieces. Nor is it safe to assume that just because there is a lot of grinding fluid (sometimes called *coolant* in the shop) flowing around the grinding area, the interface is cooled. With the grinding wheel rotating at its usual 5000 SFPM-plus rate, it creates enough of a fan effect to blow coolant away from the contact area. This creates a condition sometimes referred to as "grinding dry with water." In other words, in spite of the amount of coolant close by, the actual cutting area may be dry and hot. However, the coolant can do a good job of cooling the contact area if it is properly

FIGURE G3.11 This is a very common method of flood coolant application. For the photograph, the volume of coolant has been reduced.

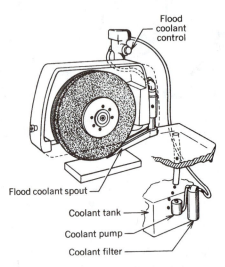

FIGURE G3.12 Fluid recirculates through the tank, piping, nozzle, and drains in a flood grinding system. (The DoAll Company)

applied. Most grinding fluids are water-based; that is, a mixture of **soluble oil** and water.

Methods of Coolant Application

Most grinding fluids are applied by a method called *flooding*. A stream of coolant under pressure is directed from a pipe, sometimes shaped but often just round, in the general direction of the grinding area (Figure G3.11). The fluid collects beneath the grinding area, is piped back to a tank where it is allowed to settle, and is cleaned; then it is pumped back around the wheel (Figure G3.12). There is always a little waste, and periodically either more coolant concentrate or water is added to the solution. Flooding is an effective method of applying coolant; provided the solution stays within the effective range, neither too rich nor too lean, it works very well. However, the fluid cannot just be in the vicinity of the grinding area; it must be right *in* that area if it is to do its job. For example, on a small surface grinder processing small parts, the nozzle can almost be pointed at the grinding wheel-workpiece contact area.

ACCESSORIES

The list of accessories for a hand and hydraulically operated toolroom-type surface grinder may be quite extensive. As mentioned before, with the proper accessories, almost any type of grinding can be done on one of these little grinders within maximum size ranges. Finally, accessories are a major point of difference between toolroom and production machines. Toolroom grinders must handle a variety of work, so accessories are needed. Production machines are used mostly for one purpose; if that purpose changes, the machine is rebuilt or modified for its new use or it is removed from service.

ATTACHMENTS

For practical purposes it probably makes very little difference whether something is called an *attachment* or an *accessory*. Both make it possible to do something with the machine that could not otherwise be done, or at least could not be done so easily or quickly.

Center-Type Cylindrical Attachment

This is basically a workholder with a headstock and a tailstock (Figure G3.13). It is mounted crosswise on the locked table so that the cross feed of the grinder makes the wheel traverse end-to-end on the workpiece. If a flat is needed on an essentially cylindrical part, then all that is necessary is to stop the

FIGURE G3.13 Center-type cylindrical attachment mounted on a surface grinder. The attachment can be tilted for grinding a taper, as shown here, or set level for grinding a straight cylinder. (Harig Manufacturing Corporation)

rotation of the work, reciprocate the table just a little, and cross feed as for any flat surface.

High-Speed Attachment

For incidental internal grinding, this attachment (Figure G3.14), driven by a belt from the grinder spindle, provides the high speed that is needed to make the small mounted wheels run at the high RPMs needed to make them grind efficiently. Of course, it is usually essential to provide an attachment for mounting the workpiece also.

FIGURE G3.14 High-speed spindle adds capability for internal grinding. (Whitnon Spindle, Division of Mite Corporation)

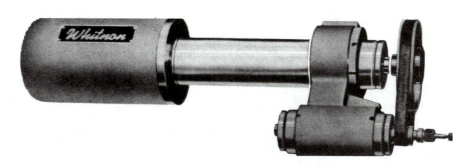

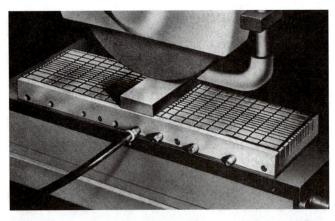

FIGURE G3.15 Vacuum chucks such as this one hold practically anything and are considered good for thin work. (Hitachi Magna-Lock Corp., Big Rapids, Mich.)

Vacuum Chuck

This replacement for an electromagnetic chuck (Figure G3.15) holds the work by exhausting the air from under it. Thus, it makes no difference whether the workpiece is magnetic or not. It is also recommended by some experts for holding pieces as thin as only a few thousandths of an inch.

SURFACE GRINDING A VEE-BLOCK

Demonstrating any skill is, of course, the final test of whether you have learned it. The workpieces selected for this unit on using the surface grinder are two matching vee-blocks of SAE 4140 or a similar alloy steel. A series of grinding steps are given in the text to detail the process for making these vee-blocks. Some of these steps will include grinding operations on both blocks in the same setup at the same time. The specifications for surface grinding are often much closer than those required on other machine tools. These grinding machines, however, are capable of holding tolerances within one-tenth of a thousandth of an inch.

Selecting the Wheel

A SAE 4140 steel in the Rockwell C hardness range of 48 to 52 is regarded as being not too difficult to grind. It can be ground satisfactorily with a number of wheel specifications that are likely to be on hand in practically any grinding shop. Wheels for machine shop use have to be able to grind a wide range of materials, because the number of parts to be ground in any single lot is not likely to be large enough to warrant the trials necessary to find *exactly* the best wheel specification. Here, however, are some guidelines in the selection.

The abrasive will be aluminum oxide because the material is steel, and the bond will be vitrified because this is precision grinding. Given the requirements for this particular workpiece, the following, giving a range of four possible selections of wheels, are the recommendations of several specialists in the grinding field:

1. 9A46-H8V
2. 9A60-K8V
3. 32A46-I8V
4. DA46-J9V

All of these may be regarded as general-purpose specifications for a part of this configuration, material, and hardness.

Looking at these recommendations in order, the abrasive recommendation of the first two is a white aluminum oxide, which is the most friable (brittle). The third and fourth are for a mixture of white and regular (gray) aluminum oxide, which is slightly tougher and perhaps wears a little longer. In wet grinding, as this is, probably the tougher abrasive would be preferred. However, in dry grinding, there would probably be no question about the use of white abrasive; it does not tend to burn the work as much as a tougher abrasive.

A grit size of 46 would be indicated for grinding efficiency, but 60 would provide a slightly better finish. Wheel grade provides the widest range (H to K), which could be interpreted to mean that any grade within this range would do the job. However, the wheel with the finest grit (60) also has the hardest grade (K). Such a wheel would wear a little longer than the others. Grit size and abrasive types are both likely to be fairly constant from various manufacturers.

The range of structure is only from 8 to 9, which is hardly significant. Grade and structure, however, result from manufacturing procedures and, hence, are at least comparable from one wheel supplier to another.

Vitrified bond is indicated. Vitrified bonds are somewhat varied, although all wheel manufacturers have one or two general-purpose or standard bonds for machine shop work.

Steps in Grinding

(*Note:* This exercise in mounting the wheel on a surface grinder may not be necessary in your shop.)

Step 1. *Select the wheel.* The previous information would help you do this.

FIGURE G3.16 Cleaning the wheel spindle with a soft cloth.

FIGURE G3.18 Mounting the wheel, flange, and nut.

FIGURE G3.17 Ring-testing the wheel.

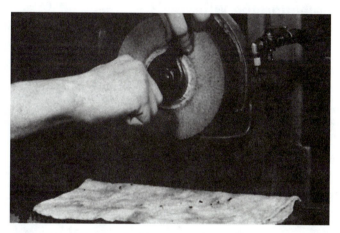

FIGURE G3.19 Tightening the nut with spanner wrenches.

Step 2. *Clean the spindle* (Figure G3.16). Use a soft cloth to remove any grit or dirt from the spindle. Note that the chuck is protected by a cloth to prevent nicks or burrs from tools laid out on it.

Step 3. *Ring-test the wheel* (Figure G3.17). As indicated in Unit 2, this is a safety precaution any time a wheel is mounted.

Step 4. *Mount the wheel* (Figure G3.18). The wheel should fit snugly on the spindle, and the outside flange must be of the same size as the inner flange, as shown in Figure G3.19 This wheel has blotters attached, but if they were damaged or there were none attached, it would be necessary to get some new blotters. Flanges should also be checked occasionally for burrs or nicks and flatness. The flange is held by a nut.

Step 5. *Tighten the nut* (Figure G3.19). The nut should be tightened snugly. The blotters will take

up a little extra force, but if either flange is warped or otherwise out of flat, it is possible to crack a wheel. Overtightening can also crack a wheel.

Step 6. *Replace the safety guard* (Figure G3.20). This is a necessary precaution for safety.

GRINDING VEE-BLOCKS

The vee-block (Figure G3.21) is a common and useful machine shop tool. Precision types are finish machined by grinding after they have been rough machined oversize (about .015 in.) and heat treated. Finish grinding of vee-blocks will provide you with a well-rounded surface grinding experience.

Refer to a working drawing (Figure G3.22) to determine dimensions and to plan the sequence of grinding operations.

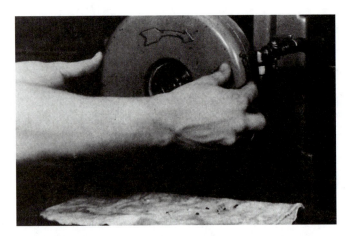

FIGURE G3.20 The safety guard being replaced.

Selecting the Wheel

For these vee-blocks, the best abrasive will be a friable grade of aluminum oxide. A vitrified wheel of 46 grit, with an I-bond hardness would be a reasonable selection.

Grinding Machine Setup

The following procedure may be used in setting up the grinding machine for vee-block grinding:

Step 1. Select a suitable wheel.

FIGURE G3.21 Finished, hardened, and ground precision vee-block.

Step 2. Clean the spindle. Use a cloth to remove any grit or dirt from the spindle. If you lay tools and wheels on the chuck, cover it with a cloth for protection.

Step 3. Ring-test the grinding wheel.

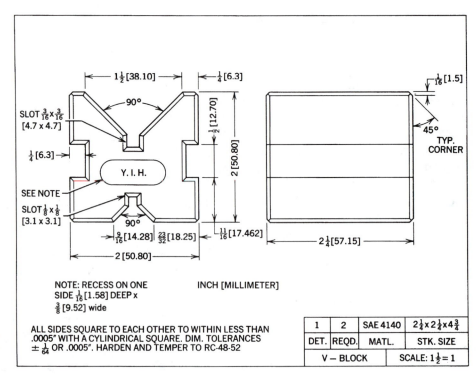

FIGURE G3.22 Dimensions and information for grinding the vee-block.

Step 4. Mount the wheel. The wheel should fit snugly and the flanges should be the same size. Blotters will probably be attached to the wheel. If not, place a blotter between each of the flanges and the wheel.

Step 5. Install the spindle nut and tighten firmly. Do not overtighten.

Step 6. Replace or close the wheel guard.

Step 7. Place the diamond dresser on the magnetic chuck in the proper position (Figure G3.23). (Or use the built-in dresser if you have one.)

Step 8. Dress the wheel, using fluid and a rapid cross feed (Figure G3.24). (Use fluid only if you can be sure of flooding the diamond.)

Step 9. Remove the dresser and clean the chucking surface (Figure G3.25).

Step 10. Check the chucking surface for nicks and burrs (Figure G3.26). Use a deburring stone if necessary.

To avoid the necessity of having to protect the chuck surface with paper, prepare the first part surfaces by removing any heat-treatment scale with an abrasive cloth on a flat surface (Figure G3.27).

Side and End Grinding Procedure

The vee-blocks should be match ground; that is, ground together in pairs so that they are exactly the same dimensions when completed. The following procedural steps and illustrations will describe the process for side and end grinding.

FIGURE G3.23 Diamond dresser in the correct position.

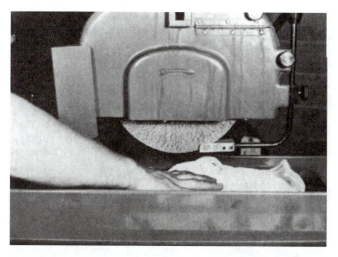

FIGURE G3.25 Cleaning the magnetic chuck with a cloth. The wheel must be completely stopped before this is done.

FIGURE G3.24 Dressing the wheel using grinding fluid.

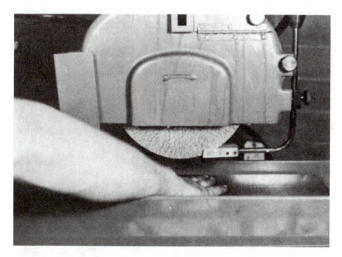

FIGURE G3.26 Checking for burrs on the magnetic chuck.

FIGURE G3.27 Preparing the part to put on the magnetic chuck.

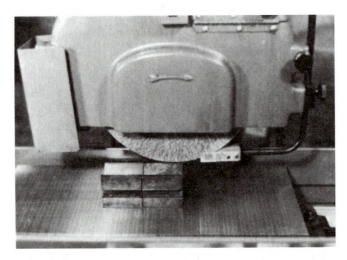

FIGURE G3.28 The rough blocks in place with the large vee-side up ready to be ground.

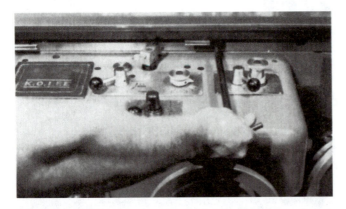

FIGURE G3.29 Setting the table trip dogs (stop dogs). (Lane Community College)

Step 1. Place the blocks with the large vee-side up on the grinder chuck (Figure G3.28).

Step 2. Magnetize the chuck.

Step 3. Lower the wheelhead until it is about an inch above the workpiece. Adjust table and saddle position so that blocks are centered.

Step 4. Adjust table-feed reverse trips so that the workpiece has about one inch of overtravel at each end of the table stroke (Figure G3.29).

Step 5. Use a feeler gage or piece of paper and lower the wheelhead until it is a few thousandths of an inch above the surface to be ground.

Step 6. Start the wheel and grinding fluid. Start table cross feed and table travel. Carefully lower the wheel until it just begins to contact the high point of the workpiece. Set the downfeed micrometer collar to zero at this point.

Step 7. The amount to grind from the workpiece will depend on the amount of extra material left from the original machining. Down feed about .001 in. per pass and grind to a cleanup condition (Figure G3.30). About .003 to .005 in. should be left for finish grinding on each surface.

Step 8. Turn the vee-blocks over and surface grind the small-vee side (Figure G3.31).

Step 9. The end of the block may be ground by clamping it to a precision angle plate (Figure G3.32). Adjust the block into alignment using a test indicator. The end of the block should extend slightly beyond the angle plate.

FIGURE G3.30 Grinding the first surface until the surface is "cleaned up."

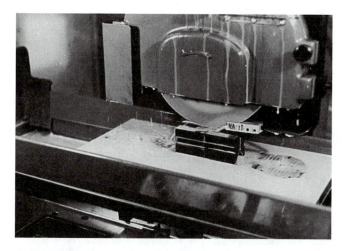

FIGURE G3.31 The blocks are turned over and the opposite sides (small-vee) are ground.

FIGURE G3.33 The precision angle plate and vee-block setup is turned with the vee-block end up on the magnetic chuck. The end of the vee-block is ground square to a ground side.

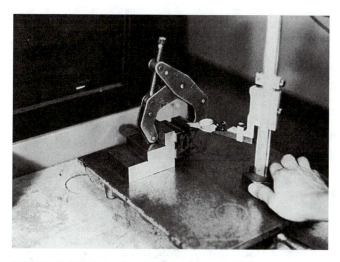

FIGURE G3.32 A ground side of the vee-block is clamped to a precision angle plate, and the part is made parallel, in preparation for grinding the end square.

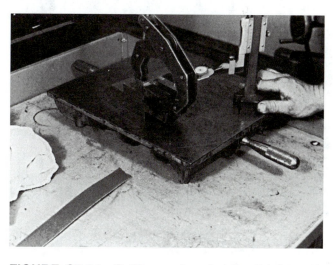

FIGURE G3.34 Setting up to grind the third square surface.

Step 10. Set the angle plate on the chuck and end grind the vee-block (Figure G3.33).

Step 11. For grinding the remaining side, clamp the ground end to the angle plate leaving the side surface projecting above the angle plate (Figure G3.34). Grind one remaining side square to the end and square to the first surface ground in Step 7.

Step 12. End grind opposite ends (Figure G3.35). The vee-blocks may be set up on the magnetic chuck without further support.

Step 13. Grind the remaining sides leaving .003 to .005 in. for finishing to the dimension (Figure G3.36).

Step 14. Re-dress the wheel with a slow pass (or passes) of the diamond dresser for finish grinding. Use a light cut of about .0002 in.

Step 15. Check all sides and ends for squareness (Figure G3.37). Use a precision cylindrical square and dial test indicator. Small errors can be corrected by further grinding if necessary. Tissue paper shims (.001 in.) may be used to achieve squareness. Take as little material as possible if corrections are needed.

Step 16. Check all dimensions with a vernier micrometer (.0001 in. discrimination) or dial test indicator (Figure G3.38).

FIGURE G3.35 Grinding the opposite ends of both blocks in one setup to make them parallel.

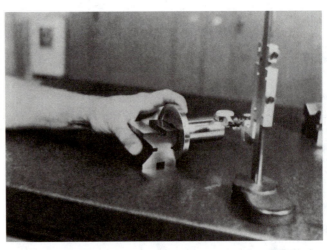

FIGURE G3.37 A vee-block now being checked for square on all sides using a precision cylindrical square, a dial indicator, and a height gage on a surface plate.

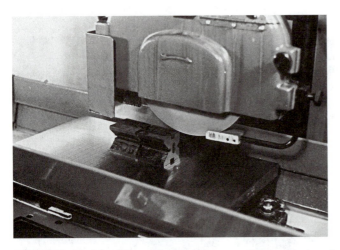

FIGURE G3.36 The sides that have not been ground are also being ground in one setup to make them parallel to the other sides.

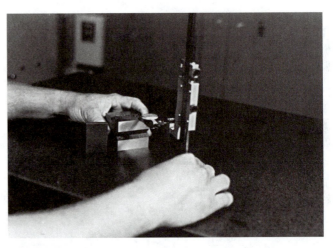

FIGURE G3.38 Dimensions being checked using a .0001-in. reading dial indicator on a height gage. Precision gage blocks are used for comparison measurement for this operation.

Step 17. Finish grind all sides and ends to final dimensions. Both blocks should be match ground.

Vee-Grinding

Step 1. Grind one side of the large vee (Figure G3.39). Set the blocks in a magnetic vee-block and carefully align with table travel. *(Do not make contact with the side of the wheel.)*

Step 2. After rough grinding one side of the vee, note the number on the downfeed micrometer collar or set it to zero index. Raise the wheel about $\frac{1}{2}$ in. Reverse the blocks in the magnetic vee-block and grind

the other side. Bring the wheel down to contact the work and make grinding passes until the micrometer collar is at the same position as it was when the first vee side was roughed. Now dress the wheel for finishing and repeat the procedure for both vees, removing only enough material necessary to obtain a finish. This procedure will insure that the vee is accurately centered on the blocks.

Step 3. Repeat the procedure for the remaining vee grooves.

Step 4. Set up the parts again in the magnetic vee-blocks to grind the external angular surfaces (chamfers) (Figure G3.40). This operation must be

FIGURE G3.39 Setting up the magnetic vee-block to grind the angular surfaces on the large vees.

FIGURE G3.40 Setting up the parts again in the magnetic vee-block to grind the external angular surfaces.

done gently, as workholding from the accessory magnetic vee-blocks is much less firm than from the surface of the magnetic chuck itself.

SELF-TEST

1. The part of the horizontal-spindle surface grinder that holds the wheel is called the _____.
2. Workholding on a surface grinder is almost always done on a _____ _____.
3. How can nonparallel (angular) surfaces be ground?
4. What do the finest divisions (marks) on the downfeed handwheel usually represent in inches?
5. What do the marks on the crossfeed handwheel usually represent?
6. On a small surface grinder, the maximum down feeds for roughing and finishing are how much?
7. When dressing or trueing the wheel with a diamond, what should the down feed be per pass?
8. The faster the traverse with diamond dressings, the sharper the wheel will be; a slower dress dulls the abrasive. What effect will this have on the surface of the piece to be ground?
9. When using flood coolant application, why is it not sufficient to just have coolant lying on top of the workpiece?
10. Precision grinding machines, such as cylindrical grinders, centerless grinders, and tool and cutter grinders, are built for special work. The horizontal-spindle surface grinder can be considered as the most versatile of grinding machines. Why is this so?

Problems and Solutions in Surface Grinding

Producing quality work on a surface grinder bears some resemblance to driving a car. It is not too difficult when everything is going right. It is knowing what to do when something is not going right that separates the skilled from the unskilled. This unit is a discussion of some of the common problems of grinding, how to recognize them, and what to do about them. It should also give you some insight into whether you should try to do something about the problem yourself or whether you should report it to your supervisor.

OBJECTIVES

After completing this unit, you should be able to:

1. Recognize common surface defects resulting from surface grinding and suggest ways of correcting them.
2. Suggest ways of correcting situations that show up in postinspection such as work that is not flat, parallel, or to form.

Conditions causing problems in surface grinding can conveniently be divided into two groups:

1. *Mechanical.* Those problems having to do with the condition of the machine; for example, worn bearings. Or those problems having to do with the electrical or hydraulic systems of the machine.
2. *Operational.* Those problems having to do with the operation of the machine; for example, the selection and condition of the grinding wheel, selection of coolant, wheel dressing, and other similar responsibilities.

Obviously, there is not a clear line that can be drawn between these two types of problems. The division of responsibilities varies among shops; in general, the mechanical condition of the machine, aside from routine daily lubrication, is a shop maintenance responsibility, while the operation of the machine is your responsibility as its operator. Still, there are some close decisions to be made; they are usually decided on the basis of shop policy or practice. In a school shop, it is generally the responsibility of the person using the machine to clean and lubricate it before and after use. If the machine ways are not lubricated, a problem called **stick-slip** develops as the table traverses, in which the table moves at a varying speed and has a jerky movement. If you are not sure of your responsibilities while using these machines, ask your instructor.

Two other general observations are in order. One is that any surface grinder is limited in the degree of precision and the quality of surface finish that it will produce. A lightly built, inexpensive grinder simply will not produce the quality of finish nor the precision of a more heavily built and more expensive machine.

The second is that any machined surface, even the finest, most mirrorlike surface, is a series of scratches. It is true that on the finer finishes the scratches are finer, closer together, and follow a definite pattern, but there are scratches nonetheless, and they will show up if the surface is sufficiently magnified.

Although you have not studied surface finish in any detail in this section, it is easy to understand that on a reciprocating-table surface grinder, the scratches will be parallel and running in the direction of table travel. Anything that differs significantly from this pattern can be considered a surface defect and a problem.

Dirt, heat, faulty wheel dressing, and vibration can cause problems, as indicated earlier. The condition of the wheel's grinding surface can also cause problems.

A wheel whose surface is either loaded or glazed will not cut well or produce a good flat surface. **Loading** means that bits of the work material have become embedded in the wheel's face. This usually means that the wheel's structure should be more open. **Glazing** means that the wheel face has worn too smooth to cut. It may result from using a wheel with a grit that is too fine, a structure that is too dense, or a grade that is too hard.

OPERATOR'S RESPONSIBILITIES

In general, it can be said that you, as the operator, are responsible for the daily checking and running of the machine. This includes such things as selecting and dressing the wheel, selecting the coolant, checking to see that the coolant tank is full and the filters are working as they should, and that the coolant is flowing in sufficient volume. You should observe the lubrication of the machine to see whether there is too little or too much. You should be alert for signs that the wheel is not secure or if the bearings are beginning to wear too much. You should not only check the work as instructed but you should also observe it for surface irregularities. Of course, you will not be working entirely on your own; much of this will be done with the advice and agreement of your instructor or supervisor, especially when you first begin to use the machine.

GOOD MACHINE CONDITION

It's an old axiom of grinding that you cannot produce quality work on a machine in poor condition. It is true that you can compensate for worn bearings to some degree by substituting a softer wheel, but this is a trap to avoid if at all possible. Not that machine condition is the only factor to be considered, but it is probably the most important single factor in preventing problems.

One frequently hears of unusual causes of trouble, such as the story about a machine that had been performing satisfactorily for years on a given wheel specification. Then suddenly, with the same specification, the operator has nothing but problems. When the manufacturer checked, he found that the customer had moved the machine from the ground floor to the sixth floor, and the added vibration required a new specification. With a wheel that suited the new conditions, there were no further problems. Many problems are not so straightforward. Indeed, the ones that are most difficult are those where there is more than one condition causing the problem. The discussion that follows indi-

cates when there is more than one condition that may cause a particular problem.

Before discussing specific problems, many of which are shown in Table G4.1, it is worth noting that good dressing practice is probably the best single way an operator can avoid problems. For example, if you cross feed the dresser too fast, you run the risk of causing distinctive spiral scratches on the surface of the workpiece. If you cross feed the diamond too slowly, the wheel is dressed too fine and the effect is similar to that resulting from grit size that is too fine or a wheel that is too hard. A wheel that is too hard also causes burning or burnishing of the work or hollow spots because an area has become too hot, has expanded, been ground off, and then has contracted below the surface as it cooled.

Surface finish problems can occur if you have not dressed a small radius on the corner of the wheel, which can be done by touching the rotating wheel with a dressing stick. The first corrective measure you should take, if you have surface finish problems, is to re-dress the wheel.

SURFACE FINISH DEFECTS

Surface defects usually show up as unwanted scratches in the finish. Figure G4.1 shows four slightly oversize (1.25X) pieces that have been selected to illustrate common problems of this type. Not all of the problems discussed in the following paragraphs lend themselves to illustration.

Chatter Marks

These are sometimes referred to as "vibration marks" (Figure G4.2) because vibration is usually the cause. The reason may be some outside source such as a punch press operating nearby and transmitting vibration through the machine to the wheel so that the wheel "slips" instead of cutting for a moment. It may also come from within the machine—a wheel that is not balanced or a wheel with one side soaked with coolant. It may be from worn wheel bearings or even from a wheel that loads and/or glazes and alternately slips, drops its load, and resumes cutting. This slipping and cutting alternation usually produces chatter marks that are close together; the more irregular and wider-spaced marks are likely to come from some other source. The remedy may be just redressing, but it could also require a change of grinding wheel, a check on wheel bearings, or even, if none of these work, a relocation of the machine.

TABLE G4.1   Summary of surface grinding defects and possible causes

Causes	Burning or Checking	Burnishing of Work	Chatter Marks	Scratches on Work	Wheel Glazing	Wheel Loading	Work Not Flat	Work Out-of-Parallel	Work Sliding on Chuck
Machine Operation									
Dirty coolant				x		x			
Insufficient coolant	x						x	x	
Wrong coolant					x	x			
Dirty or burred chuck				x			x	x	
Inadequate blocking									x
Poor chuck loading							x	x	x
Sliding work off chuck				x					
Dull diamond					x				
Too fine dress	x				x	x	x		
Too long a grinding stroke								x	
Loose dirt under guard				x					
Grinding Wheel									
Too fine grain size	x				x	x			
Too dense structure					x	x			
Too hard grade	x	x	x		x	x	x		
Too soft grade			x	x					
Machine Adjustment									
Chuck out of line								x	
Loose or cracked diamond				x			x	x	
No magnetism									x
Vibration			x						
Condition of Work									
Heat treat stresses							x		
Thin							x		

SOURCE: *Precision Surface Grinding.* Data courtesy of DoAll Company, 1964.

Irregular Scratches

Random scratches (Figure G4.3)—often called, for obvious reasons, *fishtails*—are usually caused by the recirculation of dirt, bits of abrasive or "tramp" metal in the coolant, or dirt falling from under the wheel guard. If there is not enough coolant, for instance, it may be recirculating too fast for the settling of swarf (grinding particles) to take place. A similar result can occur if you slide a workpiece off a dirty chuck instead of lifting it off, which may be more difficult.

The first thing to do to remedy these problems is to clean out the inside of the wheel guard; then replace or clean the filters in the coolant tank and make sure there is enough coolant in the tank. It should be worthwhile to take some extra time to insure that the chuck is clean before you load it again.

Discoloration

Discoloration, which is also known as *burning* or **checking,** results when a workpiece becomes overheated. It can be caused by insufficient coolant, improperly applied coolant, a wheel that is too hard or too fine, or by removing too much metal too rapidly from a small area (Figure G4.4). If carried on too long, it can result in expansion of the metal, probably in the middle of the surface being ground. Then, if the hump is ground off, there will be a low spot when the workpiece cools to normal temperature. However, the burning may be considered in the first

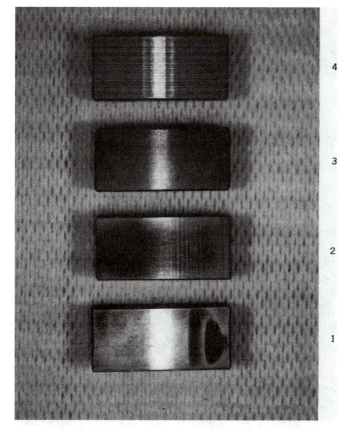

FIGURE G4.1   Four specimens of oil-hardening steel, approximately 60 RC, with specific surface defects. (Mark Drzewiecki, Surface Finishes Inc.)

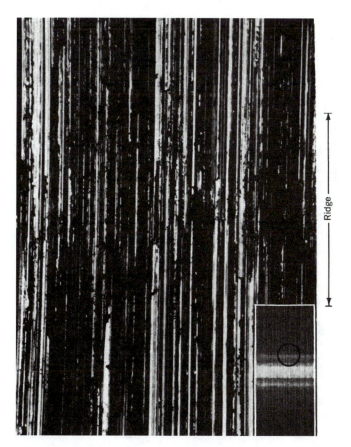

FIGURE G4.2   Chatter marks enlarged 80X. The marked inset (1.25X) shows enlarged area. (Mark Drzewiecki, Surface Finishes Inc.)

place as a surface defect that may or may not be a problem, depending on the final use of the workpiece.

Probably the first remedy is to try to get the wheel to *act* softer by speeding up table traverse, re-dressing the wheel rougher, or taking lighter cuts and, of course, by checking the supply and the application of the coolant. Whenever there is not enough coolant, it is recirculated before it has had a chance to cool off as it should, and so it simply becomes warmer and warmer.

Burnishing

A burnished surface (Figure G4.5) is one that is smoothed by abrasive rubbing instead of by cutting. It often looks good, and it may indeed not be a problem if the surface is not subject to wear. Technically, what happens when a surface is burnished is that the hills of the surface are heated enough so that they can be pushed into the valleys of the scratches. But with any wear at all, the displaced metal breaks loose, and the surface suddenly becomes much rougher. It usu-

ally results from using a wheel that is too hard. For that reason, the usual remedy is to get a softer wheel or to change grinding conditions to make the wheel act softer by increasing work speed, re-dressing the wheel, or taking lighter cuts.

Miscellaneous Surface Defects

As mentioned earlier in this unit, any ground surface is a planned series of scratches, preferably of uniform depth and direction. On the grinder you have been studying, the scratches are parallel and at right angles to the direction of work traverse. Any other pattern, scratches without a pattern, or any discoloration of the work surface can be considered a defect. Some of the causes of these defects are that the wheel may be loose, the bearings may be worn, there may be vibration from some unsuspected source, the wheel may have been dressed too fast or too slow, or the wheel may be too rough or too fine. Sometimes the causes are so remote or obscure as to puzzle even experienced troubleshooters.

FIGURE G4.3 Grinding marks or "fishtails" also enlarged 80X. The inset (enlarged 1.25X) shows the damaged area. (Mark Drzewiecki, Surface Finishes Inc.)

FIGURE G4.4 Discoloration or burning also enlarged 80X, with the damaged area marked on the inset. (Mark Drzewiecki, Surface Finishes Inc.)

Work Not Parallel

It has been said repeatedly that the grinding line of the wheel is parallel to the top of the chuck, and this is true if the grinder is in good condition. However, if the chuck is out of line, dirty, or burred, this parallelism may no longer exist, and any workpiece may be out of parallel. Of the three conditions mentioned above, dirt and burring of the chuck are definitely your responsibility, but the chuck alignment may or may not be. The remedies are usually obvious, as are those of other possible causes, as shown in Table G4.1. As you progress, you will develop a sort of routine to be followed for a given machine with this condition.

Work Out-of-Flat

Lack of flatness in a thick workpiece is likely to be the result of some local overheating, which causes an area of the surface to bulge. Then when the bulge is ground off and the work cools, there is a low spot.

Most flatness problems arise with thin work and for very obvious reasons. Thin work does not have the bulk to absorb grinding heat without distortion. If it has been rolled, then stresses caused by the passage of the metal through the rolls may have been created, and the grinding may release these stresses on one side, causing the metal to warp or bow out-of-flat.

Correcting warpage in a thin workpiece is a matter of patience, a right start, and the minimum chuck power required to hold the workpiece in place. The procedure is as follows:

1. Using the least practical amount of chuck power, place the work on the chuck with the bowed side up so that the work rests on the ends.
2. Take a light cut with minimum down feed. This should grind only the high spots on the work. Cutting should begin near the center.
3. Turn the workpiece over, shim the ends with paper, and take another light cut. This time cutting begins near the ends.
4. Repeat these steps, reducing the shims gradually, until the part is flat within specifications.

FIGURE G4.5 Burnished area enlarged 80X. The inset shows the damaged area. (Mark Drzewiecki, Surface Finishes Inc.)

Finally, it should only have to be mentioned, as a reminder, that it is impossible to grind work flat if the chuck surface is not flat. When the chuck surface is between .0001 and .0002 in. out-of-flat, it is time to regrind it.

Perhaps the last word on solving the problems that arise in surface grinding is that care in avoiding the causes of trouble—such as careful and thorough cleaning of the chuck, frequent checking of the coolant level, the condition of the filters, and care in placing and removing workpieces from the chuck—can prevent many of the problems before they happen. It might be called preventive operation of the grinder.

SELF-TEST

1. List at least three actions of an operator that will help prevent problems in surface grinding.
2. Name at least three general causes of surface grinding problems.
3. What is a surface defect?
4. What is the principal cause of chatter marks? What are some possible remedies?
5. "Fishtails" are another common problem. What are they, what causes them, and what should you do to get rid of them?
6. Name two of the problems that can result from overheating work. What two or three things might you do if you suspect that a workpiece is getting too hot?
7. What is a burnished surface? Why is it objectionable? What are some remedies?
8. List at least two conditions that could produce out-of-parallel work.
9. List at least two conditions that could cause out-of-flat work.
10. Why is it difficult to grind thin work flat? How do you correct the condition?

Cylindrical Grinders

The center-type cylindrical grinder, as its name implies, is used to grind the outside diameters of cylindrical (or conical) workpieces mounted on centers. Using accessory equipment, inside diameters can also be ground. Roundness of the ground surface is one of the criteria for success of cylindrical grinding. The cylindrical grinder is a versatile machine tool capable of finish machining a cylindrical or a conical part to a high degree of dimensional accuracy. Various setups on the basic machine permit many different grinding tasks to be accomplished. The purpose of this unit is to familiarize you with the major parts of this machine and its general capabilities.

OBJECTIVES

After completing this unit, you should be able to:

1. Identify the major parts of the cylindrical grinder.
2. Describe the various movements of the major parts.
3. Describe the general capabilities of this machine.

CYLINDRICAL GRINDING

The cylindrical grinder grinds the outside or inside diameter of a cylindrical (or conical) part. The rotating abrasive wheel, typically moving from 4000 to 6500 SFPM, is brought into contact with the rotating workpiece moving in the opposite direction between 50 and 200 SFPM. The workpiece is traversed lengthwise against the wheel to reduce the outside diameter in the case of external grinding. In the case of internal grinding, the workpiece diameter would be increasing. In cylindrical plunge grinding, the wheel is brought into contact with the work but without traverse. In all cylindrical grinding, the workpiece is rotated opposite to the rotation of the abrasive wheel.

IDENTIFYING MACHINE PARTS AND THEIR FUNCTIONS

On the center-type cylindrical grinder (Figure G5.1), the workpiece is mounted between centers much as it would be in the lathe. The plain cylindrical grinder (Figure G5.2) has a fixed wheelhead that cannot be swiveled, only moved toward or away from the center axis of the workpiece. The table can be swiveled to permit the grinding of tapered workpieces. On the universal cylindrical grinder (Figure G5.3), both the wheelhead and table may be swiveled for taper grinding. All possible motions are illustrated in Figure G5.4.

Major Parts of the Universal Center-Type Cylindrical Grinder

Major parts of the machine include the bed, slide, swivel table, headstock, footstock, and wheelhead.

BED The bed is the main structural component and is responsible for the rigidity of the machine tool. The bed supports the slide, which in turn supports the swivel table.

SLIDE AND SWIVEL TABLE The slide carries the swivel table and provides the traverse motion to carry the workpiece past the wheel. The swivel table is mounted on the slide and supports the head- and

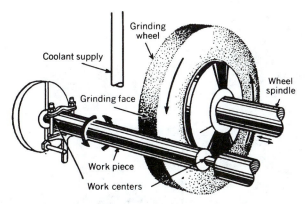

FIGURE G5.1 Sketch of center-type cylindrical grinder set up for traverse grinding. Note particularly the direction of travel of the grinding wheel and the workpiece and the method of rotating the workpiece. (Bay State Abrasives, Dresser Industries, Inc.)

FIGURE G5.3 Universal cylindrical grinder. (Landis Tool Co., Division of Litton Industries)

FIGURE G5.2 Plain cylindrical grinder. (Landis Tool Co., Division of Litton Industries)

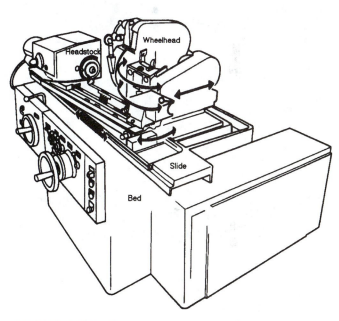

FIGURE G5.4 View of universal cylindrical grinder with arrows indicating the swiveling capabilities of various major components. (Cincinnati Milacron)

footstocks. The swivel table has graduations for establishing taper angles.

HEADSTOCK The headstock (Figure G5.5) mounts on the swivel table and is used to support one end of the workpiece. The headstock also provides the rotating motion for the workpiece. The headstock spindle is typically designed to accept a chuck or face plate. The headstock center is used when workpieces

are mounted between centers. Variable headstock spindle-speed selection is also available. For the most precise cylindrical grinding, the headstock center is held stationary while the driving plate that rotates the part rotates concentric to the dead center. This procedure eliminates the possibility of duplicating headstock bearing irregularities into the workpiece. For parts that can tolerate minor errors, it is preferable to have the center turn with the driving plate.

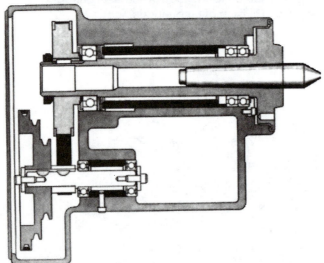

FIGURE G5.5 Typical headstock of a center-type cylindrical grinder with a cutaway sketch. (Landis Tool Co., Division of Litton Industries)

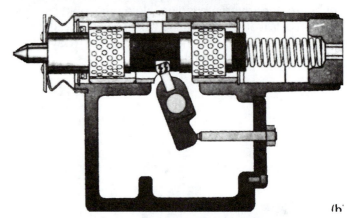

(a)

(b)

FIGURE G5.6 *(a)* Typical footstock. The lever on top of the footstock retracts the work center so that the workpiece can be mounted on the grinder. *(b)* The spring (right end of sketch) provides tension to hold the workpiece in place. (Landis Tool Co., Division of Litton Industries)

FOOTSTOCK The footstock (Figure G5.6) is also mounted on the swivel table and supports the opposite end of a workpiece mounted between centers. The footstock center does not rotate. It is spring loaded and is retracted by lever. This permits easy installation and removal of the workpiece. Compression loading on the spring is adjustable, and the footstock spindle typically can be locked after it is adjusted. The footstock assembly is positioned on the swivel table at whatever location is needed to accommodate the length of the workpiece.

WHEELHEAD The wheelhead, located at the back of the machine, contains the spindle, bearings, drive, and main motor.

WORK HOLDING AND CENTER-HOLE GRINDING

The workpiece in *on-centers* cylindrical grinding is mounted between the head and footstock centers. This effectively provides a single-point mounting on each end of the workpiece permitting maximum accuracy to be achieved in the grinding operation.

The angle of center holes in the workpiece is extremely important. The results of a cylindrical grinding job depend on this factor. Center holes in the workpiece are often precision ground to the correct geometry by a center-hole grinder (Figure G5.7). Locating the center holes (Figure G5.8) and producing the correct angle (Figure G5.9) are critical operations

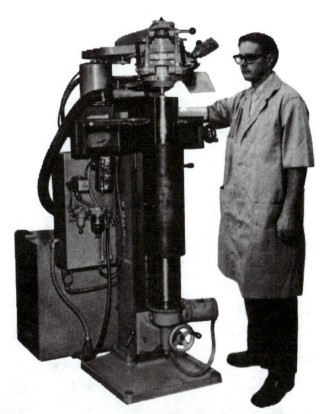

FIGURE G5.7 Center-hole grinding machine. (Bryant Grinder Corp.)

FIGURE G5.8 Closeup of center-hole locating setup. The exact location of the center is a most critical step in the operation. (Bryant Grinder Corp.)

in center-hole grinding. Since a large proportion of cylindrical grinding is done on heat-treated parts, the center-hole grinder is especially useful for removing heat-treatment scale and leaving a precise locating surface.

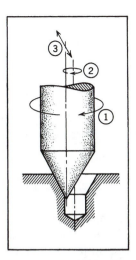

FIGURE G5.9 Sketch showing the motions of the grinding wheel in center-hole grinding. (Bryant Grinder Corp.)

FIGURE G5.10 Traverse grinding, which is probably the most common type of manually controlled cylindrical grinding. (Cincinnati Milacron)

CAPABILITIES OF THE CENTER-TYPE CYLINDRICAL GRINDER

Most common is simple traverse grinding (Figure G5.10). This may be done on interrupted surfaces (Figure G5.11) where the wheel face is wide enough to span two or more surfaces.

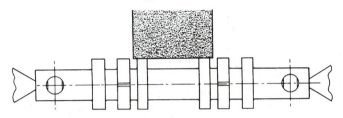

FIGURE G5.11 Sketch of traverse grinding with interrupted surfaces. The wheel should always be wide enough to span two surfaces or more at once. (Cincinnati Milacron)

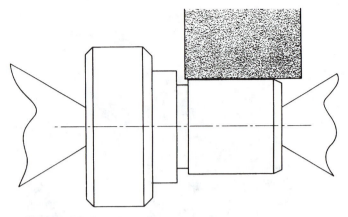

FIGURE G5.13 Straight plunge grinding, where the wheel is usually wider than the length of the workpiece feature. (Cincinnati Milacron)

Multiple-diameter form grinding (Figure G5.12) is a type of plunge grinding where the wheel may be *form dressed* to a desired shape. In straight plunge grinding, the wheel is brought into contact with the workpiece, but the workpiece is not traversed (Figure G5.13).

OD taper grinding (Figures G5.14 and G5.15) is done by swiveling the table. Steeper tapers may require that both table and wheelhead be swiveled in combination (Figure G5.16).

FIGURE G5.14 *OD* taper grinding, which may be done to produce a tapered (conical) finished workpiece or to correct a workpiece that was previously tapered. (Cincinnati Milacron)

FIGURE G5.12 Multiple-diameter or form grinding. This is usually a plunge grinding operation with the form dressed in the wheel face. (Cincinnati Milacron)

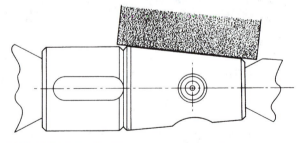

FIGURE G5.15 Taper grinding with the workpiece swiveled to the desired angle. For a steeper taper, the wheel might also have to be dressed at an angle less than 90 degrees. (Cincinnati Milacron)

FIGURE G5.16 Steep taper grinding. Here the wheel-head has been swiveled to grind the workpiece taper. (Cincinnati Milacron)

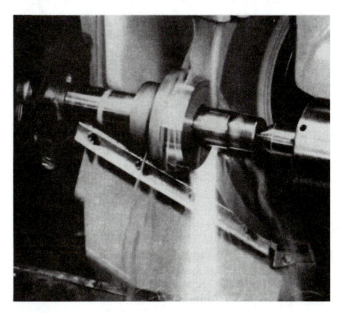

FIGURE G5.17 Angular shoulder grinding. This is very often a production-type operation, but it is shown here on a universal grinder. (Cincinnati Milacron)

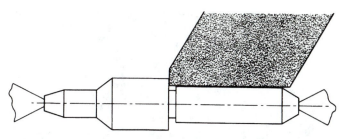

FIGURE G5.18 Angular plunge grinding with shoulder grinding. Note the dressing of the grinding wheel. (Cincinnati Milacron)

FIGURE G5.19 Internal grinding, which requires a special high-speed attachment that is mounted on the wheelhead so that it can be swung up out of the way when not in use. It also requires either a chuck or a face plate on the headstock. (Cincinnati Milacron)

Angular shoulder grinding (Figure G5.17) and angular plunge grinding may require that the wheel be dressed at an appropriate angle (Figure G5.18).

Internal cylindrical grinding can be straight (Figure G5.19) or tapered (Figure G5.20). The work-piece is held in a chuck or fixture so that the inside diameter can be accessed by the grinding wheel.

ACCESSORIES FOR CENTER-TYPE GRINDERS

Accessories for center-type cylindrical grinders include the high-speed internal attachment, headstock chucks, dressers, and the back or steady rest for grinding slender workpieces (Figure G5.21).

FIGURE G5.20 *ID* taper grinding. This is the same sort of operation as shown in Figure G5.18 but with the workhead swiveled at the required angle. (Cincinnati Milacron)

FIGURE G5.21 Back or steady rest used to support a thin piece for grinding. If only one is used, as here, it is placed *near* the center of the workpiece. If more are used, usually an odd number, they should be *nearly* equally spaced along the workpiece. Using slightly unequal spacing will sometimes help to reduce harmonic vibration of the workpiece. (Cincinnati Milacron)

SELF-TEST

1. Explain the differences in construction between plain and universal cylindrical grinders.
2. Why is the most critical cylindrical grinding done on nonrotating centers?
3. Why should the footstock adjustment be locked after the workpiece has been positioned?
4. Can the "plain" cylindrical grinder produce a tapered (conical) surface?
5. Since many cylindrically ground workpieces are hardened, what can be done for center-hole preparation to prevent heat-treatment scale from causing inaccurate cylindrical grinding?
6. What is *traverse grinding*?
7. What is *plunge grinding*?
8. Describe the basic methods by which a workpiece is held on a center-type cylindrical grinder.
9. If you are traverse-grinding a workpiece with interrupted surfaces, what is a major limitation that must be observed in the workpiece?
10. What two additional components are required to do internal grinding on a universal cylindrical grinder?

Using the Cylindrical Grinder

Using the cylindrical grinder involves many of the same steps that you learned in surface grinding. The cylindrical grinder is a versatile and precise machine tool. Used correctly, it will finish-machine your workpiece to a high degree of dimensional accuracy. The purpose of this unit is to familiarize you with the preparation of the machine and the grinding process required to finish-machine a typical cylindrical grinding project.

OBJECTIVES

After completing this unit, you should be able to:

1. Prepare the cylindrical grinder for a typical grinding job.
2. Finish-grind a lathe mandrel to required specifications.

CYLINDRICAL GRINDING A LATHE MANDREL

A lathe mandrel is a common and useful machine shop tool. Precision mandrels are finished by cylindrical grinding after they have been machined and heat treated. Finish grinding the mandrel will provide you with a well-rounded experience in cylindrical grinding. Refer to a working drawing (Figure G6.1) to determine the required dimensions.

Selecting the Wheel

For a hardened steel mandrel, the best abrasive will be aluminum oxide of 60 grit and a J bond hardness with a somewhat dense structure.

Grinding Machine Setup

Step 1. Set up the diamond dresser (Figure G6.2). Diamond dressers have a height to match the wheel centerline. On many machines the dresser is built into the footstock assembly.

Step 2. Dress the wheel using a full flow of cutting fluid. If you cannot obtain continuous fluid coverage, then dress dry. Use a rapid traverse of the diamond, with about .001 in. infeed per pass, until the wheel is true and sharp (Figure G6.3).

Step 3. Place a parallel test bar between centers (Figure G6.4). Mount a dial indicator on the wheelhead and set to zero at one end of the test bar.

Step 4. Move the table 12 inches and read the indicator at the other end of the bar. The indicator should read .003 or a total of .006 in. per foot of taper (Figure G6.5).

Step 5. Adjust the swivel table to obtain the correct amount of taper (Figure G6.6).

Step 6. Lubricate the center hole in each end of the mandrel, using a high-pressure lubricant specially prepared for use with centers. (Avoid using lubricants that have lead as a component.)

Step 7. Place the dog on the mandrel and insert it between the headstock and footstock centers. Move the footstock up so that some tension is on the footstock center (Figure G6.7).

Step 8. The work must have no detectable axial movement but be free to rotate on the centers. The dog must be clamped firmly to the work and contact the drive pin (Figure G6.8). If your machine has a slotted driver, make sure that the dog does not touch the bottom of the slot, or it could force the workpiece off of the center.

Step 9. Set the table stops (Figure G6.9) so that the wheel can be traversed some distance beyond the ground surface but has at least $\frac{1}{4}$-in. clearance from contacting the driving dog. If there is adequate clearance on the footstock center, set the stop to permit at least one-third of the wheel width to go beyond the surface that is being ground. This over-travel will insure that the ground surface of the mandrel gets completely finished.

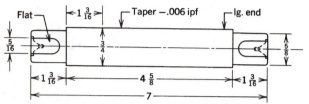

FIGURE G6.1   Sketch of tapered lathe mandrel.

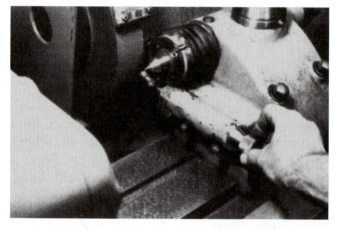

FIGURE G6.2   Setting up the diamond for wheel dressing.

FIGURE G6.3   Dressing the wheel with the grinding fluid on.

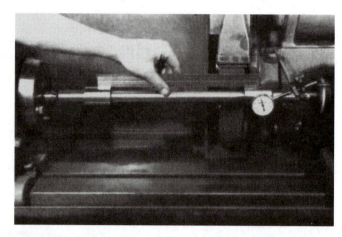

FIGURE G6.4   Setting up the parallel test bar and taking the first reading.

FIGURE G6.5   Second reading of the dial indicator.

FIGURE G6.6   Adjusting the swivel table.

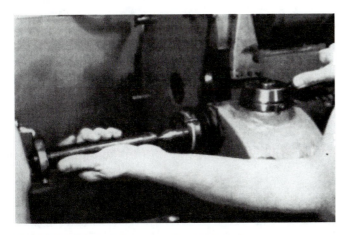

FIGURE G6.7 Clamping the footstock.

FIGURE G6.9 Adjusting the wheel overrun.

FIGURE G6.8 Checking the work dog clearance and drive pin contact.

FIGURE G6.10 Turning on the wheel and work rotation.

Mandrel Grinding Procedure

Step 1. Move the wheel close to what will become the smaller *OD* of the workpiece. With the wheel stopped, you may use a paper strip (about .003 in. thick) as a feeler gage to position the wheel close to the mandrel.

Step 2. Turn on the wheel and headstock spindle. Adjust the work speed to provide 70 to 100 SFPM.

Step 3. Start the table traverse. The table traverse rate should be about one-fourth of the wheel width for each revolution of the workpiece. (On unhardened work a more rapid traverse would be suitable, up to one-half the wheel width.) If your machine is

equipped with a Tarry control, set the adjustment for a slight dwell at the end of the traverse so that the table does not "bounce" on the table stops.

Step 4. Infeed the wheel until it contacts the rotating mandrel. Sparks will begin to show (Figure G6.10). Now start the grinding fluid and infeed about .001 in. per traverse until the surface of the part is cleaned up. At this point, set the infeed dial to read zero. Then retract the wheel.

Step 5. Stop the machine completely. Measure the mandrel in two places, 4 in. apart (Figures G6.11 and G6.12) to verify that your taper setting was correctly done. There should be a .002-in. difference in diameter between the larger and smaller diameters. Make

FIGURE G6.11 Confirming the required taper.

FIGURE G6.13 Grinding the mandrel to dimension.

FIGURE G6.12 Measuring taper 4 in. down the part.

FIGURE G6.14 The finished mandrel.

an adjustment if necessary. Be certain that the grinding head is retracted from the work if you make this adjustment.

Step 6. Rough and finish grind the mandrel to final size. The grinder may have an automatic infeed stop that will stop the infeed when final diameter is reached. On a manual machine, infeed the wheelhead to remove about .001 in. from the diameter per complete traverse cycle. The finishing passes should remove about .0002 in. from the diameter.

Step 7. Check for both taper and diameter periodically while making the finishing passes. The table traverse rate should be reduced to about one-eighth the width of the wheel per workpiece revolution. When the final size is reached, allow the work to spark out by traversing through several cycles without additional infeed (Figure G6.13).

Only after careful measurement should the workpiece be removed from between the centers (Figure G6.14), as returning the work to the centers to remove very small amounts of material is usually not successful. This is because of the minor irregularities that occur. The smallest amount of grit can cause large differences.

SELF-TEST

1. Why is the wheel used on a cylindrical grinder typically of a more dense structure than a surface grinding wheel for the same workpiece material condition?
2. What advantage is there in having the diamond wheel dresser integral with the footstock?

3. What is the purpose of a parallel test bar on a cylindrical grinder?
4. What amount of end play should there be when mounting workpieces on centers?
5. What special care must be taken with the driver dog in cylindrical grinding?
6. When setting the traversing stops, how much of the wheel width should be permitted to over-travel the ground part length?
7. What purpose is served by the Tarry control?
8. For rough cylindrical grinding, about how much of the wheel width should be traversed for each rotation of the workpiece?
9. For finish grinding, about how much of the wheel width should be traversed for each rotation of the workpiece?
10. Finishing passes on the cylindrical grinder should be about _____ in. from the diameter.

WORKSHEET: CYLINDRICAL GRINDING

Objectives

1. Learn to use the cylindrical grinding machine to make a straight precision cylinder.
2. Be able to produce a precision taper.
3. Be able to grind an internal surface.

Outline for Study

Prior to starting each procedure for this project, study and complete Post-Tests for:

1. Grinding a cylindrical shape: Section G, Units 5 and 6.
2. Grinding external tapers.

Procedures

Begin the procedures for this project by completing the following worksheets:

Worksheet: Making a Lathe Mandrel; Grinding a Cylinder

Worksheet: Grinding a Taper

WORKSHEET: MAKING A LATHE MANDREL; GRINDING A CYLINDER

Material

$7\frac{1}{8}$ in. of 1-in. SAE 4140 steel for a $\frac{3}{4}$-in. dia. mandrel or sufficient material for a selected mandrel size. Standard mandrel dimensions may be found in *Machinery's Handbook*.

Procedure

1. To see if a different size or another part is to be made.
2. Make the turned diameter 0.015 to 0.030 in. oversize (depending on size).
3. Turn a mandrel between centers on a lathe as shown in Figure G6.15. Check with your instructor. Mill the flats on the ends.
4. Harden and temper to about RC 50. (This operation is optional.)
5. Follow the instructions in your textbook in Section G, Units 5 and 6, for grinding a mandrel. Before beginning the mandrel grinding procedure for grinding the taper, make a straight grind.

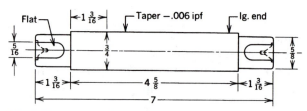

FIGURE G6.15 Sketch of tapered lathe mandrel.

6. When you have made the setup, have your instructor inspect it before turning on the wheel.
7. With the wheel and work rotating and the table traverse set, feed the wheel to the work. Infeed the wheel until it just begins to produce sparks.
8. Turn on the coolant.
9. Make several passes, infeeding about 0.001 to 0.002 in. per pass until the surface scale is all cleaned up. Allow the table to traverse without infeed until it sparks out. Retract the wheel and stop the rotation and table traverse.
10. Check the two ends of the mandrel with a "tenth" micrometer. It should be the same within one "tenth."
11. If there is taper, move the table to compensate and regrind until the mandrel is the same diameter within plus or minus 0.0001 in. The mandrel should still be over the final finish diameter.

WORKSHEET: GRINDING A TAPER

Material

The mandrel.

Procedure

1. Follow the steps given in your textbook for grinding a taper. When adjusting the swivel table for the required taper, a dial indicator may be used in a position near the end of the table at a given distance from the pivot point of the table. This distance and the dial indicator reading is a proportional ratio to the ratio of the amount to be removed from the mandrel *on one side* in a given length. For example, in 6 inches the mandrel should taper 0.003 in. in diameter. The table must be rotated so that 0.0015 in. more is removed from one end *on one side* than is removed on the other. If the table pivot point is 30 inches from the

dial indicator, the ratio would be 30 in. to 6 in. or 5:1. Therefore, the dial indicator movement should be 5 × 0.0015 or 0.0075 in.

2. The final dimension of the taper should be determined by making the nominal diameter about one-fourth of taper length away from the large end. For example, the $\frac{3}{4}$-in. mandrel should be 0.750 in. dia. at $\frac{13}{16}$ in.

3. When you are near the final size, you should have an acceptable finish. If you do not, check with your instructor.

4. Clean up your finished mandrel.

Turn in the completed mandrel to your instructor for evaluation.

TABLE 1 Hardness and tensile strength comparison table

Hardness Conversion Table									
Brinell		Rockwell			Brinell		Rockwell		
Indentation diameter (mm)	No.[a]	B	C	Tensile Strength (1000 PSI approximately)	Indentation diameter (mm)	No.[a]	B	C	Tensile Strength (1000 PSI approximately
2.25	745		65.3		3.75	262	(103.0)	26.6	127
2.30	712		—		3.80	255	(102.0)	25.4	123
2.35	682		61.7		3.85	248	(101.0)	24.2	120
2.40	653		60.0		3.90	241	100.0	22.8	116
2.45	627		58.7		3.95	235	99.0	21.7	114
2.50	601		57.3		4.00	229	98.2	20.5	111
2.55	578		56.0		4.05	223	97.3	(18.8)	—
2.60	555		54.7	298	4.10	217	96.4	(17.5)	105
2.65	534		53.5	288	4.15	212	95.5	(16.0)	102
2.70	514		52.1	274	4.20	207	94.6	(15.2)	100
2.75	495		51.6	269	4.25	201	93.8	(13.8)	98
2.80	477		50.3	258	4.30	197	92.8	(12.7)	95
2.85	461		48.8	244	4.35	192	91.9	(11.5)	93
2.90	444		47.2	231	4.40	187	90.7	(10.0)	90
2.95	429		45.7	219	4.45	183	90.0	(9.0)	89
3.00	415		44.5	212	4.50	179	89.0	(8.0)	87
3.05	401		43.1	202	4.55	174	87.8	(6.4)	85
3.10	388		41.8	193	4.60	170	86.8	(5.4)	83
3.15	375		40.4	184	4.65	167	86.0	(4.4)	81
3.20	363		39.1	177	4.70	163	85.0	(3.3)	79
3.25	352	(110.0)	37.9	171	4.80	156	82.9	(0.9)	76
3.30	341	(109.0)	36.6	164	4.90	149	80.8		73
3.35	331	(108.5)	35.5	159	5.00	143	78.7		71
3.40	321	(108.0)	34.3	154	5.10	137	76.4		67
3.45	311	(107.5)	33.1	149	5.20	131	74.0		65
3.50	302	(107.0)	32.1	146	5.30	126	72.0		63
3.55	293	(106.0)	30.9	141	5.40	121	69.8		60
3.60	285	(105.5)	29.9	138	5.50	116	67.6		58
3.65	277	(104.5)	28.8	134	5.60	111	65.7		56
3.70	269	(104.0)	27.6	130					

[a]Values above 500 are for tungsten carbide ball; below 500 for standard ball.

NOTE 1 This is a condensation of Table 2, Report J417b, SAE 1971 Handbook. Values in ( ) are beyond normal range and are presented for information only.

NOTE 2 The following is a formula to approximate tensile strength when the Brinell hardness is known:

Tensile strength = $BHN \times 500$

SOURCE Bethlehem Steel Corporation, *Modern Steels and Their Properties*, Handbook 3310, 1980.

TABLE 2   Flat bar steel—Weight per linear foot, pounds

Thickness (in.)	Width in. $\frac{3}{8}$	$\frac{1}{2}$	$\frac{5}{8}$	$\frac{3}{4}$	$\frac{7}{8}$	1	$1\frac{1}{4}$	$1\frac{1}{2}$	$1\frac{3}{4}$	2	$2\frac{1}{4}$	$2\frac{1}{2}$
$\frac{1}{8}$	.1594	.2125	.2656	.3188	.3719	.4250	.5313	.638	.744	.850	.956	1.063
$\frac{1}{4}$	.3188	.4250	.5313	.6375	.7438	.8500	1.0625	1.275	1.488	1.700	1.913	2.125
$\frac{3}{8}$	.4781	.6375	.7969	.9563	1.1156	1.2750	1.5938	1.913	2.231	2.550	2.869	3.188
$\frac{1}{2}$	.6375	.8500	1.0625	1.2750	1.4875	1.7000	2.1250	2.550	2.975	3.400	3.825	4.250
$\frac{5}{8}$	.7969	1.0625	1.3281	1.5938	1.8594	2.1250	2.6563	3.188	3.719	4.250	4.781	5.313
$\frac{3}{4}$	.9563	1.2750	1.5938	1.9125	2.2313	2.5500	3.1875	3.825	4.463	5.100	5.738	6.375
$\frac{7}{8}$	1.1156	1.4875	1.8594	2.2313	2.6031	2.9750	3.7188	4.463	5.206	5.950	6.694	7.438
1	1.2750	1.7000	2.1250	2.5500	2.9750	3.4000	4.2500	5.100	5.950	6.800	7.650	8.500
$1\frac{1}{8}$	1.4344	1.9125	2.3906	2.8688	3.3469	3.8250	4.7813	5.738	6.694	7.650	8.606	9.563
$1\frac{1}{4}$	1.5938	2.1250	2.6563	3.1875	3.7188	4.2500	5.3125	6.375	7.438	8.500	9.563	10.625
$1\frac{3}{8}$	1.7531	2.3375	2.9219	3.5063	4.0906	4.6750	5.8438	7.013	8.181	9.350	10.519	11.688
$1\frac{1}{2}$	1.9125	2.5500	3.1875	3.8250	4.4625	5.1000	6.3750	7.650	8.925	10.200	11.475	12.113
$1\frac{5}{8}$	2.0719	2.7625	3.4531	4.1438	4.8344	5.5250	6.9063	8.288	9.669	11.050	12.431	13.122
$1\frac{3}{4}$	2.2313	2.9750	3.7188	4.4625	5.2063	5.9500	7.4375	8.925	10.413	11.900	13.388	14.131
$1\frac{7}{8}$	2.3906	3.1875	3.9844	4.7813	5.5781	6.3750	7.9688	9.563	11.156	12.750	14.344	15.141
2	2.5500	3.4000	4.2500	5.1000	5.9500	6.8000	8.5000	10.200	11.900	13.600	15.300	16.150

2¾	3	3¼	3½	3¾	4	4¼	4½	4¾	5	5¼	5½	5¾	6
						Width in.							
1.169	1.275	1.381	1.488	1.594	1.700	1.806	1.913	2.019	2.125	2.231	2.338	2.444	2.550
2.338	2.550	2.763	2.975	3.188	3.400	3.613	3.825	4.038	4.250	4.463	4.675	4.888	5.100
3.506	3.825	4.144	4.463	4.781	5.100	5.419	5.738	6.056	6.375	6.694	7.013	7.331	7.650
4.675	5.100	5.525	5.950	6.375	6.800	7.225	7.650	8.075	8.500	8.925	9.350	9.775	10.200
5.844	6.375	6.906	7.438	7.969	8.500	9.031	9.563	10.094	10.625	11.156	11.688	12.219	12.750
7.013	7.650	8.288	8.925	9.563	10.200	10.838	11.475	12.113	12.750	13.388	14.025	14.663	15.300
8.181	8.925	9.669	10.413	11.156	11.900	12.644	13.388	14.131	14.875	15.619	16.363	17.106	17.850
9.350	10.200	11.050	11.900	12.750	13.600	14.450	15.300	16.150	17.000	17.850	18.700	19.550	20.400
10.519	11.475	12.431	13.388	14.344	15.300	16.256	17.213	18.169	19.125	20.081	21.038	21.994	22.950
11.688	12.750	13.813	14.875	15.938	17.000	18.063	19.125	20.188	21.250	22.313	23.375	24.438	25.500
12.856	14.025	15.194	16.363	17.531	18.700	19.869	21.038	22.206	23.375	24.544	25.713	26.881	28.050
14.025	15.300	16.575	17.850	19.125	20.400	21.675	22.950	24.225	25.500	26.775	28.050	29.325	30.600
15.194	16.575	17.956	19.338	20.719	22.100	23.481	24.863	26.244	27.625	29.006	30.388	31.769	33.150
16.363	17.850	19.338	20.825	22.313	23.800	25.288	26.775	28.263	29.750	31.238	32.725	34.213	35.700
17.531	19.125	20.719	22.313	23.906	25.500	27.094	28.688	30.281	31.875	33.469	35.063	36.656	38.250
18.700	20.400	22.100	23.800	25.500	27.200	28.900	30.600	32.300	34.000	35.700	37.400	39.100	40.800

TABLE 3   Weights of round bars per linear foot

Size (diameter in in.)	Weight (pounds per foot)	Size (diameter in in.)	Weight (pounds per foot)	Size (diameter in in.)	Weight (pounds per foot)
$\frac{1}{8}$	.042	$1\frac{3}{4}$	8.178	$3\frac{7}{16}$	31.554
$\frac{3}{16}$	.094	$1\frac{13}{16}$	8.773	$3\frac{1}{2}$	32.712
$\frac{1}{4}$	.167	$1\frac{7}{8}$	9.388	$3\frac{9}{16}$	33.891
$\frac{5}{16}$	.261	$1\frac{15}{16}$	10.024	$3\frac{5}{8}$	35.090
$\frac{3}{8}$	.376	2	10.681	$3\frac{11}{16}$	36.311
$\frac{7}{16}$	.511	$2\frac{1}{16}$	11.359	$3\frac{3}{4}$	37.552
$\frac{1}{2}$	.668	$2\frac{1}{8}$	12.058	$3\frac{13}{16}$	38.814
$\frac{9}{16}$	.845	$2\frac{3}{16}$	12.778	$3\frac{7}{8}$	40.097
$\frac{5}{8}$	1.043	$2\frac{1}{4}$	13.519	$3\frac{15}{16}$	41.401
$\frac{11}{16}$	1.262	$2\frac{5}{16}$	14.280	4	42.726
$\frac{3}{4}$	1.502	$2\frac{3}{8}$	15.062	$4\frac{1}{16}$	44.071
$\frac{13}{16}$	1.763	$2\frac{7}{16}$	15.866	$4\frac{1}{8}$	45.438
$\frac{7}{8}$	2.044	$2\frac{1}{2}$	16.690	$4\frac{3}{16}$	46.825
$\frac{15}{16}$	2.347	$2\frac{9}{16}$	17.535	$4\frac{1}{4}$	48.233
1	2.670	$2\frac{5}{8}$	18.400	$4\frac{5}{16}$	49.662
$1\frac{1}{16}$	3.766	$2\frac{11}{16}$	19.287	$4\frac{3}{8}$	51.112
$1\frac{1}{8}$	3.379	$2\frac{3}{4}$	20.195	$4\frac{7}{16}$	52.583
$1\frac{3}{16}$	3.766	$2\frac{13}{16}$	21.123	$4\frac{1}{2}$	54.075
$1\frac{1}{4}$	4.173	$2\frac{7}{8}$	22.072	$4\frac{9}{16}$	55.587
$1\frac{5}{16}$	4.600	$2\frac{15}{16}$	23.042	$4\frac{5}{8}$	57.121
$1\frac{3}{8}$	5.049	3	24.033	$4\frac{11}{16}$	58.675
$1\frac{7}{16}$	5.518	$3\frac{1}{16}$	25.045	$4\frac{3}{4}$	60.250
$1\frac{1}{2}$	6.008	$3\frac{1}{8}$	26.078	$4\frac{13}{16}$	61.846
$1\frac{9}{16}$	6.520	$3\frac{3}{16}$	27.131	$4\frac{7}{8}$	63.463
$1\frac{5}{8}$	7.051	$3\frac{1}{4}$	28.206	$4\frac{15}{16}$	65.100
$1\frac{11}{16}$	7.604	$3\frac{5}{16}$	29.301	5	66.759
		$3\frac{3}{8}$	30.417		

TABLE 4 Table of wire gages

No. of Wire Gage	Stubs' Steel Wire	Birmingham or Stubs' Iron Wire	American or B. & S.	New Am. S. & W. Co's Music Wire Gage	Imperial Wire Gage	Washburn & Moen, Worcester, Mass.	W. & M. Steel Music Wire
00000000	—	—	—	—	—	—	0.0083
0000000	—	—	—	—	—	0.490	0.0087
000000	—	—	—	0.004	0.464	0.4615	0.0095
00000	—	—	—	0.005	0.432	0.4305	0.010
0000	—	0.454	0.460	0.006	0.400	0.3938	0.011
000	—	0.425	0.40964	0.007	0.372	0.3625	0.012
00	—	0.380	0.3648	0.008	0.348	0.3310	0.0133
0	—	0.340	0.32486	0.009	0.324	0.3065	0.0144
1	0.227	0.300	0.2893	0.010	0.300	0.2830	0.0156
2	0.219	0.284	0.25763	0.011	0.276	0.2625	0.0166
3	0.212	0.259	0.22942	0.012	0.252	0.2437	0.0178
4	0.207	0.238	0.20431	0.013	0.232	0.2253	0.0188
5	0.204	0.220	0.18194	0.014	0.212	0.2070	0.0202
6	0.201	0.203	0.16202	0.016	0.192	0.1920	0.0215
7	0.199	0.180	0.14428	0.018	0.176	0.1770	0.023
8	0.197	0.165	0.12849	0.020	0.160	0.1620	0.0243
9	0.194	0.148	0.11443	0.022	0.144	0.1483	0.0256
10	0.191	0.134	0.10189	0.024	0.128	0.1350	0.027
11	0.188	0.120	0.090742	0.026	0.116	0.1205	0.0284
12	0.185	0.109	0.080808	0.029	0.104	0.1055	0.0296
13	0.182	0.095	0.071961	0.031	0.092	0.0915	0.0314
14	0.180	0.083	0.064804	0.033	0.080	0.0800	0.0326
15	0.178	0.072	0.057068	0.035	0.072	0.0720	0.0345
16	0.175	0.065	0.05082	0.037	0.064	0.0625	0.036
17	0.172	0.058	0.045257	0.039	0.056	0.0540	0.0377
18	0.168	0.049	0.040303	0.041	0.048	0.0475	0.0395
19	0.164	0.042	0.03589	0.043	0.040	0.0410	0.0414
20	0.161	0.035	0.031961	0.045	0.036	0.0348	0.0434
21	0.157	0.032	0.028462	0.047	0.032	0.03175	0.046
22	0.155	0.028	0.025347	0.049	0.028	0.0286	0.0483
23	0.153	0.025	0.022571	0.051	0.024	0.0258	0.051
24	0.151	0.022	0.0201	0.055	0.022	0.0230	0.055
25	0.148	0.020	0.0179	0.059	0.020	0.0204	0.0586
26	0.146	0.018	0.01594	0.063	0.018	0.0181	0.0626
27	0.143	0.016	0.014195	0.067	0.0164	0.0173	0.0658
28	0.139	0.014	0.012641	0.071	0.0149	0.0162	0.072
29	0.134	0.013	0.011257	0.075	0.0136	0.0150	0.076
30	0.127	0.012	0.010025	0.080	0.0124	0.0140	0.080
31	0.120	0.010	0.008928	0.085	0.0116	0.0132	—
32	0.115	0.009	0.00795	0.090	0.0108	0.0128	—
33	0.112	0.008	0.00708	0.095	0.0100	0.0118	—
34	0.110	0.007	0.006304	—	0.0092	0.0104	—
35	0.108	0.005	0.005614	—	0.0084	0.0095	—
36	0.106	0.004	0.005	—	0.0076	0.0090	—
37	0.103	—	0.004453	—	0.0068	—	—
38	0.101	—	0.003965	—	0.0060	—	—
39	0.099	—	0.003531	—	0.0052	—	—
40	0.097	—	0.003144	—	0.0048	—	—

NOTE Dimensions are in inches.

TABLE 5   Useful formulas for finding areas and dimensions of geometric figures

A = Area; S = Side; D = Diagonal of circumscribed circle; d = Height of circular segment (distance cut into round stock to make 3 flats); h = Height of triangle.

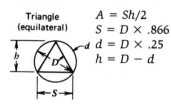

Triangle (equilateral)

$A = Sh/2$
$S = D \times .866$
$d = D \times .25$
$h = D - d$

A = Area; S = Side; D = Diagonal; d = Height of circular segment (distance cut into round stock to make 4 flats).

Square

$A = S^2 = \frac{1}{2}d^2$
$S = D \times .7071$
$D = S \times 1.4142$
$d = D \times .14645$

A = Area, S = Side; D = Diagonal; R = Radius of circumscribed circle; r = Radius of inscribed circle; d = Height of circular segment (distance cut into round stock to make 6 flats).

Hexagon

$A = 2.598S^2 = 2.598R^2 = 3.464r^2$
$S = D \times .5 = R = 1.155r$
$d = D \times .067$
$r = .866S$

A = Area; S = Side; D = Diagonal; R = Radius of circumscribed circle; r = Radius of inscribed circle; d = Height of circular segment (distance cut into round stock to make 8 flats).

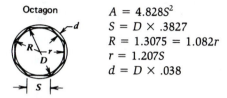

Octagon

$A = 4.828S^2$
$S = D \times .3827$
$R = 1.3075 = 1.082r$
$r = 1.207S$
$d = D \times .038$

A = Area; D = Diameter; R = Radius; C = Circumference

Circle

$A = \pi R^2 = 3.1416R^2 = .7854D^2$
$C = 2\pi R^2 = 6.2832R = 3.1416D$
$R = C \div 6.2832$

A = Area; l = Length of arc; C = Chord length; R = Radius; λ = angle (in degrees); d = Height of circular segment (distance cut into round stock to produce a flat of a given width).

Circular segment

$A = \frac{1}{2}[Rl - C(R - d)]$
$C = 2\sqrt{d(2R - d)}$
$d = R - \frac{1}{2}\sqrt{4R^2 - C^2}$
$l = .01745R\lambda$
$\lambda = \dfrac{57.296l}{R}$

NOTE To find the weight of a metal bar of any cross-sectional shape, find the area of cross section, multiply by the length of the bar in inches and by the weight in pounds per cubic inch of the material. The weights of some metals in pounds per cubic inch are as follows: steel, .284; aluminum, .0975; bronze, .317; copper, .321; lead, .409; and silver, .376.

TABLE 6   Inch-metric measures

Linear Measures	
English	**Metric**
1 mile = 1760 yards = 5280 feet	10 millimeters (mm) = 1 centimeter (cm)
1 yard = 3 feet = 36 inches	10 centimeters = 1 decimeter (dm)
1 foot = 12 inches	10 decimeters = 1 meter (m)
1 inch = 1000 mils	1000 meters = 1 kilometer (km)
Conversion Factors	
1 inch = 2.54 cm = 25.4 mm	1 millimeter = .03937 inch
1 foot = .3048 meter	10 millimeters (cm) = .3937 inch
1 yard = .9144 meter	1 meter = 39.37 inches
	3.2808 feet
	1.0936 yards
1 mile = 1.6093 km	1 kilometer = 1093.6 yards or .62137 mile

TABLE 7 Decimal and metric equivalents of fractions of an inch

Fractional Inch	Decimal Inch	Milli-meters	Fractional Inch	Decimal Inch	Milli-meters
$\frac{1}{64}$	0.015625	0.3969	$\frac{33}{64}$	0.515625	13.0969
$\frac{1}{32}$	0.03125	0.7937	$\frac{17}{32}$	0.53125	13.4937
$\frac{3}{64}$	0.046875	1.1906	$\frac{35}{64}$	0.546875	13.8906
$\frac{1}{16}$	0.0625	1.5875	$\frac{9}{16}$	0.5625	14.2875
$\frac{5}{64}$	0.078125	1.9844	$\frac{37}{64}$	0.578125	14.6844
$\frac{3}{32}$	0.09375	2.3812	$\frac{19}{32}$	0.59375	15.0812
$\frac{7}{64}$	0.109375	2.7781	$\frac{39}{64}$	0.609375	15.4781
$\frac{1}{8}$	0.125	3.1750	$\frac{5}{8}$	0.625	15.8750
$\frac{9}{64}$	0.140625	3.5719	$\frac{41}{64}$	0.640625	16.2719
$\frac{5}{32}$	0.15625	3.9687	$\frac{21}{32}$	0.65625	16.6687
$\frac{11}{64}$	0.171875	4.3656	$\frac{43}{64}$	0.671875	17.0656
$\frac{3}{16}$	0.1875	4.7625	$\frac{11}{16}$	0.6875	17.4625
$\frac{13}{64}$	0.203125	5.1594	$\frac{45}{64}$	0.703125	17.8594
$\frac{7}{32}$	0.21875	5.5562	$\frac{23}{32}$	0.71875	18.2562
$\frac{15}{64}$	0.234375	5.9531	$\frac{47}{64}$	0.734375	18.6531
$\frac{1}{4}$	0.25	6.3500	$\frac{3}{4}$	0.75	19.0500
$\frac{17}{64}$	0.265625	6.7469	$\frac{49}{64}$	0.765625	19.4469
$\frac{9}{32}$	0.28125	7.1437	$\frac{25}{32}$	0.78125	19.8437
$\frac{19}{64}$	0.296875	7.5406	$\frac{51}{64}$	0.796875	20.2406
$\frac{5}{16}$	0.3125	7.9375	$\frac{13}{16}$	0.8125	20.6375
$\frac{21}{64}$	0.328125	8.3344	$\frac{53}{64}$	0.828125	21.0344
$\frac{11}{32}$	0.34375	8.7312	$\frac{27}{32}$	0.84375	21.4312
$\frac{23}{64}$	0.359375	9.1281	$\frac{55}{64}$	0.859375	21.8281
$\frac{3}{8}$	0.375	9.5250	$\frac{7}{8}$	0.875	22.2250
$\frac{25}{64}$	0.390625	9.9219	$\frac{57}{64}$	0.890625	22.6219
$\frac{13}{32}$	0.40625	10.3187	$\frac{29}{32}$	0.90625	23.0187
$\frac{27}{64}$	0.421875	10.7156	$\frac{59}{64}$	0.921875	23.4156
$\frac{7}{16}$	0.4375	11.1125	$\frac{15}{16}$	0.9375	23.8125
$\frac{29}{64}$	0.453125	11.5094	$\frac{61}{64}$	0.953125	24.2094
$\frac{15}{32}$	0.46875	11.9062	$\frac{31}{32}$	0.96875	24.6062
$\frac{31}{64}$	0.484375	12.3031	$\frac{63}{64}$	0.984375	25.0031
$\frac{1}{2}$	0.50	12.7000	1	1.000000	25.4000

TABLE 8   Metric-inch conversion table

Milli-meters	Decimal Inches	Milli-meters	Decimal Inches	Milli-meters	Decimal Inches	Milli-meters	Decimal Inches	Milli-meters	Decimal Inches	Milli-meters	Decimal Inches
.1	.00394	11	.4331	29	1.1417	47	1.8504	65	2.5590	83	3.2677
.2	.00787	12	.4724	30	1.1811	48	1.8898	66	2.5984	84	3.3071
.3	.01181	13	.5118	31	1.2205	49	1.9291	67	2.6378	85	3.3464
.4	.01575	14	.5512	32	1.2598	50	1.9685	68	2.6772	86	3.3858
.5	.01968	15	.5905	33	1.2992	51	2.0079	69	2.7165	87	3.4252
.6	.02362	16	.6299	34	1.3386	52	2.0472	70	2.7559	88	3.4646
.7	.0275	17	.6693	35	1.3779	53	2.0866	71	2.7953	89	3.5039
.8	.0315	18	.7087	36	1.4173	54	2.1260	72	2.8346	90	3.5433
.9	.03541	19	.7480	37	1.4567	55	2.1653	73	2.8740	91	3.5827
1	.0394	20	.7874	38	1.4961	56	2.2047	74	2.9134	92	3.6220
2	.0787	21	.8268	39	1.5354	57	2.2441	75	2.9527	93	3.6614
3	.1181	22	.8661	40	1.5748	58	2.2835	76	2.9921	94	3.7008
4	.1575	23	.9055	41	1.6142	59	2.3228	77	3.0315	95	3.7401
5	.1968	24	.9449	42	1.6535	60	2.3622	78	3.0709	96	3.7795
6	.2362	25	.9842	43	1.6929	61	2.4016	79	3.1102	97	3.8189
7	.2756	26	1.0236	44	1.7323	62	2.4409	80	3.1496	98	3.8583
8	.3150	27	1.0630	45	1.7716	63	2.4803	81	3.1890	99	3.8976
9	.3543	28	1.1024	46	1.8110	64	2.5197	82	3.2283	100	3.9370
10	.3937										

TABLE 9   Allowances for fits of bores in inches

Diameter (in.)	Running and Sliding Fits (free to rotate and free to slide)	Standard Fits (readily assembled)	Driving Fits (permanent assembly, light drive)	Forced Fits (permanent assembly with hydraulic press)
Up to $\frac{1}{2}$	+.0005 to +.001	+.00025 to +.0005	−.0005	−.00075
$\frac{1}{2}$ to 1	+.001 to +.0015	+.0003 to +.001	−.0075	−.0015
1 to 2	+.0015 to +.0025	+.0004 to +.0015	−.001	−.0025
2 to $3\frac{1}{2}$	+.002 to +.003	+.0005 to +.002	−.0015	−.0035
$3\frac{1}{2}$ to 6	+.003 to +.004	+.00075 to +.003	−.002	−.0045

NOTE In this table, shaft sizes are considered as nominal and bore sizes are varied for fits, thus a negative fit is an interference (press) fit and a positive fit is a free or loose fit.

TABLE 10 Metric tap drill sizes

Metric Tap Size	Recommended Metric Drill				Closest Recommended Inch Drill			
	Drill Size (mm)	Inch Equivalent	Probable Hole Size (in.)	Probable Percent of Thread	Drill Size	Inch Equivalent	Probable Hole Size (in.)	Probable Percent of Thread
M1.6 × .35	1.25	.0492	.0507	69	—	—	—	—
M1.8 × .35	1.45	.0571	.0586	69	—	—	—	—
M2 × .4	1.60	.0630	.0647	69	#52	.0635	.0652	66
M2.2 × .45	1.75	.0689	.0706	70	—	—	—	—
M2.5 × .45	2.05	.0807	.0826	69	#46	.0810	.0829	67
*M3 × .5	2.50	.0984	.1007	68	#40	.0980	.1003	70
M3.5 × .6	2.90	.1142	.1168	68	#33	.1130	.1156	72
*M4 × .7	3.30	.1299	.1328	69	#30	.1285	.1314	73
M4.5 × .75	3.70	.1457	.1489	74	#26	.1470	.1502	70
*M5 × .8	4.20	.1654	.1686	69	#19	.1660	.1692	68
*M6 × 1	5.00	.1968	.2006	70	#9	.1960	.1998	71
M7 × 1	6.00	.2362	.2400	70	15/64	.2344	.2382	73
*M8 × 1.25	6.70	.2638	.2679	74	17/64	.2656	.2697	71
M8 × 1	7.00	.2756	.2797	69	J	.2770	.2811	66
*M10 × 1.5	8.50	.3346	.3390	71	Q	.3320	.3364	75
M10 × 1.25	8.70	.3425	.3471	73	11/32	.3438	.3483	71
*M12 × 1.75	10.20	.4016	.4063	74	Y	.4040	.4087	71
M12 × 1.25	10.80	.4252	.4299	67	27/64	.4219	.4266	72
M14 × 2	12.00	.4724	.4772	72	15/32	.4688	.4736	76
M14 × 1.5	12.50	.4921	.4969	71	—	—	—	—
*M16 × 2	14.00	.5512	.5561	72	35/64	.5469	.5518	76
M16 × 1.5	14.50	.5709	.5758	71	—	—	—	—
M18 × 2.5	15.50	.6102	.6152	73	39/64	.6094	.6144	74
M18 × 1.5	16.50	.6496	.6546	70	—	—	—	—
*M20 × 2.5	17.50	.6890	.6942	73	11/16	.6875	.6925	74
M20 × 1.5	18.50	.7283	.7335	70	—	—	—	—
M22 × 2.5	19.50	.7677	.7729	73	49/64	.7656	.7708	75
M22 × 1.5	20.50	.8071	.8123	70	—	—	—	—
*M24 × 3	21.00	.8268	.8327	73	53/64	.8281	.8340	72
M24 × 2	22.00	.8661	.8720	71	—	—	—	—
M27 × 3	24.00	.9449	.9511	73	15/16	.9375	.9435	78
M27 × 2	25.00	.9843	.9913	70	63/64	.9844	.9914	70
*M30 × 3.5	26.50	1.0433						
M30 × 2	28.00	1.1024						
M33 × 3.5	29.50	1.1614						
M33 × 2	31.00	1.2205		Reaming recommended to the drill size shown				
M36 × 4	32.00	1.2598						
M36 × 3	33.00	1.2992						
M39 × 4	35.00	1.3780						
M39 × 3	36.00	1.4173						

Formula for metric tap drill size: Basic major diameter (mm) $- \dfrac{\% \text{ Thread} \times \text{Pitch (mm)}}{76.980}$ = Drilled hole size (mm)

Formula for percent of thread: $\dfrac{76.980}{\text{Pitch (mm)}} \times \left[\underset{\text{(mm)}}{\text{Basic major diameter}} - \underset{\text{(mm)}}{\text{Drilled hole size}} \right]$ = Percent of thread

SOURCE Material courtesy of TRW Inc., *New Greenfield Geometric ISO Metric Screw Thread Manual*, 1973.

BASIC DESIGNATIONS

ISO metric threads are designated by the letter *M* followed by the *nominal size* in millimeters, and the *pitch* in millimeters, separated by the sign "✕."

Example: M16 ✕ 1.5

Those numbers in Table 10 marked with an asterisk are the commercially available sizes in the United States.

TOLERANCE SYMBOLS

3   4   5   6

7   8   9

Numbers are used to define the amount of product tolerance permitted on either internal or external threads. Smaller grade numbers carry smaller tolerances, that is, grade 4 tolerances are smaller than grade 6 tolerances, and grade 8 tolerances are larger than grade 6 tolerances.

e   H   G   g

Letters are used to designate the "position" of the product thread tolerances relative to basic diameters. Lower-case letters are used for external threads, and capital letters for internal threads.

In some cases the "position" of the tolerance establishes an allowance (a definite clearance) between external and internal threads.

By combining the tolerance amount number and the tolerance position letter, the *tolerance symbol* is established that identifies the actual maximum and minimum product limits for external or internal threads. Generally, the first number and letter refer to the pitch diameter symbol. The second number and letter refer to the **crest** diameter symbol (minor diameter of internal threads or major diameter of external threads).

Example:

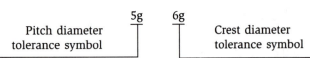

Pitch diameter tolerance symbol / Crest diameter tolerance symbol (5g / 6g)

Where the pitch diameter and crest diameter tolerance symbols are the same, the symbol need only be given once.

Example:

6g

Pitch diameter and crest diameter tolerance symbol

It is recommended that the *coarse series* be selected whenever possible and that *general purpose grade* 6 be used for both internal and external threads.

Tolerance positions g for external threads and H for internal threads are preferred.

Other product information may also be conveyed by the ISO metric thread designations. Complete specifications and product limits may be found in the ISO Recommendations or in the B1 report "ISO Metric Screw Threads."

Some examples of ISO Metric Thread designations are as follows:

> M10
> M18 ✕ 1.5
> M6 — 6H
> M4 — 6g
> M12 ✕ 1.25 — 6H
> M20 ✕ 2 — 6H/6g
> M6 ✕ 0.75 — 7g 6g

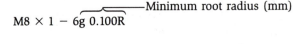

M8 ✕ 1 — 6g 0.100R — Minimum root radius (mm)

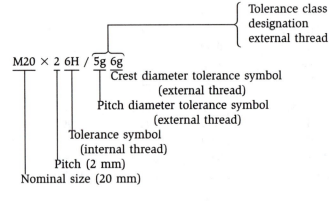

M20 ✕ 2 6H / 5g 6g

Tolerance class designation external thread

Crest diameter tolerance symbol (external thread)

Pitch diameter tolerance symbol (external thread)

Tolerance symbol (internal thread)

Pitch (2 mm)

Nominal size (20 mm)

---

source Material courtesy of TRW Inc., *New Greenfield Geometric Screw ISO Metric Thread Manual*, 1973.

TABLE 11 Tap drill sizes for unified and American standard series screw threads

Thread Size	Threads Per Inch	Series	Tap Drill Diameter		Percent of Full Thread	Thread Size	Threads Per Inch	Series	Tap Drill Diameter		Percent of Full Thread
			Size	Inches					Size	Inches	
0	80	NF	1.25 mm	.0492	66	6	40	NF	31	.1200	55
			1.2 mm	.0472	79				32	.1160	68
			$\frac{3}{64}''$	.0469	81				33	.1130	77
1	64	NC	$\frac{1}{16}''$	.0625	51				34	.1110	83
			53	.0595	66	8	32	NC	28	.1405	58
			54	.0550	88				29	.1360	69
1	72	NF	$\frac{1}{16}''$	.0625	58				3.4 mm	.1338	74
			1.55 mm	.0610	66				3.3 mm	.1299	84
			53	.0595	75	8	36	NF	27	.1440	55
2	56	NC	49	.0730	56				28	.1405	65
			50	.0700	69				29	.1360	77
			1.75 mm	.0689	74				3.4 mm	.1338	83
			51	.0670	82	10	24	NC	22	.1570	61
2	64	NF	1.9 mm	.0748	55				24	.1520	70
			49	.0730	64				25	.1495	75
			1.8 mm	.0709	74				26	.1470	79
			50	.0700	79	10	32	NF	19	.1660	59
3	48	NC	45	.0820	63				20	.1610	71
			46	.0810	66				21	.1590	76
			47	.0785	76				22	.1570	81
			48	.0760	85	12	28	NF	14	.1820	63
3	56	NF	44	.0860	56				15	.1800	66
			2.15 mm	.0846	62				16	.1770	72
			45	.0820	73				17	.1730	79
			46	.0810	77	12	28	NF	12	.1890	58
4	40	NC	42	.0935	57				14	.1820	73
			2.3 mm	.0905	66				15	.1800	77
			43	.0890	71				16	.1770	84
			44	.0860	80	12	32	NEF	$\frac{3}{16}''$	.1875	70
4	48	NF	41	.0960	59				13	.1850	76
			42	.0935	68				14	.1820	84
			2.3 mm	.0905	79	$\frac{1}{4}''$	20	UNC	5	.2055	67
			43	.0890	85			NC	6	.2040	71
5	40	NC	37	.1040	65				7	.2010	75
			38	.1015	72				8	.1990	78
			39	.0995	78	$\frac{1}{4}''$	28	UNF	2	.2210	62
			40	.0980	83			NF	$\frac{7}{32}''$	.2187	67
5	44	NF	36	.1065	62				5.5 mm	.2165	72
			37	.1040	71				3	.2130	80
			38	.1015	79	$\frac{1}{4}''$	32	NEF	5.7 mm	.2244	63
			39	.0995	86				2	.2210	71
6	32	NC	34	.1110	66				$\frac{7}{32}''$	.2187	77
			35	.1100	69				5.5 mm	.2165	82
			36	.1065	77	$\frac{5}{16}''$	18	UNC	$\frac{17}{64}''$	.2656	65
			37	.1040	83			NC	G	.2610	71

TABLE 11 *(continued)*

Thread Size	Threads Per Inch	Series	Tap Drill Diameter Size	Inches	Percent of Full Thread
			F	.2570	77
			6.4 mm	.2520	84
5/16"	24	UNF NF	J	.2770	66
			I	.2720	75
			H	.2660	85
5/16"	32	NEF	7.3 mm	.2874	62
			7.2 mm	.2835	71
			9/32"	.2812	77
			J	.2770	87
3/8"	16	UNC NC	P	.3230	64
			O	.3160	72
			5/16"	.3120	77
			7.8 mm	.3071	83
3/8"	24	UNF NF	R	.3390	67
			8.5 mm	.3346	74
			Q	.3320	79
			21/64"	.3281	86
7/16"	14	UNC NC	3/8"	.3750	67
			U	.3680	75
			23/64"	.3594	84
7/16"	20	UNF NF	X	.3970	62
			25/64"	.3906	72
			W	.3860	79
1/2"	13	UNC NC	7/16"	.4375	62
			27/64"	.4219	78
			Z	.4130	87
1/2"	20	UNF NF	11.75 mm	.4626	57
			29/64"	.4531	72
9/16"	12	UNC NC	1/2"	.5000	58
			31/64"	.4844	72
			15/32"	.4687	86
9/16"	18	UNF NF	33/64"	.5156	65
			13 mm	.5118	70
			1/2"	.5000	86
5/8"	11	UNC NC	14 mm	.5512	63
			35/64"	.5469	66
			17/32"	.5312	79
5/8"	18	UNF NF	37/64"	.5781	65
			14.5 mm	.5709	75
			9/16"	.5625	87
3/4"	10	UNC NC	17 mm	.6693	62
			21/32"	.6562	72
			41/64"	.6406	84
3/4"	16	UNF NF	45/64"	.7031	58
			11/16"	.6875	77
			43/64"	.6719	96
7/8"	9	UNC NC	25/32"	.7812	65
			49/64"	.7656	76
			3/4"	.7500	87
7/8"	14	UNF NF	13/16"	.8125	67
			20.5 mm	.8071	73
			51/64"	.7969	84
1"	8	UNC NC	57/64"	.8906	67
			7/8"	.8750	77
			55/64"	.8593	87
1"	12	UNF N	15/16"	.9375	58
			59/64"	.9218	72
			29/32"	.9062	87

Some symbols used for American threads are:

Symbol	Reference
NC	American National Coarse Thread Series
NF	American National Fine Thread Series
NEF	American National Extra Fine Thread Series
NS	Special Threads of American National Form
NH	Am. Natl. Hose Coupling and Fire Hose Coupling Thread
NPT	American Standard Taper Pipe Thread
NPTF	American Standard Taper Pipe Thread (Dryseal)
NPS	American Standard Straight Pipe Thread
ACME	Acme Threads—(Acme-C) Centralizing—(Acme-G) General Purpose
STUB ACME	Stub Acme Threads
V	A 60° V thread with truncated crests and roots. The theoretical V form is usually flattened several thousandths of an inch
SB	Manufacturers Stovebolt Standard Thread

Symbols used for Unified threads are:

Symbol	Reference
UNC	Unified Coarse Thread Series
UNF	Unified Fine Thread Series
UNEF	Unified Extra Fine Thread Series

Answers to Self-Tests

SECTION A / UNIT 1 / SHOP SAFETY

Self-Test Answers

1. Eye protection equipment.
2. Wear a safety goggle or full face shield. Prescription glasses may be made as safety glasses.
3. Shoes, short sleeves, short or properly secured hair, no rings and wristwatches, shop apron or shop coat with short sleeves.
4. Use of cutting fluids and vacuum dust collectors.
5. They may cause skin rashes or infections.
6. Bend knees and squat, lift with your legs, keeping your back straight.
7. Compressed air can propel chips through the air, implant dirt into skin, and possibly injure ear drums.
8. Good housekeeping includes cleaning oil spills, keeping material off the floor, and keeping aisles clear of obstructions.
9. In the horizontal position with a person on each end.
10. Do I know how to operate this machine?
    What are the potential hazards involved?
    Are all guards in place?
    Are my procedures safe?
    Am I doing something I probably should not do?
    Have I made all proper adjustments and tightened all locking bolts and clamps?
    Is the workpiece secured properly?
    Do I have proper safety equipment?
    Do I know where the stop switch is?
    Do I think about safety in everything I do?
11. No. You should use a hoist or crane for more than 20 or 30 lbs because a lathe chuck is in an awkward location.
12. (b) They are CO_2, dry chemical, and dry chemical multiuse (all purpose).

SECTION A / UNIT 2 / SELECTION AND IDENTIFICATION OF METALS

Self-Test Answers

1. Carbon and alloy steels are designated by the numerical SAE or AISI system.

2. The three basic types of stainless steels are: martensitic (hardenable) and ferritic (nonhardenable)—both magnetic and of the 400 series—and austenitic (nonmagnetic and nonhardenable, except by work hardening) of the 300 series.
3. The identification for each piece would be as follows:
    a. AISI C1020 CF is a soft, low-carbon steel with a dull metallic luster surface finish. Use the observation test, spark test, and file test for hardness.
    b. AISI B1140 (G and P) is a medium-carbon, resulfurized, free-machining steel with a shiny finish. Use the observation test, spark test, and machinability test.
    c. AISI C4140 (G and P) is a chromium-molybdenum alloy, medium-carbon content with a shiny finish. Since an alloy steel would be harder than a similar carbon or low-carbon content steel, a hardness test should be used such as the file or scratch test to compare with known samples. The machinability test would be useful as a comparison test.
    d. AISI 8620 HR is a tough low-carbon steel used for carburizing purposes. A hardness test and a machinability test will immediately show the difference from low-carbon hot-rolled steel.
    e. AISI B1140 (ebony) is the same as the resulfurized steel in (b), only the finish is different. The test would be the same as for (b).
    f. AISI C1040 is a medium-carbon steel. The spark test would be useful here as well as the hardness and machinability tests.
4. A magnetic test can quickly determine whether it is a ferrous metal or perhaps nickel since they will both be attracted by a magnet. If the metal is white in color, a spark test will be needed to determine whether it is a nickel casting or one of white cast iron, since they are similar in appearance. If a small piece can be broken off, the fracture will show whether it is white or gray cast iron. Gray cast iron will leave a black smudge on the finger. If it is cast

steel, it will be more ductile than cast iron and a spark test should reveal a smaller carbon content.

5. O1 refers to an alloy-type oil hardening (oil quench) tool steel. W1 refers to a water-hardening (water quench) tool steel.

6. (a) No. (b) Hardened tool steel or case-hardened steel.

7. Austenitic (having a face-centered cubic unit cell in its lattice structure). Examples are aluminum, nickel, 300 series stainless steel, and high manganese alloy steel.

8. Nickel is a nonferrous metal that has magnetic properties. Some alloy combinations of nonferrous metals make strong permanent magnets; for example, the well-known Alnico magnet—an alloy of aluminum, nickel, and cobalt.

9. Some properties of steel to be kept in mind when ordering or planning for a job would be: strength, machinability, hardenability, weldability (if welding is involved), fatigue resistance, and corrosion resistance (especially if the piece is to be exposed to a corrosive atmosphere).

10. *Advantages:* Since aluminum is about one-third lighter than steel, it is used extensively in aircraft. It also forms an oxide on the surface that resists further corrosion. *Disadvantages:* The initial cost is much greater. Higher-strength aluminum alloys cannot be welded.

11. The letter *H* following the four-digit number always designates strain or work hardening. The letter *T* refers to heat treatment.

12. Magnesium weighs approximately one-third less than aluminum and is approximately one-quarter the weight of steel. Magnesium will burn in air when finely divided.

13. Copper is most extensively used in the electrical industry because of its low resistance to the passage of current when it is unalloyed with other metals. Copper can be hardened or work hardened and certain alloys may be hardened by a solution heat treatment and aging process.

14. Bronze is basically copper and tin. Brass is basically copper and zinc.

15. Nickel is used to electroplate surfaces of metals for corrosion resistance and as an alloying element with steels and nonferrous metals.

16. Both resist deterioration from corrosion.

17. Alloy.

18. Tin, lead, and cadmium.

19. Die cast metals, sometimes called "pot metal."

20. Wrought aluminum is stronger.

21. Large rake angles (12 to 20 degrees back rake). Use of a lubricant. Proper cutting speeds.

22. No. If a fire should start in the chips, the water-based coolant will intensify the burning.

23. The rake should be zero on all cutting tools for brasses or bronzes.

24. Tungsten is combined with carbon to form a tungsten carbide powder that is compressed into briquettes and sintered in a furnace. The resultant tungsten carbide cutting tool is extremely hard and will resist the high temperatures of machining.

SECTION A / UNIT 3 / CUTOFF MACHINES

Self-Test Answers

1. Reciprocating, horizontal band, and abrasive.

2. Reciprocating saws.

3. Wear eye protection, make sure all guards are in place, make sure the frame will not strike the workpiece before you turn on the machine. Keep coolant from spilling onto the floor.

4. Wear eye protection, avoid overspeeding blades, and do not use a defective blade that has cracks.

5. Pinch points. Fingers can get mashed from heavy stock moving or rolling sideways on the roller table.

6. The abrasive blade can fly apart, throwing out bits of the blade at a high velocity. Even the guard will not prevent some pieces from flying out.

SECTION B / UNIT 1 / SYSTEMS OF MEASUREMENT

Self-Test Answers

1. To find in. knowing mm, multiply mm by .03937; $35 \times .03937 = 1.378$ in.

2. To find mm knowing in., multiply in. by 25.4; $.125 \times 25.4 = 3.17$ mm.

3. To find cm knowing in., multiply in. by 2.54; $6.273 \times 2.54 = 15.933$ cm.

4. Length.

5. Degrees, minutes, and seconds.

6. It is checked with an acceptable standard and adjusted if necessary.

7. To find mm knowing in., multiply in. by 25.4; $.050 \times 25.4 = 1.27$ mm.

8. 10 mm = 1 cm; therefore, to find cm knowing mm, divide by 10.

9. To find in. knowing mm, multiply mm by .03937; .02 × .03937 = .0008 in. The tolerance would be ±.0008 in.

10. Yes, by the use of appropriate conversion dials.

SECTION B / UNIT 2 / MICROMETERS AND CALIPERS

Self-Test Answers

1. People who take pride in their tools usually take pride in their workmanship. The quality of a product produced depends to a large extent on the accuracy of the measuring tools used. Skilled craftsmen protect their tools because they guarantee the product.

2. Moisture between the contact faces can cause corrosion.

3. Even small dust particles will change a dimension. Oil or grease attract small chips and dirt. All of these can cause incorrect readings.

4. A measuring tool is no more discriminatory than the smallest division marked on it. This means that a standard micrometer can discriminate to the nearest thousandth. A vernier scale on a micrometer will make it possible to discriminate a reading to one ten-thousandths of an inch under controlled conditions.

5. The reliability of a micrometer depends on the inherent qualities built into it by its maker. Reliability also depends upon the skill of the user and the care the tool receives.

6. The sleeve is stationary in relation to the frame and is engraved with the main scale, which is divided into 40 equal spaces each equal to .025 in. The thimble is attached to the spindle and rotates with it. The thimble circumference is graduated with 25 equal divisions, each representing a value of .001 in.

7. There is less chance of accidentally moving the thimble when reading a micrometer while it is still in contact with the workpiece.

8. Measurement should be made at least twice. On critical measurements, checking the dimensions additional times will assure that the size measurement is correct.

9. As the temperature of a part is increased, the size of the part will increase. When a part is heated by the machining process, it should be permitted to cool down to room temperature before being measured. Holding a micrometer by the frame for an extended period of time will transfer body heat through the hand and affect the accuracy of the measurement taken.

10. The purpose of the ratchet stop or friction thimble is to enable equal pressure to be repeatedly applied between the measuring faces and the object being measured. Use of the ratchet stop or friction thimble will minimize individual differences in measuring pressure applied by different persons using the same micrometer.

EXERCISE ANSWERS

Outside micrometer readings (Figures B2.30a to B2.30e):

Figure B2.30a   .669 in.
Figure B2.30b   .787 in.
Figure B2.30c   .237 in.
Figure B2.30d   .994 in.
Figure B2.30e   .072 in.

Inside micrometer readings (Figures B2.31a to B2.31e):

Figure B2.31a   1.617
Figure B2.31b   2.000
Figure B2.31c   2.254
Figure B2.31d   2.562
Figure B2.31e   2.784

Depth micrometer readings (Figures B2.32a to B2.32e):

Figure B2.32a   .535 in.
Figure B2.32b   .815 in.
Figure B2.32c   .732 in.
Figure B2.32d   .535 in.
Figure B2.32e   .647 in.

SECTION B / UNIT 3 / COMPARISON AND ANGULAR MEASURING INSTRUMENTS

Self-Test Answers

1. Comparison measurement is measurement by which an unknown dimension is compared to a known dimension. This often involves a transfer device that represents the unknown and is then transferred to the known where the reading can be determined.

2. Most comparison instruments do not have the capability to show measurement directly. Direct-reading measuring instruments are more reliable than transfer-type tools since the measuring "feel" with its potential for error must be made twice with transfer measuring devices.

3. If the axis of measurement is not the same as the axis of the measuring instrument, an error will result. Only dial indicators should be used to make direct measurements; dial test indicators will always have an error in measurement due to the difference between the arc and chord length in the tip travel.
4. Error can be reduced by making sure that the axis of the measuring instrument is exactly in line with the axis of measurement.
5. (c) Adjustable parallel.
6. (i) and (j) Dial test indicator and vernier height gage.
7. (d) Radius gage.
8. (g) Combination square.
9. (b) Telescope gage.
10. (e) Thickness (feeler) gage.

SECTION B / UNIT 4 / TOLERANCES, FITS, GEOMETRIC DIMENSIONS, AND STATISTICAL PROCESS CONTROL (SPC)

Self-Test Answers

1. Tolerances are important because they control the size and therefore the ability of parts to fit together in complex assemblies.
2. Fractional dimensions $\pm \frac{1}{64}$ in.
   Two-place decimals $\pm$ .010 in.
   Three-place decimals $\pm$ .005 in.
   Four-place decimals $\pm$ .0005 in.
   Angles $\pm \frac{1}{2}$ degree
3. Straightness, roundness, flatness, perpendicularity, parallelism, concentricity, runout.
4. Typical press fit allowance is calculated by the formula .0015 in. $\times$ the diameter of the part.
5. Shrink and expansion fits are accomplished by cooling (shrinking) or heating (expanding) the parts to be fitted together. After fitting in the shrunk or expanded state, the parts will securely hold together upon cooling or warming to ambient temperature.
6. SPC tools include all the electronic digital varieties of most of the measuring tools including calipers, micrometers, and dial indicators. These are coupled to a microcomputer which is used to generate the desired statistical information. SPC activities involve the inspection of parts where dimensions are recorded, analyzed, and graphed by the computer. Various statistics may be generated, and various graphs such as histograms prepared.

SECTION B / UNIT 5 / READING SHOP DRAWINGS

Self-Test Answers

1. See Figure A2.B5.1.
2. The tolerance of the hole is specified as $\pm$.005. Therefore, the minimum size of the hole would be .750 $-$ .005 or .745.

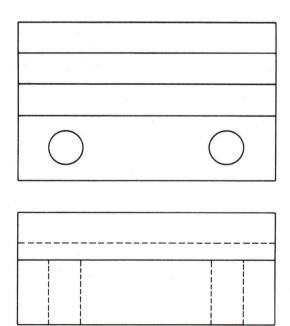
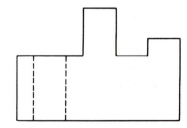

FIGURE A2.B5.1

3. $2\frac{1}{4}$ in.
4. The slot conforms to standard tolerances, width is $\pm\frac{1}{64}$ in.
5. $\frac{1}{8}$ in.
6. 6 in.
7. 1 in.
8. Drilling and reaming.
9. $\frac{1}{8}$ in. by 45 degrees.
10. Note 1 indicates that all sharp corners are to be broken, which means that all burrs and sharp edges left from machining are to be smoothed off.

SECTION C / UNIT 1 / MACHINABILITY AND CHIP FORMATION

Self-Test Answers

1. At lower speeds, negative-rake tools usually produce a poorer surface finish than do positive rakes.
2. It is called a *built-up edge* and causes a rough, ragged cutting action that produces a poor surface finish.
3. Thin, uniform chips indicate the least surface disruption.
4. No. It slides ahead of the tool on a shear plane, elongating and altering the grain structure of the metal.
5. Negative rake.
6. Higher cutting speeds produce better surface finishes. There also is less disturbance and disruption of the grain structure at higher speeds.
7. Surface irregularities such as scratches, tool marks, microcracks, and poor radii can shorten the working life of the part considerably by causing stress concentration that can develop into a metal fatigue failure.
8. The property of hardness is related to machinability. Machinists sometimes use a file to determine hardness.
9. The "9"-shaped chip.
10. Machinability is the relative difficulty of a machining operation with regard to tool life, surface finish, and power consumption.

SECTION C / UNIT 2 / SPEEDS AND FEEDS FOR MACHINE TOOLS

Self-Test Answers

1. The RPM for the $\frac{1}{8}$ in. twist drill should be 2880, and for the $\frac{3}{4}$ in. drill, 480.
2. They would be the same.

3. Use the next-lower speed setting or the highest setting on the machine if the calculated speed is higher than the top speed of the machine.
4. 100 RPM.
5. Inches per revolution (IPR).
6. .010 in. per revolution.
7. Milling machines, both vertical and horizontal.
8. About $\frac{1}{4}$ in. per table stroke.

SECTION C / UNIT 3 / CUTTING FLUIDS

Self-Test Answers

1. Cooling and lubrication.
2. Synthetic and semisynthetic fluids, emulsions, and cutting oils.
3. Remove the cutting fluid, clean the tank, and replace with new cutting fluid.
4. Because the tramp oil will contaminate the soluble oil-water mix.
5. They tend to remove skin oils. The "oilier" types do not cause this problem.
6. Using a pump oiler can contain the appropriate cutting fluid.
7. A coolant tank, pump, hose, and nozzle.
8. Flooding the workpiece-tool area with cutting fluid.
9. Because the rapidly spinning grinding wheel tends to blow it away.
10. Because only a very small amount of liquid which contains the fatty particles ever reaches the cutting area.

SECTION D / UNIT 1 / THE DRILL PRESS

Self-Test Answers

1. Loose clothing of any kind that could get caught in a machine would be hazardous. Gloves are a danger around rotating machinery. Long loose hair is extremely unsafe around rotating machinery. Loose hair must be tied or restrained when working in the shop. Watches or rings should not be worn.
2. Workpieces that are not clamped can become a danger to the operator. Even work clamped in a drill press vise that is not itself bolted or clamped to the table, can begin to spin if the drill jams. Often this results in a flying workpiece of great mass. The only corrective measure needed is to securely clamp everything that is being drilled.
3. Most machine tools are capable of throwing metal chips at high velocities. Those who have

had painful eye injuries can attest to the fact that you cannot get out of the way quickly enough. *Always* wear safety glasses in the machine shop at *all* times.

4. Never clean the spindle taper with the machine running, nor should you change belts, lubricate, or make any other adjustments. The result could be sprained or broken fingers or worse.

5. No. It is not acceptable to leave the chuck key in for even a moment, as this is just the situation in which you are most likely to forget it and turn on the machine. Make it a practice never to let the key leave your hand and you won't have the problem of remembering to remove it from the chuck.

6. No! Slowing or stopping a spindle with the hand can lead to injury. Chips can accumulate around the chuck. The open jaws in an empty chuck could break a finger. Let the machine slow down and stop by itself.

7. This problem can be eliminated by placing a block of wood under the drill on which it can drop. This also protects the cutting edge of the drill.

8. A brush or stick will safely remove chips. A machine table may be wiped down with a cloth only after all of the chips have been removed.

9. The oil can cause anyone walking past or carrying a heavy object to fall. The operator could slip and fall into the machine or against a switch or lever, thus causing an additional accident.

10. Burrs and sharp edges are extremely effective cutting tools for fingers and hands. The sooner the burrs are removed, the less chance there is for you to be cut. Ends of threads, both internal and external, are particularly bad as they are an unexpected hazard.

11. (a) The sensitive, upright, and radial arm drill presses are the three basic types. The sensitive drill press is made for light-duty work; it provides the operator with a sense or "feel" of the feed on the drill. The upright is a similar, but heavy-duty, drill press equipped with power feed. The radial arm drill press allows the operator to position the drill over the work where it is needed rather than to position the work under the drill as with other drill presses.

  (b) The sensitive, upright, and radial arm drill presses all perform much the same functions of drilling, reaming, counterboring, countersinking, spot facing, and tapping, but the upright and radial machines do heavier and larger jobs. The radial arm drill can support large, heavy castings and work can be done on them without the workpiece being moved.

12. Sensitive drill press
  __A__ Spindle
  __G__ Quill lock handle
  __E__ Column
  __L__ Switch
  __B__ Depth stop
  __N__ Head
  __O__ Table
  __H__ Table lock
  __F__ Base
  __C__ Power feed
  __J__ Motor
  __D__ Variable speed control
  __K__ Table lift crank
  __I__ Quill return spring
  __M__ Guard

13. Radial drill press
  __B__ Column
  __C__ Radial arm
  __A__ Spindle
  __D__ Base
  __E__ Drill head

SECTION D / UNIT 2 / DRILLING TOOLS

Self-Test Answers

1. __T__ Web
  __U__ Margin
  __D__ Drill point angle
  __P__ Cutting lip
  __K__ Flute
  __O__ Body
  __E__ Lip relief angle
  __G__ Land
  __B__ Chisel edge angle
  __J__ Body clearance
  __I__ Helix angle
  __Y__ Axis of drill
  __N__ Shank length
  __C__ Tang
  __X__ Taper shank
  __W__ Straight shank

2.

	Decimal Diameter	Fractional Size	Number Size	Letter Size	Metric Size
a	.0781	$\frac{5}{64}$			
b	.1495		25		
c	.272			I	
d	.159		21		
e	.1969				5

	Decimal Diameter	Fractional Size	Number Size	Letter Size	Metric Size
f	.323			P	
g	.3125	$\frac{5}{16}$			
h	.4375	$\frac{7}{16}$			
i	.201		7		
j	.1875	$\frac{3}{16}$			

3. The basic advantage of drill-point sharpening is precision. The drill requires no spotting drill or pilot hole, and it doesn't wobble and go off center.

4. Production drilling with CNC machines requires closer tolerances than that of conventional drilling on drill presses.

5. There is no chisel end on a drill when split-point sharpened on a grinding machine; therefore the drill will not wander but drill the hole precisely where it is placed.

6. Split point drills as small as .008 in. can be sharpened on precision drill grinders.

SECTION D / UNIT 3 / WORK LOCATING AND HOLDING DEVICES ON DRILLING MACHINES

Self-Test Answers

1. The purpose for using workholding devices is to keep the workpiece rigid, from turning with the drill, and for operator safety.

2. Included in a list of workholding devices would be strap clamps, T-bolts, and step blocks. Also used are C-clamps, vee-blocks, vises, jigs and fixtures, and angle plates.

3. Parallels are mostly used to raise workpieces off the drill press table or to lift the workpiece higher in a vise, thus providing a space for the drill breakthrough. They are made of hardened steel, so care should be exercised in their use.

4. Thin, limber materials tend to be sprung downward from the force of drilling until drill breakthrough begins. The drill then "grabs" as the material springs upward: a broken drill is often the result. Place the support or parallels as near the drill as possible to avoid this.

5. Angle drilling is done by tilting a drill press table (not all types tilt) or by using an angular vise. If no means of setting the exact angle is provided, use a protractor head with level.

6. Vee-blocks are suited to hold round stock for drilling. The most frequent use of vee-blocks is for cross drilling holes in shafts, although many other setups are used.

7. The wiggler is used for locating a center punch mark under the center axis of a drill spindle.

8. Some odd-shaped workpieces, such as gears with extending hubs that need holes drilled for set screws, might be difficult to set up without an angle plate.

9. One of the difficulties with hand tapping is the tendency for taps to start crooked or be misaligned with the tap-drilled hole. Starting a tap by hand in a drill press with the same setup as used for the tap drilling assures a perfect alignment.

10. Since jigs and fixtures are mostly used for production manufacturing, small machine shops rarely have a use for them.

SECTION D / UNIT 4 / OPERATING DRILLING MACHINES

Self-Test Answers

1. The three considerations would be speeds, feeds, and coolants.

2. The RPMs of the drills would be
   a. $\frac{1}{4}$-in. diameter is 1440 RPM.
   b. 2-in. diameter is 180 RPM.
   c. $\frac{3}{4}$-in. diameter is 480 RPM.
   d. $\frac{3}{8}$-in. diameter is 960 RPM.
   e. $1\frac{1}{2}$-in. diameter is 240 RPM.

3. Worn margins and outer corners broken down. The drill can be ground back to its full size and resharpened.

4. The operator will increase the feed in order to produce a chip. This increased feed is often greater than the drill can stand without breaking.

5. The feed is about right when the chip rolls into a tight helix. Long, stringy chips can indicate too much feed.

6. Feeds are designated by a small measured advance movement of the drill for each revolution. A .001-in. feed for a $\frac{1}{8}$-in. diameter drill, for example, would move the drill .001 in. into the work for every turn of the drill.

7. The water soluble oil types and the cutting oils, both animal and mineral.

8. Besides having the correct cutting speed, a sulfurized oil-based cutting fluid helps to reduce friction and cool the cutting edge.

9. Drill "jamming" can be avoided by a "pecking" procedure. The operator drills a small amount and pulls out the drill to remove the chips. This is repeated until the hole is finished.

10. The depth stop is used to limit the travel of the drill so that it will not go on into the table or vise. The depth of blind holes is preset and

drilled. Countersink and counterbore depths are set so that several can be easily made the same.

SECTION D / UNIT 5 / COUNTERSINKING AND COUNTERBORING

Self-Test Answers

1. Countersinks are used to chamfer holes and to provide tapered holes for flathead fasteners such as screws and rivets.
2. Countersink angles vary to match the angles of different flathead fasteners or different taper hole requirements.
3. A center drill is used to make a 60-degree countersink hole in workpieces for lathes and grinders.
4. A counterbore makes a cylindrical recess concentric with a smaller hole so that a hex head bolt or socket head capscrew can be flush mounted with the surface of a workpiece.
5. The pilot diameter should always be a few thousandths of an inch smaller than the hole but not more than .005 in.
6. Lubrication of the pilot prevents metal-to-metal contact between it and the hole. It will also prevent the scoring of the hole surface.
7. A general rule is to use approximately one-third of the cutting speed when counterboring as when using a twist drill with the same diameter.
8. Feeds and speeds when counterboring are controlled to a large extent by the condition of the equipment, the available power, and the material being counterbored.
9. Spot facing is performed with a counterbore. It makes a flat bearing surface, square with a hole, to seat a nut, washer, or bolt head.
10. Counterboring requires a rigid setup with the workpiece being securely fastened and provisions made for the pilot to protrude below the bottom surface of the workpiece.

SECTION D / UNIT 6 / REAMING IN THE DRILL PRESS

Self-Test Answers

1. Machine reamers are identified by the design of the shank, either a straight or tapered shank and usually a 45-degree chamfer on the cutting end.
2. A chucking reamer is a finishing reamer, the fluted part is cylindrical, and the lands are re-

lieved. A rose reamer is a roughing reamer. It can remove a considerable amount of material. The body has a slight back taper and no relief on the lands. All cutting takes place on the chamfered end.
3. A jobbers reamer is a finishing reamer like a chucking reamer but it has a longer fluted body.
4. Shell reamers are more economical to produce than solid reamers, especially in larger sizes.
5. An accurate hole size cannot be obtained without a high quality surface finish.
6. As a general rule the cutting speed used to ream a hole is about one-third to one-half of the speed used to drill a hole of the same size in the same material.
7. The feed rate, when reaming as compared to drilling the same material, is approximately two to three times greater. As an example, for a 1-in. drill the feed rate is about .010 to .015 in. per revolution. A 1 in. reamer would have a feed rate of between .020 and .030 in.
8. The reaming allowance of a $\frac{1}{2}$-in. diameter hole would be $\frac{1}{64}$ in.
9. Coolants cool the tool and workpiece and act as lubricants.
10. Chatter may be eliminated by reducing the speed, increasing the feed, or using a piloted reamer.
11. Oversize holes may be caused by a bent reamer shank or buildup on the cutting edges. Check also if there is a sufficient amount and if the correct kind of coolant is being used.
12. Bell-mouthed holes are usually caused by a misaligned reamer and workpiece setup. Piloted reamers, bushings, or a floating holder may correct this problem.
13. Surface finish can be improved by decreasing the feed and checking the reaming allowance. Too much or not enough material will cause poor finish. Use a large volume of coolant.
14. Carbide-tipped reamers are recommended for long production runs where highly abrasive materials are reamed.
15. Cemented carbides are very hard but also very brittle. The slightest amount of chatter or vibration may chip the cutting edges.

SECTION E / UNIT 2 / TURNING MACHINE SAFETY

Self-Test Answers

1. Both the employer and employee bear the responsibility for worker safety. Also responsible

are the manufacturer, reconstructor, and modifier of machinery.

2. A pinch point is where associated parts of a lathe move close together, thus making a hazard to the operator. An example would be a finger caught in gears.

3. When a spindle that has a threaded nose is suddenly reversed, the chuck that was screwed on can unscrew and come out of the lathe. A chuck wrench left in a lathe chuck may fly out when the power is turned on.

4. A lathe operator can get burned from hot chips that fly and strike his body where skin is exposed or by handling hot chips with bare hands. He can also get burned from a part just turned.

5. Sharp edges, projections, stringy chips, and burred edges on workpieces all produce cuts. When measuring a bored hole, cover the tool and back off the carriage.

6. On special setups that are hazardous, signs and verbal instructions are used. The operator should always rotate the setup slowly by hand before turning on the power. He should stand aside when he starts the spindle. A geared scroll chuck should not be rotated with power when empty. Workpieces should be securely gripped in chuck jaws.

7. Do not turn on the switch yourself. Get the mechanic to do it.

8. The rod will begin to whip immediately and will bend so that its end will describe about a 4-ft diameter circle. At that RPM there would be no time to get out of the way. A trough or tube can be used to guard the extended rod.

9. The bolt that is overstressed may be almost ready to break and when a sudden workload is placed on it, it does break—suddenly. Understressed bolts usually just allow slight slippage of parts out of position.

10. It is probably the result of a wrong attitude. One who acts without thinking, who is careless about work habits, and who does not do the safe thing at all times is a sure prospect for an accident.

SECTION E / UNIT 3 / TOOLHOLDERS AND TOOLHOLDING FOR THE LATHE

Self-Test Answers

1. A toolholder is needed to rigidly support and hold a cutting tool during the actual cutting operation. The cutting tool is often only a small piece of high-speed steel or other cutting material that has to be clamped in a much larger toolholder in order to be usable.

2. Quick-change toolholders are adjusted for height with a micrometer collar.

3. Tool height on turret toolholders is adjusted by placing shims under the tool.

4. Toolholder overhang affects the rigidity of a setup; too much overhang may cause chatter.

5. Drilling machine tools are used in a lathe tailstock.

6. The lathe tailstock is bored with a Morse taper hole to hold Morse taper shank tools.

7. When a series of repeat tailstock operations are to be performed on several different workpieces, a tailstock turret should be used.

SECTION E / UNIT 4 / CUTTING TOOLS FOR THE LATHE

Self-Test Answers

1. High-speed steel is easily shaped into the desired shape of cutting tool. It produces better finishes on low-speed machines and on soft metals.

2. Its geometrical form: the side and back rake, front and side relief angles, and chip breakers.

3. Unlike single-point tools, form tools produce their shape by plunging directly into the work.

4. When a chip trap is formed by improper grinding on a tool, the chip is not able to clear the tool; this prevents a smooth flow across the face of the tool. The result is tearing of the surface on the workpiece and a possible broken tool.

5. Some toolholders provide a built-in back rake of about 16 degrees; adding any back rake to the tool would make an excessive total back rake.

6. A zero rake should be used for threading tools. A zero to slightly negative back rake should be used for plastics and brass because tools tend to "dig in" to these materials.

7. The side relief angle allows the tool to feed into the work material. The end relief angle keeps the tool end from rubbing on the work.

8. The side rake directs the chip flow away from the cut and it also provides for a keen cutting edge. The back rake promotes smooth chip flow and good finishes.

9. The angles can be checked with a tool grinding gage, a protractor, or an optical comparator.

10. Long, stringy chips or those that become snarled on workpieces, tool post, chuck or lathe

dog are hazardous to the operator. Chip breakers and correct feeds can produce an ideal chip that does not fly off but will simply drop to the chip pan and is easily handled.

11. Chips can be broken up by using coarse feeds and maximum depth for roughing cuts and by using tools with chip breakers on them.

12. Overheating a tool causes small cracks to form on the edge. When a stress is applied, as in a roughing cut, the tool end may break off.

13. Ceramic tools are made of cemented oxide, usually aluminum oxide.

14. They should be used on relatively hard materials at high speeds with no intermittent cuts.

15. Carbide tools have greater hardness at high and low temperatures than high-speed steel tools and are widely used in industry for cutting tough alloy steels.

16. When diamond tools are used, the removal rate is small and the finishes are extremely good.

17. In such intermittent cutting, ceramic or diamond tools should not be used. A tough grade of carbide is best for such cutting operations.

18. Polycrystalline diamond tools are useful for cutting abrasive materials.

19. The advantage of CBN tools is that they are next to diamond in hardness and can therefore cut materials that are too hard for tungsten carbides.

20. A right-hand tool cuts toward the headstock.

SECTION E / UNIT 5 / LATHE SPINDLE TOOLING

Self-Test Answers

1. The lathe spindle is a hollow shaft that can have one of three mounting devices machined on the spindle nose. It has an internal Morse taper that will accommodate centers or collets.

2. The spindle nose types are the threaded, long-taper key drive, and the camlock.

3. The independent chuck is a four-jaw chuck in which each jaw can move separately and independently of the others. It is used to hold odd-shaped workpieces.

4. The universal chuck is most often a three-jaw chuck although they are made with more or less jaws. Each jaw moves in or out by the same amount when the chuck wrench is turned. They are used to hold and quickly center round stock.

5. Combination chucks and Adjust-Tru three-jaw chucks.

6. A drive plate.

7. The live center is made of soft steel, so it can be turned to true it up if necessary. It is made with a Morse taper to fit the spindle taper or special sleeve if needed.

8. A hardened drive center is serrated, so it will turn the work between centers but only light cuts can be taken. Face drives use a number of driving pins that dig into the work and are hydraulically compensated for irregularities.

9. Workpieces and fixtures are mounted on face plates. These are identified by their heavy construction and the T-slots. Drive plates have only slots.

10. Collet chucks are very accurate workholding devices. Spring collets are limited to smaller material and to specific sizes.

SECTION E / UNIT 6 / OPERATING THE MACHINE CONTROLS

Self-Test Answers

1. The spindle must not be turning when headstock gears are shifted. Both drive clutch and motor must be in the off position.

2. The vari-drive is changed while the motor is running, but the back gear lever is only shifted with the motor off.

3. Levers that are located on the headstock can be shifted in various arrangements to select speeds.

4. The feed reverse lever.

5. These levers are used for selecting feeds or threads per inch for threading.

6. The quick approach and return of the tool for delicate work and when approaching a shoulder or chuck jaw.

7. Because the cross feed is geared differently (about one-third of the longitudinal feed), the outside diameter would have a coarser finish than the face.

8. The half-nut lever is used only for threading.

9. They are graduated in English units. Some metric conversion collars are being made and used that read in both English and metric units at the same time.

10. You can test with a rule and a given slide movement such as .125 or .250 in. If the slide moves one-half that distance, the lathe is calibrated for double depth and reads the same amount as what is taken off the diameter of the workpiece.

SECTION E / UNIT 7 / FACING AND CENTER DRILLING

Self-Test Answers

1. A lathe board is placed on the ways under the chuck and the chuck is removed because it is the wrong chuck to hold rectangular work. The mating parts of an independent chuck and the lathe spindle are cleaned and the chuck is mounted. The part is roughly centered in the jaws and adjusted to center by using the back of a toolholder or a dial indicator.

2. The tool should be on the center of the lathe axis.

3. No. The resultant facing feed would be approximately 0.3 to 0.5 × .012 in., which would be a finishing feed.

4. The compound must be swung to either 30 or 90 degrees so that the tool can be fed into the face of the work by a measured amount. A depth micrometer or a micrometer caliper can be used to check the trial finish cut.

5. A right-hand facing tool is used for shaft ends. It is different from a turning tool in that its point is only a 58-degree included angle to fit in the narrow space between the shaft face and the center.

6. $\text{RPM} = \dfrac{300 \times 4}{4} = 300$

7. Center drilling is done to prepare work for turning between centers and for spotting workpieces for drilling in the lathe.

8. Round stock is laid out with a center head and square or rectangular with diagonal lines. Stock can be held vertically in a vise or angle plates and vee blocks for center drilling in a drill press.

9. Center drills are broken as a result of feeding the drill too fast and having the lathe speed too slow. Other causes result from having the tailstock off center, the work off center in a steady rest, or lack of cutting oil.

10. The sharp edge provides a poor bearing surface and soon wears out-of-round, causing machining problems such as chatter.

SECTION E / UNIT 8 / TURNING BETWEEN CENTERS

Self-Test Answers

1. A shaft between centers can be turned end for end without loss of concentricity, and it can be removed from the lathe and returned without loss of synchronization between thread and tool. Cutting off between centers is not done, as it would break the parting tool and ruin the work. Steady rest work is not done with work mounted in a center in the headstock spindle.

2. The other method is holding the workpiece in a chuck on one end and in the tailstock center on the other end.

3. Coarser feeds, deeper cuts, and smaller rake angles all tend to increase chip curl. Chip breakers also make the chip curl.

4. Dead centers are hardened 60-degree centers that do not rotate with the work but require high pressure lubricant. Ball-bearing centers turn with work and do not require lubricant. Pipe centers turn with the work and are used to support tubular material.

5. With no end play in the workpiece and the bent tail of the lathe dog free to click against the sides of the slot.

6. Because of expansion of the workpiece from the heat of machining, it tightens on the center, thus causing more friction and more heat. This could ruin the center.

7. Excess overhang promotes lack of rigidity. This causes chatter and tool breakage.

8. $\text{RPM} = \dfrac{90 \times 4}{1\frac{1}{2}} = 360 \div \dfrac{3}{2} = 360 \times \dfrac{2}{3} = 240$

   or $\dfrac{360}{1.5} = 240$

9. The spacing would be .010 in., as the tool moves that amount for each revolution of the spindle.

10. The feed rate should be limited to what the tool, workpiece, or machine can stand without undue stress.

11. For most purposes where liberal tolerances are allowed, .015 to .030 in. can be left for finishing. When closer tolerances are required, two finish cuts are taken with .005 to .010 in. left for the last finish cut.

12. After roughing is completed, .015 to .030 in. is left for finishing. The diameter of the workpiece is checked with a micrometer, and the remaining amount is dialed on the cross-feed micrometer collar. A short trial cut is taken and the lathe is stopped. This diameter is again checked. If the diameter is within tolerance, the finish cut is taken.

SECTION E / UNIT 9 / ALIGNMENT OF THE LATHE CENTERS

Self-Test Answers

1. The workpiece becomes tapered.
2. The workpiece is tapered with the small end at the tailstock.
3. By the witness mark on the tailstock, by using a test bar, and by taking a light cut on a workpiece and measuring.
4. The dial indicator.
5. With a micrometer. The tailstock is set over with a dial indicator.

SECTION E / UNIT 10 / LATHE OPERATIONS

Self-Test Answers

1. Drilled holes are not sufficiently accurate for bores in machine parts because they would be loose on the shaft and would not run true.
2. The workpiece is center spotted with a center drill at the correct RPM and, if the hole is to be more than a $\frac{3}{8}$-in. diameter, a pilot drill is put through. Cutting oil is used. The final-size drill is put through at a slower speed.
3. The chief advantage of boring in the lathe is the bore runs true with the centerline of the lathe and the outside of the workpiece, if the workpiece has been set up to run true (with no runout). This is not always possible when reaming bores that have been drilled, since the reamer follows the eccentricity or runout of the bore.
4. Ways to eliminate chatter in a boring bar are:
   a. Shorten the bar overhang, if possible.
   b. Reduce the spindle speed.
   c. Make sure the tool is on center.
   d. Use as large a diameter bar as possible without binding in the bore.
   e. Reduce the nose radius on the tool.
   f. Apply cutting oil to the bore.
   g. Use tuned or solid-carbide boring bars.
5. Through boring is making a bore the same diameter all the way through the part. Counterboring is making two or more diameters in the same bore, usually with 90-degree or square internal shoulders. Blind holes are bores that do not go all the way through.
6. Grooves and thread relief are made in bores by means of specially shaped or ground tools in a boring bar.

7. A floating reamer holder will help to eliminate the bell mouth, but it does not remove the runout.
8. Hand reamers produce a better finish than machine reaming.
9. Cutting speeds for reaming are *one-half* that used for drilling; feeds used for reaming are *twice* that used for drilling.
10. Large internal threads are produced with a boring tool. Heavy forces are needed to turn large hand taps, so it is not advisable to use large taps in a lathe.
11. A tap drill can be used as a reamer by first drilling with a drill that is $\frac{1}{32}$ to $\frac{1}{16}$ in. undersized. This procedure assures a more accurate hole size by drilling.
12. A spiral-point tap works best for power tapping.
13. The variations in pitch of the hand-cut threads would cause the micrometer collar to give erroneous readings. A screw used for this purpose and for most machine parts must be threaded with a tool guided by the lead screw on the lathe.
14. Thread relief and external grooves are produced by specially ground tools that are similar to internal grooving tools except that they have less end relief. Parting tools are often used for making external grooves.
15. Parting tools tend to seize in the work, especially with deep cuts or heavy feeds. Without cutting oil, seizing is almost sure to follow with the possibility of a broken parting tool and misaligned or damaged work.
16. You can avoid chatter when cutting off with a parting tool by maintaining a rigid setup and keeping enough feed to produce a continuous chip, if possible.
17. Knurling is used to improve the appearance of a part, to provide a good gripping surface, and to increase the diameter of a part for press fits.
18. Ordinary knurls make a straight or diamond pattern impression by displacing the metal with high pressures.
19. When knurls produce a double impression, they can be readjusted up or down and moved to a new position. Angling the toolholder 5 degrees may help.
20. You can avoid producing a flaking knurled surface by stopping the knurling process when the diamond points are almost sharp. Also use a lubricant while knurling.

SECTION E / UNIT 11 / CUTTING EXTERNAL AND INTERNAL THREADS

Self-Test Answers

1. A series of cuts are made in the same groove with a single-point tool by keeping the same ratio and relative position of the tool on each pass. The quick-change gearbox allows choices of various pitches or leads.
2. The chips are less likely to bind and tear off when feeding in with the compound set at 29 degrees, and the tool is less likely to break.
3. The 60-degree angle on the tool is checked with a center gage or optical comparator.
4. The number of threads per inch can be checked with a screw pitch gage or by using a rule and counting the threads in one inch.
5. A center gage is used to align the tool to the work.
6. No. The carriage is moved by the thread on the lead screw when the half-nuts are engaging it.
7. Even-numbered threads may be engaged on the half-nuts at any line and odd-numbered threads at any numbered line. It would be best to use the same line every time for fractional-numbered threads.
8. The spindle should be turning slowly enough for the operator to maintain control of the threading operation—about one-fourth the usual turning speed.
9. The lead screw rotation is reversed, which causes the cut to be made from the left to the right. The compound is set at 29 degrees to the left. The threading tool and lathe settings are set up in the same way as for cutting right-hand threads.
10. Picking up the thread or resetting the tool is a procedure that is used to position a tool to existing threads.
11. The most convenient method of finding the tap drill size is to consult a tap drill table, which is usually based on 75 percent thread. If other percentages of threads are desired, the following formula may be used:

Outside diameter of thread $-$
$$\frac{.01299 \times \text{percentage of thread}}{\text{Number of threads per inch}} = \text{Hole size}$$

12. When available, the best tap to use in soft aluminum, especially in a blind hole, would be a fluteless forming tap.
13. The minor diameter of the thread.
14. By varying the bore size, usually larger than the minor diameter. This is done to make tapping easier.
15. 75 percent.
16. The compound can be set in the same position as in external threading, and the threading operation can be observed by the operator.
17. Large internal threads of various forms can be made, and the threads are concentric to the axis of the work.
18. To the left of the operator.
19. A screw pitch gage should be used.
20. Boring bar and tool deflection cause the threads to be undersize from the calculations and settings on the micrometer collars.
21. The minor diameter equals $D - (P \times .65 \times 2)$.

$$P = \tfrac{1}{8} \text{ in.} = .125 \text{ in.}$$
$$D = 1 \text{ in.} - (.125 \times .65 \times 2) = .8375 \text{ or } .838 \text{ in.}$$

22. A thread plug gage, a shop-made plug gage, or the mating part.

SECTION E / UNIT 12 / TAPER TURNING, TAPER BORING, AND FORMING

Self-Test Answers

1. Steep tapers are quick-release tapers and slight tapers are self-holding tapers.
2. Tapers are expressed in taper per foot, taper per inch, and by angles.
3. Tapers are turned by hand feeding the compound slide, by offsetting the tailstock and turning between centers, or by using a taper attachment. A fourth method is to use a tool that is set to the desired angle and to form cut the taper.
4. No. The angle on the workpiece would be the included angle, which is twice that on the compound setting. The angle on the compound swivel base is the angle with the work centerline.
5. The reading at the lathe centerline index would be 55 degrees, which is the complementary angle.
6. Offset $=$

$$\frac{10 \times (1.125 - .75)}{2 \times 3} = \frac{3.75}{6} = .625 \text{ in.}$$

7. Four methods of measuring the offset on the tailstock are with centers and a rule, witness

marks and a rule, the dial indicator, and tool-holder-micrometer dial.

8. The two types of taper attachments are the plain and the telescopic. Internal and external and slight to fairly steep tapers can be made. Centers remain in line, and power feed is used for good finishes.

9. The taper plug gage and the taper ring gage are the simplest and most practical means to check a taper. Four methods of measuring tapers are the plug and ring gages; using a micrometer on layout lines; using a micrometer with precision parallels and drill rod on a surface plate; and using a sine bar, gage block, and a dial indicator.

10. Chamfers, V-grooves, and very short tapers may be made by the form tool method.

SECTION E / UNIT 13 / LATHE ACCESSORIES

Self-Test Answers

1. When workpieces extend from the chuck more than four or five workpiece diameters and are unsupported by a dead center; when workpieces are long and slender.

2. Since steady rests are useful for supporting long workpieces, heavier cuts can be taken or operations such as turning, threading, and grooving may be performed without chattering. Internal operations such as boring may be done on long workpieces.

3. The steady rest is placed near the tailstock end of the shaft, which is supported in a dead center. The steady rest is clamped to the lathe bed and the lower jaws are adjusted to the shaft finger tight. The upper half of the frame is closed and the top jaw is adjusted with some clearance. The jaws are locked and lubricant is applied.

4. The jaws should be readjusted when the shaft heats up from friction in order to avoid scoring. Also, soft materials are sometimes used on the jaws to protect finishes.

5. A center punch mark is placed in the center of the end of the shaft. The lower two jaws on the steady rest are adjusted until the center punch mark aligns with the point of the dead center.

6. No.

7. No. When the surface is rough, a bearing spot must be turned for the steady rest jaws.

8. By using a cat head.

9. A follower rest.

10. The shaft is purposely made one or two inches longer and an undercut is machined on the end to clear the follower rest jaws.

SECTION F / UNIT 1 / VERTICAL MILLING MACHINES

Self-Test Answers

1. Chips should never be handled with bare hands because they are sharp, possibly hot, and often contaminated from cutting fluids.

2. Eye protection is always worn in a machining facility.

3. Machine guards are there for safety. Guards not in place leave operators or bystanders exposed to dangerous moving parts of machines.

4. A clean work area prevents one from stumbling over parts on the floor and slipping on spilled oil or cutting fluids.

5. The cutting edges of tools are sharp and can cause cuts. Use a rag when handling tools.

6. The motor on top of the workhead is heavy and will flip down if the clamping bolts are loosened completely. The clamping bolts should be kept snug to provide enough drag so that the head only swivels when pushed against.

7. Measurements should only be made when the machine spindle has stopped.

8. When viewed from the cutting end, a right-hand-cut end mill will rotate counterclockwise.

9. An end mill has to have center cutting teeth to be used for plunge cutting.

10. End mills for aluminum usually have a fast helix angle and also highly polished flutes and cutting edges.

11. Carbide end mills are very effective when milling abrasive or hard materials.

12. The six major components of a vertical milling machine are the column, knee, saddle, table, ram, and toolhead.

SECTION F / UNIT 2 / USING THE VERTICAL MILLING MACHINE

Self-Test Answers

1. Tool material (HS or carbide), workpiece material, condition of the machine, rigidity of the setup, and use of coolant.

2. Thick chips are formed with high feedrates and they often cause a built-up edge on the cutter

tooth, leading to rough finishes and high stress on the tool edge. The result is often tool breakage.

3. About one-half the diameter of the end mill.
4. High-speed steel tools.
5. No. The previous operator may have been careless in his alignment procedure or perhaps deliberately set an angle for a specific job.
6. An edge finder or dial indicator.
7. By using a paper strip for a feeler gage.
8. All machines have backlash unless they have special antibacklash devices on the feedscrew. A worn feedscrew nut may allow anywhere from a .010- to .050-in. error when reversing the machine. Backlash may be eliminated by resetting the micrometer dial.
9. By tilting the toolhead or by setting the work at an angle in a vise.
10. Dividing heads, rotary tables, shaping attachments, and right-angle milling attachments.

SECTION F / UNIT 3 / HORIZONTAL MILLING MACHINES

Self-Test Answers

1. Profile-sharpened cutters and form-relieved cutters.
2. Light-duty plain milling cutters have many teeth and are used for finishing operations. Heavy-duty plain mills have few but coarse teeth, designed for heavy cuts.
3. Plain milling cutters do not have side cutting teeth. This would cause extreme rubbing if used to mill steps or grooves. Plain milling cutters should be wider than the flat surface they are machining.
4. Side milling cutters having side cutting teeth are used when grooves are machined.
5. Straight-tooth side mills are used only to mill shallow grooves because of their limited chip space between the teeth and their tendency to chatter. Staggered-tooth mills have a smoother cutting action because of the alternate helical teeth; more chip clearance allows deeper cuts.
6. Metal slitting saws are used in slotting or cutoff operations.
7. Gear-tooth cutters and corner-rounding cutters.
8. To mill V-notches, dovetails, or chamfers.
9. Face mills over 6 in. in diameter.
10. The two classes of taper are self-holding, with a small included angle, and self-releasing, with a steep taper.
11. $3\frac{1}{2}$ IPF.

12. Any small nick or chip between shank and socket or between spacers will cause the cutter to run out and will mar these contact surfaces.
13. Where small diameter cutters are used on light cuts, and where little clearance is available.
14. As close to the cutter as the workpiece and workholding device permit.
15. A style C arbor is a shell end mill arbor.
16. Tightening or loosening an arbor nut without the arbor support in place will bend or spring the arbor.

SECTION F / UNIT 4 / USING THE HORIZONTAL MILLING MACHINE

Self-Test Answers

1. Flying chips can cause eye injuries, loose-fitting clothing or jewelry can be caught in a cutter, and long or loose hair can be caught.
2. No.
3. $\text{RPM} = \dfrac{CS \times 4}{D} = \dfrac{90 \times 4}{4} = 90 \text{ RPM.}$
4. Feedrate = .003 × 16 × 90 = 4.32 IPM.
5. The machine vise.
6. The solid jaw.
7. In a milling vise or clamped to a T-slot.
8. Conventional mode.
9. When climb milling is used on a machine that has backlash.
10. When insufficient coolant is available to flood the cutter and work. Intermittent heating and cooling caused by insufficient coolant causes thermal cracking of carbide-insert tools.

SECTION G / UNIT 1 / SELECTION AND IDENTIFICATION OF GRINDING WHEELS

Self-Test Answers

1. Use silicon carbide for the bronze, aluminum oxide for the steel, either manufactured or natural diamond for the tungsten carbide inserts, and cubic boron nitride or aluminum oxide for the high-speed steel. Cubic boron nitride could be eliminated.
2. For a straight wheel the third dimension is the hole size. On a cylinder wheel the third dimension is the wall thickness.
3. Side grinding: cylindrical (Type 2), straight cup (Type 6), flaring cup (Type 11), and dish (Type

12). Peripheral grinding: straight (Type 1) and dish (Type 12). The saucer or dish wheel is the only shape on which both peripheral and side grinding are rated safe.

4. These cutoff wheels would be rubber bonded, because they are the only ones that can be made so thin.

5. Ball grinding requires the hardest wheels, then cylindrical grinding (cylinder ground with a peripheral wheel), then flat surface grinding with a peripheral wheel, then internal grinding and, finally, the softest, flat surfacing with a side grinding wheel.

6. Wheel 1 is a resinoid-bonded (B) aluminum oxide (A) wheel, very coarse grit (14), very hard (Z) and dense (3) structure. It is a typical specification for grinding castings. Wheel 2 is a vitrified-bonded (V) silicon carbide (C) wheel, same grit size but with medium (J) grade and structure (6), for grinding some soft metal like copper.

7. Wheel 1 is the peripheral grinding wheel, because it is a harder (K) and denser (8) wheel. Both wheels have the same abrasive, silicon carbide, and the same bond, vitrified. The abrasive in wheel 2 is coarser, 24 grit as against 36, and may be used for side grinding.

8. This is a vitrified aluminum oxide wheel, H grade, 8 structure, in a 46 grit size. The 32 indicates a particular kind of aluminum oxide, and the "BE" a particular vitrified bond.

9. In a straight wheel for flat grinding, you would use a softer (J or I) and more open structure (7 or 8) wheel. With a cup or segmental wheel, you would use a still softer grade, like H. For flat grinding, probably a size coarser grit, 46, would be used. The important thing is to go softer.

10. The material to be ground usually determines the abrasive to be used, although it is not practical to stock cubic boron nitride or diamond wheels unless you have a good deal of work to be done with them.

   Hard materials require fine grit sizes and soft grades. With soft, ductile materials, coarser grit sizes and harder, longer-wearing wheels are needed.

   Except for very hard materials, the rule is that grit sizes are coarser as the amount of stock to be removed becomes greater.

   Generally, the finer the grit, the better the finish.

SECTION G / UNIT 2 / GRINDING WHEEL SAFETY

Self-Test Answers

1. A 12-in. wheel can be safe on a 2000 RPM grinder. The calculation is either:

$$12 \times 3.1416 \div 12 \times 2000 = 6283 \text{ SFPM}$$

or

$$\frac{6500}{12 \times 3.1416 \div 12} = 2069 \text{ RPM}$$

2. a. Tap about 45 degrees either side of the vertical centerline.
   b. Use a wooden tapper or mallet.
   c. Suspend the wheel, or place it on a clean, hard floor. Make sure there is nothing to muffle the sound.

3. a. On a receipt from supplier, to insure that the wheels are sound.
   b. Before first mounting, to catch cracks developed in toolroom storage.
   c. Before each remounting, to catch cracks developed in storage at machine.

4. a. Do not bump or drop wheels.
   b. Do not roll wheel like a hoop.
   c. If you cannot carry a wheel, use a hand truck, but support and guard the wheel to prevent cracking.
   d. Never stack anything, particularly metal tools, on top of wheels.
   e. In storage, separate wheels from each other with cardboard or other protection.

5. Flanges must be equal in diameter, clean and flat on the mounting rim, but hollow inside the rim so there is no pressure at the wheel hole. Flanges must be checked to see that there are no burrs and no nicks. They must be at least one-third the wheel diameter.

6. a. Ring test the wheel.
   b. Check machine RPM against RPM of wheel. Wheel must be equal or higher.
   c. See that wheel fits snugly on spindle or mounting flange. It should not be so tight that it must be forced on, nor should it be loose and sloppy.
   d. Check flanges as above. Blotters should be larger than flanges.
   e. Stand to one side while starting the wheel.
   f. Run the wheel at least a minute without load before starting to grind.

7. a. Always wear approved safety glasses or other face protection.

b. Grind on the periphery, except for Type 2 cylinder wheels, cup wheels, and other wheels designed for side grinding.

c. Never jam work into the wheel.

d. Always make sure safety guard is in place and properly adjusted.

e. Never use a wheel that has been dropped.

f. If using coolant, turn coolant off a minute or so before stopping the wheel. This keeps the wheel from becoming unbalanced.

8. If wheel diameter remains the same, then wheel speed (SFPM) increases as RPM increases. Or to put it another and more practical way, as wheel diameter decreases through wear and dressing, the RPM must be increased to maintain the same speed in surface feet.

9. The fact that a wheel will run at 150 percent of safe speed in a factory test room, under extremely safe conditions with extra protection, and *without any load,* is no guarantee that it would be equally safe *with load under operating* conditions. The 50 percent overspeed has been determined by research to be a reasonable assurance of safe operation at the listed safe speed. It is common manufacturing practice to test products under conditions much more severe than the suppliers ever expect the product to meet in actual use.

10. Long sleeves, neckties, or anything that could be caught in a machine are not acceptable in a grinding shop or in any other shop. This includes finger rings and wristwatches. In fact, you should not wear any watch at all around a surface grinder, where the magnetic chuck could magnetize it and ruin it.

SECTION G / UNIT 3 / USING THE SURFACE GRINDER

Self-Test Answers

1. Wheelhead.
2. Magnetic chuck.
3. By mounting the workpiece either in a prepared fixture or on a magnetic sine chuck.
4. One ten-thousandths of an inch (.0001 in.), called a "tenth."
5. One-thousandth of an inch (.001 in.).
6. .002 in. for roughing and .0005 in. for finishing.
7. About .002 in.
8. The sharp wheel will remove stock more rapidly without burning but will give a coarse finish. The duller wheel produces a good finish but tends to burn the work if too much stock is removed per pass.

9. The wheel simply blows it away and the contact area is not cooled. The coolant ideally should be applied to the wheel so that it will be carried to the contact area.

10. Many attachments are available for surface grinders making it possible to perform grinding operations other than flat grinding.

SECTION G / UNIT 4 / PROBLEMS AND SOLUTIONS IN SURFACE GRINDING

Self-Test Answers

1. Careful dressing of the wheel, keeping coolant tank full, checking coolant filters, thorough cleaning of chuck before each loading, checking tightness of wheel.
2. Vibration, heat, and dirt, plus poor quality wheel dressing.
3. A surface defect is anything that is out of pattern; any kind of unwanted scratches or discoloration.
4. Vibration is the principal cause. Off-grade wheels and wheels principally too hard or soaked on one side with coolant.
5. Fishtails are random scratches caused most often by dirt and grit in the coolant. Sliding the work off the chuck might also be considered a cause, and, less often, a wheel that is very much too soft. Usually, the starting point in getting rid of them is to check the coolant level and filters.
6. Burning, checking, or discoloration is one problem; low spots (out of flatness) is the other. Probably you would check first to see that there was enough coolant and that it was actually getting to the cutting area. Speeding up the table speed, if this is possible, could also help.
7. A burnished surface is one in which the hills of the scratches have been rubbed over into the valleys. It is bad because the surface will not resist wear. Probably the wheel is too hard and you should change it.
8. Poor chuck alignment is probably the first cause. You might also look for dirt or swarf on the chuck or insufficient coolant.
9. Overheating, particularly of thin work, is one. Dressing a wheel too fine could also be a cause. A wheel that is too hard is another cause. In fact, although work can be parallel and not flat, or flat and not parallel, the two conditions usually go together.
10. Mostly because thin work cannot absorb the heat, but also because the grinding may release stresses rolled into the workpieces.

SECTION G / UNIT 5 / CENTER-TYPE CYLINDRICAL GRINDERS

Self-Test Answers

1. On the plain cylindrical grinder the wheelhead does not swivel. The wheelhead may travel parallel to the rotating workpiece, in one design type, or the workpiece can be traversed past the grinding wheel in lighter-duty plain cylindrical grinders. On the universal cylindrical grinder, all the major components are able to swivel, and the grinding head can be equipped with a chuck for internal grinding. The plain cylindrical grinder is usually a very rigidly constructed production machine, while the construction of the universal grinder is much less rigid because of all the motions that are designed into the machine. The universal cylindrical grinder is used mainly in toolroom applications where versatility is critical.
2. This prevents the duplication of headstock bearing irregularities into the workpiece.
3. This prevents the force of the grinding toward the footstock end from deflecting the part away from the grinding wheel.
4. Yes, the table that carries the centers is usually capable of being swiveled for this purpose.
5. A center-hole grinding machine can be used to prepare workpieces for cylindrical grinding.
6. This is grinding in which the rotating workpiece is moved across the face of the grinding wheel, which generates the surface.
7. This is cylindrical grinding in which the rotating workpiece is moved directly into the grinding wheel. This action imparts a "mirror image" of the form of the periphery of the grinding wheel into the part.
8. The workpiece is supported on two conical work centers that project into matching conical holes in the ends of the workpiece.
9. The interruption should be narrower than the width of the grinding wheel face.
10. A chuck or other workholding device, like a faceplate, to hold and rotate the part relative to the grinding spindle. A "high-speed" attachment that mounts to the wheelhead assembly and carries a mounted internal grinding wheel.

SECTION G / UNIT 6 / USING THE CYLINDRICAL GRINDER

Self-Test Answers

1. The combined curvature of the wheel and the workpiece create a very narrow line of contact from which the grinding swarf can easily escape.
2. It does not have to be mounted for each use, hence it is very convenient for the operator.
3. It is useful for aligning the swivel table to parallel or for offsetting the table for accurate conical (tapered) surfaces.
4. None. The part should be adjusted for nil end play, but be able to rotate freely on centers that have been lubricated with a special high-pressure (lead-free) center lubricant.
5. The "tail" of the driving dog must not touch the bottom of the driving slot, or the workpiece could be forced from its conical seat. This would result in poor accuracy.
6. At least one-third of the wheel width, where possible. Half the wheel width would be optimal to permit full grinding to size.
7. The Tarry control permits a dwell at the end of the traverse to stop "table bounce." (Without a Tarry control, it is difficult to obtain the highest possible dimensional accuracy.)
8. About one-fourth of the wheel width.
9. About one-eighth of the wheel width.
10. .0002 in. (or less).

Abrasive. A substance, such as finely divided aluminum oxide or silicon carbide, used for grinding (abrading), smoothing, or polishing.

Acute angle. An angle less than 90 degrees.

Aging. The process of holding metals at room temperature or at a predetermined temperature for the purpose of increasing their hardness and strength by precipitation: aging is also used to increase dimensional stability in metals such as castings.

Alignment. The proper positioning or state of adjustment of parts in relation to each other, especially in line as in axial alignment.

Aluminum oxide. Also alumina (Al_2O_3). Occurs in nature as corundum and is used extensively as an abrasive. Today, most aluminum oxide abrasives are manufactured.

Angular. Having one or more angles; measured by an angle; forming an angle.

Angularity. The quality or characteristic of being angular.

Angular measure. The means by which an arc of a circle is divided and measured. This can be in degrees (360 degrees in a full circle), minutes (60 min. in 1 degree), and seconds (60 sec. in 1 minute), or in radians. (See *Radian*.)

Anhydrous. Free from water.

Anneal. A heat treatment in which metals are heated and then cooled very slowly for the purpose of decreasing hardness. Annealing is used to improve machinability, remove stresses from weldments, forgings, and castings, to remove stresses resulting from cold work, and to refine and make uniform the microscopic internal structures of metals.

Anodizing. The formation of a relatively thick corrosion product on some metals such as aluminum to impede further corrosion and to produce a hard wearing surface. The anodized surface is formed in a similar fashion to that of electroplating but with the polarity reversed so that the part to be anodized is the anode.

Arbor. A rotating shaft upon which a cutting tool is fastened. Often used as a term for mandrel.

As rolled. When metal bars are hot rolled and allowed to cool in air, they are said to be in the "as rolled" or natural condition.

Axial. Having the characteristics of an axis (that is, centerline or center of rotation); situated around and in relation to an axis as in axial alignment.

Axial rake. An angular cutting surface that is rotated about the axial centerline or a cutting tool such as a drill or reamer.

Axis. Centerline or center of rotation of an object or part; the rotational axis of a machine spindle, which extends beyond the spindle and through the workpiece. Machining of the part imparts the machine axis to that area of metal cutting.

Babbitt metal. An alloy of soft metals such as tin, copper, and antimony, used for lining bearings.

Backlash. Looseness or slack in mechanical parts caused by normal clearance between working parts or wear. Feed and lead screws are common examples of backlash. Some machines have backlash eliminators built into them.

Bearings. Bearings are used on rotating members to give them freedom to rotate with a minimum of friction. Ball and roller-type bearings are called *antifriction bearings* because they were designed to operate with very little friction. In contrast, sleeve bearings of bronze or babbitt metal are called *friction bearings* since they develop more friction when in service.

Bell mouth. A taper extending inward a short way at the entrance to a bore.

Bezel. A rim that holds a transparent covering, as on a watch or dial indicator.

Blind hole. A hole that does not go completely through an object.

Blueprint. A photographic print, of maps, mechanical drawings, and architectural plans, in white on a bright blue background. In contrast, blue line prints are on a white background.

Boring. The process of removing metal from a hole by using a single-point tool. The workpiece can rotate with a stationary bar, or the bar can rotate on a stationary workpiece to bore a hole.

Boss. A raised portion (usually round), on a casting to provide a seat for a bolt head or nut.

Brinell hardness. The hardness of a metal or alloy measured by hydraulically pressing a hard ball (usually a 10-mm dia. ball) with a standard load into the specimen. A number is derived by measuring the indentation with a special microscope.

Brittleness. That property of a material that causes it to suddenly break at a given stress without bending or distortion of the edges of the broken surface. Glass, ceramics, and cast iron are somewhat brittle materials.

Broaching. The process of removing unwanted metal by pulling or pushing a tool on which cutting teeth project through or along the surface of a workpiece. The cutting teeth are each progressively longer by a few thousandths of an inch to give each tooth a chip load. One of the most frequent uses of broaching is for producing internal shapes such as keyseats and splines.

Burnish. To make shiny by rubbing. No surface material is removed by this finishing process. External and internal surfaces are often smoothed with high-pressure rolling.

Hardened plugs are sometimes forced through bores to finish and size them by burnishing.

Burr. (1) A small rotary file. (2) A thin edge of metal that is usually very sharp leftover from a machining operation.

Bushing. A hollow cylinder that is used as a spacer, a reducer for a bore size, or for a bearing. Bushings can be made of metals or nonmetals such as plastics or formica.

Button die. A thread-cutting die that is round and usually slightly adjustable. It is held in a diestock or holder by means of a cone point set screw that fits into a depression on the periphery of the die.

Calibration. The adjustment of a measuring instrument such as a micrometer or dial indicator so that it will measure accurately.

Celsius. A temperature scale used in the SI metric systems of measurement where the freezing point of water is 0° and the boiling point is 100°. Also called *centigrade*.

Cementite. Iron carbide, a compound of iron and carbon (Fe_3C) found in steel and cast iron.

Centerline. A reference line on a drawing or part layout from which all dimensions are located.

Chamfer. A bevel cut on a sharp edge of a part to improve resistance to damage and as a safety measure to prevent cuts.

Chasing a thread. In machine terminology, chasing a thread is making successive cuts in the same groove with a single-point tool. Also when cleaning or repairing, a damaged thread.

Chatter. Vibration of workpiece, machine, tool, or a combination of all three due to looseness or weakness in one or more of these areas. Chatter may be found in either grinding or machine operations and is usually noted as a vibratory sound and seen on the workpiece as wave marks.

Checked. A term used mostly in grinding operations indicating a surface having many small cracks (checks). The term *heat checked* or *crazed* is used in reference to friction clutch surfaces.

Chips. The particles that are removed when materials are cut. Also called *filings*.

Chip trap. A deformed end of a lathe cutting tool that prevents the chip from flowing across and away from the tool.

Circularity. The extent to which an object has the form of a circle. The measured accuracy or roundness of a circular or cylindrical object such as a shaft. A lack of circularity is referred to in shops as out-of-round, egg-shaped, or having a flat spot.

Circumference. The periphery or outer edge of a circle. Its length is calculated by multiplying pi (3.1416) times the diameter of the circle.

Coarseness. A definition of grit size in grinding or spacing of teeth on files and other cutting tools.

Cold finish. Refers to the surface finish obtained on metal by any of several means of cold working, such as rolling or drawing.

Cold working. Any process such as rolling, forging, or forming a cold metal in which the metal is stressed beyond its yield point. Grains are deformed and elongated in the process, causing the metal to have a higher hardness and lower ductility.

Complementary angles. Two angles whose sum is 90 degrees. Often referred to in machine shop work since most angular machining is done within one quadrant, or 90 degrees.

Concave. An internal arc or curve; a dent.

Concentricity. The extent to which an object has a common center or axis. Specifically, in machine work, the extent to which two or more surfaces on a shaft rotate in relation to each other; the amount of runout on a rotating member.

Conductivity. The quality and rate of conducting electricity or heat in materials.

Convex. An external arc or curve; a bulge.

Coolant. A cutting fluid used to cool the tool and workpiece, especially in grinding operations; usually water-based.

Crest of thread. Outer edge (point or flat) of a thread form.

Cutting fluid. A term referring to any of several materials used in cutting metals: cutting oils, soluble or emulsified oils (water based), and sulfurized oils.

Cyclic. To recur in cycles as on a rotating member. A rotating shaft that is slightly bent due to side stress is subject to cyclic stress reversals.

Deburr. The removal of a sharp edge or corner caused by a machining process.

Decibel. A unit for expressing the relative intensity of sounds on a scale from zero (least perceptible sound) to about 130 (the average pain level).

Deformation. Distortion; alteration of the form or shape as a result of the plastic behavior of a material under stress.

Degrees. The circle is divided into 360 degrees, four 90-degree quadrants. Each degree is divided into 60 minutes and each minute into 60 seconds. Degrees are measured with protractors, optical comparators, and sine bars, to name a few methods. Degrees are also divisions of temperature scales.

Diagonal. A straight line from corner to corner on a square, rectangle, or any parallelogram.

Diameter. Twice the radius; the length of any straight line going through the center of a figure or body; specifically, a circle in drafting and layout.

Diametral pitch. The ratio of the number of teeth on gears to the number of inches of pitch diameter.

Die. (1) Cutting tool for producing external threads. (2) A device that is mounted in a press for cutting and forming sheet metal.

Die cast metal. Metal alloys, often called *pot metals*, that are forced into a die in a molten state by hydraulic pressure. Thousands of identical parts can be produced from a single die or mold by this process of die casting.

Dimension. A measurement in one direction. One of three coordinates: length, width, and depth. Also thickness, radius, and diameter are given as dimensions on drawings.

Discrimination. The level of measurement to which an instrument is capable within a given measuring system. A .001-in. micrometer can be read to within one-thousandth of an inch. With a vernier, it can discriminate to one ten-thousandth of an inch.

Distortion. The alteration of the shape of an object that would normally affect its usefulness. Bending, twisting, and elongation are common forms of distortion in metals.

Drawings. Mechanical drawings of machines and parts as in the machining trades. Machinists use both assembly and detail drawings.

Dress. The process of trueing a grinding wheel by using a single-point diamond or other dressing tool such as a diamond impregnated form dresser.

Drift. A flat, tapered tool used to remove tapered shanks from sleeves and sockets.

Ductility. The property of a metal to be deformed permanently without rupture while under tension. A metal that can be drawn into a wire is ductile.

Ebonized. Certain cold-drawn or -rolled bars that have black stained surfaces are said to be *ebonized* (not the same as the black scaly surface of hot-rolled steel products).

Eccentricity. A rotating member whose axis of rotation is different or offset from the primary axis of the part or mechanism. Thus, when one turned section of a shaft centers on a different axis than the shaft, it is said to be *eccentric* or to have *runout*. For example, the throws or cranks on an engine crankshaft are eccentric to the main bearing axis.

Edge finder. A tool fastened in a machine spindle that locates the position of the workpiece edge in relation to the spindle axis.

E.D.M. Electro-discharge machining. With this process, a graphite or metal electrode is slowly fed into the workpiece that is immersed in oil. A pulsed electrical charge causes sparks to jump to the workpieces, each tearing out a small particle. In this way, the electrode gradually erodes its way through the workpiece, which can be a soft or an extremely hard material such as tungsten carbide.

Elasticity. The property of a material to return to its original shape when stretched or compressed.

Elastic limit. The extent to which a material can be deformed and still return to the original shape when the load is released. Deformation occurs beyond the elastic limit.

Electroplating. An electrical deposition of one metal on another in an electrolyte. An electric current is induced through the electrolyte at the cathode (part to be plated) and the anode (the plating material).

Embossing. The raising of a pattern in relief on a metal by means of a high pressure on a die plate.

Embrittle. To cause brittleness. For example, when trapped in steel from welding or plating operations, hydrogen can cause brittleness.

Emulsifying oils. An oil containing an emulsifying agent such as detergent so it will mix with water. Oil emulsions are used extensively for coolants in machining operations.

Expansion. The enlargement of an object, usually caused by an increase in temperature. Metals expand when heated and contract when cooled in varying amounts depending on the coefficient of expansion of the particular metal.

Fabricated. Manufactured; put together. In metalworking the term generally refers to the assembling of standard shapes such as angles, channels, beams, and plates into a useful device by the process of welding; a weldment.

Face. (1) The side of a metal disc or end of a shaft when turning in a lathe. A facing operation is usually at 90 degrees to the spindle axis of the lathe. (2) The periphery or outer cylindrical surface of a straight grinding wheel.

Fahrenheit. A temperature scale that is calibrated with the freezing point of water at 32 degrees and the boiling point at 212 degrees. The Fahrenheit scale is gradually being replaced with the Celsius scale used with the metric system of measurement.

Fatigue. Metal fatigue is a gradual, slow development of a crack that will in time progress through the piece to the point of failure. It is caused by cyclic stress reversals, as on a rotating shaft, and is usually initiated at a point of stress concentration such as a groove, notch, or sharp corner. Fatigue resistance is the property of certain materials to resist fatigue failures under conditions of cyclic stress.

Feed. The longitudinal or rotary movement of a cutting tool into a work piece or of a workpiece into a cutting tool to produce a chip and remove material.

Feeler gages. Strips of thin metals of specific thickness that are used to determine the dimension of small space as in between two metal parts.

Ferromagnetic. Metals or other substances that have unusually high magnetic permeability, a saturation point with some residual magnetism and high hysteresis. Iron and nickel are both ferromagnetic.

Ferrous. Iron, from the Latin word *ferrum,* meaning iron. An alloy containing a significant amount of iron.

Fiber. A threadlike material that is strong and tough. Fibers are made of glass, wood, cotton, asbestos, metals, and synthetic materials. When fibers are added to a material to give it added tensile strength, it is called a *composite.*

Fillet. A concave junction of two surfaces; an inside corner radius of a shoulder on a shaft; an inside corner weld.

Finishing (surface). The control of roughness by turning, grinding, milling, lapping, superfinishing, or a combination of any of these processes. Surface texture is designated in terms of roughness profile in microinches, waviness, and lay (direction of roughness).

Fixture. A device that holds workpieces and aligns them with the tool or machine axis with repeatable accuracy.

Flammable. Any material that will readily burn or explode when brought into contact with a spark or flame.

Flange. A rib for adding strength or for attachment to other objects. An enlargement on the end of a shaft or pipe through which a circle of fasteners (bolts or rivets) hold a mating part.

Flash. (1) Excess material that is extruded between die halves in die castings or forging dies, and also the upset material formed when welding bandsaws. (2) The brilliant light of arc welding that can damage eyes.

Flexible seals. Rubbers, elastomers, and soft plastics are used for sealing material because they are elastic materials and will continue to maintain sealing pressure between two surfaces indefinitely or until the material embrittles from age and begins to crack.

Flutes. Grooves and lands along a cylindrical tool or shaft such as the flutes of a reamer or drill. The part of a flute that does the cutting is called a *tooth.*

Flux. A solid or gaseous material that is applied to metal in order to clean and remove the oxides.

Forging. A method of metalworking in which the metal is hammered into the desired shape or is forced into a mold by pressure or hammering, usually after being heated to a more plastic state. Hot forging requires less force to form a given point than does cold forging, which is usually done at room temperature.

Forming. A method of working sheet metal into useful shapes by pressing or bending.

FPM or SFM. Surface feet per minute on a moving workpiece or tool.

FPT. Feed Per Tooth.

Fracture. A failure of a metal part causing it to separate at a given point. A fracture in metal may be caused by metal fatigue, overload, or stress corrosion.

Friction. Rubbing of one part against another; resistance to relative motion between two parts in contact.

Galling. Cold welding of two metal surfaces in intimate contact under pressure. Also called *seizing*, it is more severe and more likely to happen between two similar, soft metals, especially when they are clean and dry.

Glazing. (1) A work-hardened surface on metals resulting from using a dull tool or a too rapid cutting speed. (2) A dull grinding wheel whose surface grains have worn flat causing the workpiece to be overheated and "burned" (discolored).

Globular. Spherical; a ball-like shape.

Graduations. Division marks on a rule, measuring instrument, or machine dial.

Graphite. One of the three allotropic forms of carbon. The two others are diamond and amorphous carbon (soot, charcoal).

Grit. (1) Any small, hard particles such as sand or grinding compound. Dust from grinding operations settles on machine surfaces as grit, which can damage sliding surfaces. (2) Diamond dust, aluminum oxide, or silicon carbide particles used for grinding wheels is called *grit.*

Ground and polished (G and P). A finishing process for some steel-alloy shafts during their manufacture. The rolled, drawn, or turned shafting is placed on a centerless grinder and precision ground, after which a polishing operation produces a fine finish.

Gullet. The bottom of the space between teeth on saws and circular milling cutters.

Harmonic chatter. A harmonic frequency is a multiple of the fundamental frequency of sound. Any machine part, such as a boring bar, has a fundamental frequency and will vibrate at that frequency and also at several harmonic or multiple frequencies. Thus, chatter or vibration of a tool may be noted at several different spindle speeds.

Hazard. A situation that is dangerous to any person in the vicinity. Also a danger to property, as a fire hazard, for example.

Heat treated. Metal whose structure has been altered or modified by the application of heat.

Helix. A line traced around a cylinder at a constant oblique angle (helix angle). Often erroneously referred to as a *spiral.*

High-pressure lube. A petroleum grease or oil containing graphite or molybdenum disulfide that continues to lubricate even after the grease has been wiped off.

Hogging. When large amounts of workpiece materials are deliberately removed in roughing operations, it is called "hogging." However, the term is also used when a tool accidentally digs into the work and removes more material than is desired.

Homogenous. A material having a uniform structure throughout its mass is said to be homogenous.

Horizontal. Parallel to the horizon or base line; level.

Hot rolled. Metal flattened and shaped by rolls while at a red heat.

Hub. A thickening near the axis of a wheel, gear, pulley, sprocket, and other parts that provides a bore in its center to receive a shaft. The hub also provides extra strength to transfer power to or from the shaft by means of a key and keyseat.

Increment. A single step of a number of steps. A succession of regular additions. A minute increase.

Inert gas. A gas, such as argon or helium, that will not readily combine with other elements.

Infeed. The depth a tool is fed into the workpiece. This is in contrast to *feed* which is the rate material is removed from a workpiece.

Interface. The contact area between two materials or mechanical elements.

Interference fit. Force fit of a shaft and bore, bearings, and housings or shafts. Negative clearance in which the fitted part is very slightly larger than the bore.

Involute. A profile tooth for gears. A spiral curve that is formed when a point on a thread or string is traced as it is unwound from a circle or cylinder.

Jig. A device that guides a cutting tool and aligns it to the workpiece.

Journal. The part of a rotating shaft or axle that turns in a bearing.

Kerf. Width of saw cut.

Key. A removable metal part that, when assembled into keyseats, provides a positive drive for transmitting torque between shaft and hub.

Keyseat. An axially located rectangular groove in a shaft or hub.

Keystock. Square or rectangular cold-rolled steel bars used for making and fitting keys in keyseats.

Keyway. Same as keyseat.

Knurl. Diamond or straight impressions on a metal surface produced by rolling with pressure. The rolls used are called *knurls.*

Lead. The distance a thread or nut advances along a threaded rod in one revolution.

Loading. A grinding wheel whose voids are being filled with metal, causing the cutting action of the wheel to be diminished.

Longitudinal. Lengthwise, as the longitudinal axis of the spindle or machine.

Machinability. The relative ease of machining that is related to the hardness of the material to be cut.

Magnetic. Having the property of magnetic attraction and permeability.

Malleability. The ability of a metal to deform permanently without rupture when loaded in compression.

Mandrel. A cylindrical bar upon which the workpiece is affixed and subsequently machined between centers. Mandrels, often erroneously called *arbors,* are used in metal turning and cylindrical grinding operations.

Mechanical properties (of metals). Some mechanical properties of metals are tensile strength, ductility, malleability, elasticity, and plasticity. Mechanical properties can be measured by mechanical testing.

Metal spinning. A process in which a thin disc of metal is rapidly turned in a lathe and forced over a wooden form or mandrel to form various conical or cylindrical shapes.

Metrology. The science of weights and measures or measurement.

Micrometer. A measuring device used in determining dimensions to within .001 or .0001 in. or .01 mm.

Microstructure. Structure that is only visible at high magnification.

Miscible. Capable of being mixed in any ratio without separation of the two phases.

Mushroom head. (1) An oversize head on a fastener or tool that allows it to be easily pushed with the hand. (2) A deformed striking end of a chisel or punch that should be removed by grinding.

Neutral. In machine work, neither positive nor negative rake is a neutral or zero rake; a neutral fit is neither a clearance nor interference fit.

Nominal. Usually refers to a standard size or quantity as named in standard references.

Nonferrous. Metals other than iron or iron alloys; for example, aluminum, copper, and nickel are nonferrous metals.

Normalizing. A heat treatment consisting of heating to a temperature above the critical range of steel followed by cooling in air. Normalizing produces in steel what is called a *normal structure* consisting of free ferrite and cementite or free pearlite and cementite, depending on the carbon content.

Nose radius. Refers to the rounding of the point of a lathe cutting tool. A large radius produces a better finish and is stronger than a small one.

Obtuse angle. An angle greater than 90 degrees.

Oxide scale. At a red heat, oxygen readily combines with iron to form a black oxide scale (Fe_3O_4), also called *mill scale.* At lower temperatures, 400 to 650°F (204 to 343°C), various oxide scale colors (straw, yellow, gold, violet, blue, and gray) are produced, each color within a narrow temperature range. These colors are used by some heat treaters to determine temperatures for tempering.

Oxidize. To combine with oxygen; to burn or corrode by oxidation.

Oxyacetylene. Mixture of oxygen and acetylene gases to produce an extremely hot flame used for heating and welding.

Parallax error. An error in measurement caused by reading a measuring device, such as a rule, at an improper angle.

Parting. Also called *cutting off;* a lathe operation in which a thin-blade tool is fed into a turning workpiece to make a groove that is continued to the center to sever the material.

Pecking. A process used in drilling deep holes to remove chips before they can seize and jam the drill. The drill is fed into the hole a short distance to accumulate some chips in the flutes, and then drawn out of the hole, allowing the chips to fly off. This process is repeated until the correct depth of the hole is reached.

Periphery. The perimeter or external boundary of a surface or body.

Perpendicular. At 90 degrees to the horizontal or base line.

Pin. Straight, tapered, or cotter pins are used as fasteners of machine parts or for light drives.

Pinion. The smaller gear of a gear set.

Pitch diameter. For threads, the pitch diameter is an imaginary circle that on a perfect thread occurs at the point where the widths of the thread and groove are equal. On gears, it is the diameter of the pitch circle.

Pitting. A form of corrosion in which anodes form on a metal in many locations, causing a pitted surface.

Plating. The process of applying a thin coating of metal to another metal by electrolysis; electroplating.

Porous. Having pores or voids so that liquids may penetrate or pass through.

Pot metals. Die casting alloys, which can be zinc-, lead-, or aluminum-based (among others).

Precipitation hardening. A process of hardening an alloy by heat treatment in which a constituent or phase precipitates from a solid solution at room temperature or at a slightly elevated temperature.

Precision. A relative but higher level of accuracy within certain tolerance limits. Precision gage blocks are accurate within a few millionths of an inch, yet precision lathe work in some shops may be within a few thousandths of an inch tolerance.

Pressure. Generally expressed in units as pounds per square inch (PSI) and called *unit pressure,* while force is the total load.

Pulley. A flat-faced wheel used to transmit power by means of a flat belt. Grooved pulleys are called *sheaves.*

Quench. A rapid cooling of heated metal for the purpose of imparting certain properties, especially hardness. Quenchants are water, oil, fused salts, air, and molten lead.

Quick-change gearbox. A set of gears and selector levers by which the ratio of spindle rotation to lead screw rotation on the lathe can be quickly set. Many ratios in terms of feeds or threads per inch can be selected without the use of change gears.

Quick-change tool post. A lathe toolholding device in which preset cutting tools are clamped in toolholders that can be placed on the tool post or interchanged with others to an accurately repeatable location.

Radial rake. On cylindrical or circular cutting tools, such as milling cutters or taps, the rake angle that is off the radius is called the *radial rake.*

Radian. A unit of angular measurement that is equal to the angle at the center of a circle subtended by an arc equal in length to the radius.

Radioactivity. The property possessed by some elements such as uranium of spontaneously emitting radiation by the disintegration of the nuclei of atoms.

Rake. A tool angle that provides a keenness to the cutting edge.

Rapid traverse. A rapid travel arrangement on a machine tool used to quickly bring the workpiece or cutting tool into close proximity before the cut is started.

Recessing. Grooving.

Reciprocating. A back-and-forth movement.

Reference point. On a layout or drawing, there must be a point of reference from which all dimensions originate for a part to avoid an accumulating error. This could be a machined edge, datum, or centerline.

Relief angle. An angle that provides the cutting edge clearance that allows the cutting action is called the *relief angle.*

Right angle. A 90-degree angle.

Ring test. A means of detecting cracks in grinding wheels. The wheel is lightly struck and, if a clear tone is heard, the wheel is not cracked.

Rockwell. A hardness test that uses a penetrator and known weights. Several scales are used to cover the very soft to the very hard materials. The Rockwell "C" scale is mostly used for steel.

Root. The bottom of a thread or gear tooth.

Root truncation. The flat at the bottom of a thread groove.

Roughing. In machining operations, the rapid removal of unwanted material on a workpiece, leaving a small amount for finishing, is called *roughing.* Since coarse feeds are used, the surface is often rough.

RPM. Revolutions per minute.

Runout. An eccentricity of rotation, as that of a cylindrical part held in a lathe chuck being off center as it rotated. The amount of runout of a rotating member is often checked with a dial indicator.

Scriber. A sharp pointed tool used for making scratch marks on metal for the purpose of layout.

Semiprecision. Using a method of layout, measurement, or machining in which the tolerances are greater than that capable by the industry for convenience or economy.

Serrated. Small grooves, often in a diamond pattern, used mostly for a gripping surface.

Set. The width of a saw tooth. The set of saw teeth is wider than the blade width.

Setup. The arrangement by which the machinist fastens the workpiece to a machine table or workholding device and aligns the cutting tool for metal removal. A poor setup can cause the workpiece to move from the pressure of the cutting tool, thus damaging the workpiece or tool, or chatter from lack of rigidity.

SFPM. Surface feet per minute.

Shank. The part of a tool that is held in a workholding device or in the hand.

Shearing action. A concentration of forces in which the bending moment is virtually zero and the metal tends to tear or be cut along a transversal axis at the point of applied pressure.

Sheaves. Grooved pulleys such as those used for V-belts or cables.

Shim. A thin piece of material, usually metal, that is used as a spacer.

Shock loads. Sudden high stresses applied to a machine part far beyond the anticipated design for loading. Shock loads are often the cause of mechanical failures.

S.I. Système Internationale. The metric system of weights and measures.

Silicon carbide. A manufactured abrasive. Silicon carbide wheels are used for grinding nonferrous metals, cast iron, and tungsten carbide but are not normally used for grinding steel.

Sine bar. A small precision bar with a given length (5 or 10 in.) that remains constant at any angle. It is used with precision gage blocks to set up or to determine angles within a few seconds of a degree.

Sintering. Holding a compressed metal-powder briquette at a temperature just below its melting point until it fuses into a solid mass of metal.

Sleeve. (1) In drill press tooling, a hollow tube, tapered inside and out, for the purpose of adapting various size tapered shanks on tools to a larger size taper in a drill press or lathe spindle. (2) A straight, metal tube used to reduce the size of a bore or as an inside bearing surface.

Slot. Groove or depression, as in a keyseat slot.

Snagging. Rough grinding to remove unwanted metal from castings and other products.

Socket. A device with an internal taper to receive a tapered shank, having a smaller, external taper on the other end, thus adapting a large taper shank to a smaller taper in a machine spindle.

Sodium hydroxide (NaOH). A strong caustic used for cleaning metals.

Solder. An alloy of lead and tin that is used for joining parts requiring only low strength bonding.

Soluble oils. Oils that have been emulsified and will combine with water are called *soluble oils.*

Solution heat treating. Certain alloys, usually nonferrous metals, are raised to a predetermined temperature for a length of time that is suitable to allow a certain constituent to enter into solid solution. Then the alloy is quickly cooled to hold the constituent in solution at room temperature, causing the metal to be in an unstable supersaturated condition. This condition is often followed by age hardening. See *Precipitation hardening.* See also *Aging.*

Solvent. A material, usually liquid, that dissolves another. Dissolved material is the solute.

Spark testing. A means of determining the relative carbon content of plain carbon steels and identifying some other metals by observing the sparks given off while grinding the metal.

Specifications. Requirements and limits for a particular job.

Speeds. Machine speeds are expressed in revolutions per minute; cutting speeds are expressed in surface feet per minute.

Sphericity. The quality of being in the shape of a ball. The extent to which a true sphere can be produced with a given process.

Spiral. A path of a point in a rotating plane that is continuously receding from the center is called a *flat spiral.* The term *spiral* is often used, though incorrectly, to describe a helix.

Spotfacing. When a hole is drilled in a part, a small, round area is often machined square to the hole to make a load-bearing surface for a bolt head or nut. Often the spot is machined on a rough boss or angled workpiece, using a spotfacer.

Sprockets. Toothed wheels used with chain for drive or conveyor systems.

Squareness. The extent of accuracy that can be maintained when making a workpiece with a right angle.

Stepped shaft. A shaft having more than one diameter.

Stick-slip. A tendency of some machine parts that slide on ways to bind slightly when pressure to move them is applied, followed by a sudden release, which often causes the movement to be greater than desired.

Straightedge. A comparison measuring device used to determine flatness. A precision straightedge usually has an accuracy about plus or minus .0002 in. in 24 inches of length.

Stroke. A single movement of many, as in a forward stroke with a hacksaw.

Sulfides. Sulfur compounds such as metallic ores.

Surface plate. A cast-iron or granite surface having a precision flatness that is used for precision layout, measurement, and setup.

Symmetrical. Usually bilateral in machinery where two sides of an object are alike but usually as a mirror image.

Synthetic oils. Artificially produced oils that have been given special properties, such as resistance to high temperatures. Synthetic water-soluble oils or emulsions are replacing water-soluble petroleum oils for cutting fluids and coolants.

Tang. The part of a file on which a handle is affixed.

Taper. A gradual change in dimension along the length of a part. Taper is measured by the difference in size in one inch length (TPI), one foot length (TPF), and by angular measure (degrees).

Tapered thread. A thread made on a taper, such as a pipe thread.

Tap extractor. A tool that is sometimes effective in removing broken taps.

Tapping. A method of cutting internal threads by means of rotating a tap into a hole that is sufficiently under the nominal tap size to make a full thread.

Telescoping gage. A transfer-type tool that assumes the size of the part to be measured by expanding or telescoping. It is then measured with a micrometer.

Temper. (1) The cold-worked condition of some nonferrous metals. (2) Also called *draw;* a method of toughening hardened carbon steel by reheating it.

Temperature. The level of heat energy in a material as measured by a thermometer or thermostat and recorded with any of three temperature scales: Celsius, Fahrenheit, or Kelvin.

Template. A metal, cardboard, or wooden form used to transfer a shape or layout when it must be repeated many times.

Tensile strength. The maximum unit load that can be applied to a material before ultimate failure occurs.

Terminating threads. Methods of ending the thread, such as undercutting, drilled holes, or tool removal.

Test bar. A precision-ground bar that is placed between centers on a lathe to test for center alignment using a dial indicator.

Thermal cracking. Checking or cracking caused by heat.

Thread axis. The centerline of the cylinder on which the thread is made.

Thread chaser. A tool used to restore damaged threads.

Thread crest. The top of the thread.

Thread die. A device used to cut external threads.

Thread engagement. The distance a nut or mating part is turned onto the thread is called the *thread engagement.*

Thread fit. Systems of thread fits for various thread forms range from interference fits to very loose fits; extensive references on thread fits may be found in machinist's handbooks.

Thread lead. The distance a nut travels in one revolution. The pitch and lead are the same on single-lead threads but not on multiple-lead threads.

Thread pitch. The distance from a point on one thread to a corresponding point on the next thread.

Thread relief. Usually an internal groove that provides a terminating point for the threading tool.

Tolerance. The allowance of acceptable error within which the mechanism will still fit together and be totally functional.

Tool geometry. The proper shape of a cutting tool that makes it work effectively for a particular application.

Torque. A force that tends to produce rotation or torsion. Torque is measured by multiplying the applied force by the distance at which it is acting to the axis of the rotating part.

Toughness. Toughness in metals is usually measured with an impact test and an Izod-Charpy machine in which a specimen that is notched is struck with a swinging hammer. The amount of energy absorbed by the specimen is the measure of its toughness.

Toxic fumes. Gases resulting from heating certain materials are toxic, sometimes causing illness (as metal fume fever from zinc fumes) or permanent damage (as from lead or mercury fumes).

Transverse. Crosswise to the major or lengthwise axis.

Traverse. To move a machine table or part from one point to another, usually crosswise to the major axis of the machine.

Trueing. In machine work, the use of a dial indicator to set up work accurately. In grinding operations, to dress a wheel with a diamond.

Truncation. To remove the point of a triangle (as of a thread), cone, or pyramid.

Tungsten carbide. An extremely hard compound that is formed with cobalt and tungsten carbide powders by briquetting and sintering into tool shapes.

Turning. Machine operations in which the work is rotated against a single-point tool.

Ultimate strength. Same as tensile strength; or ultimate tensile strength.

Ultrasonic. Sound frequencies above those audible to the human ear.

Vernier. A means of dividing a unit measurement on a graduated scale with a short scale made to slide along the divisions of a graduated instrument.

Vibration. An oscillating movement caused by loose bearings or machine supports, off-center weighting on rotating elements, bent shafts, or nonrigid setups.

Viscosity. The property of a fluid or semifluid that enables it to maintain an amount of shear stress and offer resistance to flow, depending on the velocity.

Vise. A workholding device. Some types are bench, drill press, and machine vises.

Vitrified. Fired clay or porcelain.

Wedge angle. Angle of keenness; cutting edge.

Wheel dressing. Trueing the grinding surface of an abrasive wheel by means of a dressing tool such as a diamond or Desmond dresser.

Wiggler. A device used to align a machine spindle to a punch mark.

Witness marks. Index marks used to align two mating parts. Witness marks are often made with a punch, chisel, or number stamp by mechanics when disassembling parts to facilitate reassembly.

Wrought. Hot or cold worked; forged.

Zero back rake. Also *neutral rake;* neither positive nor negative; level.

Zero index. Also *zero point.* The point at which micrometer dials on a machine are set to zero and the cutting tool is located to a given reference such as a workpiece edge.

DATE DUE

Feb 2/11			